Studies in Systems, Decision and Control

Volume 532

The series "Studies in Systems, Decision and Control" (SSDC) covers both new developments and advances, as well as the state of the art, in the various areas of broadly perceived systems, decision making and control–quickly, up to date and with a high quality. The intent is to cover the theory, applications, and perspectives on the state of the art and future developments relevant to systems, decision making, control, complex processes and related areas, as embedded in the fields of engineering, computer science, physics, economics, social and life sciences, as well as the paradigms and methodologies behind them. The series contains monographs, textbooks, lecture notes and edited volumes in systems, decision making and control spanning the areas of Cyber-Physical Systems, Autonomous Systems, Sensor Networks, Control Systems, Energy Systems, Automotive Systems, Biological Systems, Vehicular Networking and Connected Vehicles, Aerospace Systems, Automation, Manufacturing, Smart Grids, Nonlinear Systems, Power Systems, Robotics, Social Systems, Economic Systems and other. Of particular value to both the contributors and the readership are the short publication timeframe and the world-wide distribution and exposure which enable both a wide and rapid dissemination of research output.

Indexed by SCOPUS, DBLP, WTI Frankfurt eG, zbMATH, SCImago.

All books published in the series are submitted for consideration in Web of Science.

Guanghui Sun · Chengwei Wu · Xiaolei Li ·
Zhiqiang Ma · Shidong Xu · Xiangyu Shao

Fractional-Order Sliding Mode Control: Methodologies and Applications

Springer

Guanghui Sun
School of Astronautics
Harbin Institute of Technology
Harbin, China

Chengwei Wu
School of Astronautics
Harbin Institute of Technology
Harbin, China

Xiaolei Li
School of Astronautics
Harbin Institute of Technology
Harbin, China

Zhiqiang Ma
School of Astronautics
Northwestern Polytechnical University
Xi An, China

Shidong Xu
College of Aerospace Engineering
Nanjing University of Aeronautics
and Astronautics
Nanjing, China

Xiangyu Shao
School of Astronautics
Harbin Institute of Technology
Harbin, China

ISSN 2198-4182 ISSN 2198-4190 (electronic)
Studies in Systems, Decision and Control
ISBN 978-3-031-60846-9 ISBN 978-3-031-60847-6 (eBook)
https://doi.org/10.1007/978-3-031-60847-6

This Springer imprint is published by the registered company Springer Nature Switzerland AG
The registered company address is: Gewerbestrasse 11, 6330 Cham, Switzerland

Preface

Fractional-order calculus, which has a historical memory capacity can be viewed as the generalization of its traditional integer-order counterpart. Due to its special properties, considerable physical systems can be modeled by fractional-order calculus, such as viscoelastic systems, polymeric chemistry systems, biomedical systems, and electrode processes, and even the electromagnetic theory can be modeled by the fractional-order calculus. Various researchers have been dedicated to fractional-order control theory in recent years. Compared with the integer-order control, fractional-order control preserves several advantages. One is that it is more suitable for flexible structures, especially for structures with viscoelastic characteristics. Another lies in that the robustness of the system can be effectively improved. The last owes to the faster response to input and the smaller overshoots, simultaneously. On the other hand, sliding mode control, an effective nonlinear control method, has attracted much attention and is widely applied to practical applications. To make full use of the advantages of fractional-order control and sliding mode control, fractional-order sliding mode control has been proposed by introducing fractional-order operation in sliding mode control. Such a control scheme has been applied to real engineering systems described by Lagrangian mechanics, such as the linear motor system and space tethered system, just to name a few. However, there still exist some problems to be solved for this kind of system. For example, since existing continuous sliding mode control strategies are not suitable for linear motor equipped with discrete digital controller and communication network, the discrete counterpart needs to be designed. Fractional-order sliding mode control schemes for the deployment of space tethered systems are still seldomly reported. To solve such problems, this monograph attends to introduce the fractional-order control theory and discusses how to design fractional-order sliding mode control schemes for linear motors and space tethered systems, respectively.

The fractional-order control and fractional-order sliding mode control problems of the linear motor and the deployment of space tethered system are thoroughly investigated in this work. Several novel fractional-order control schemes and fractional-order sliding mode control strategies including adaptive fractional-order sliding mode control, fractional-order terminal sliding mode control, and fractional-order fuzzy sliding mode control are proposed to improve the performance of the system. This monograph consists of three parts. One focuses on the problems of fractional-order control for rigid-flexible coupling space structures and space tethered systems. Another is concerned with the problems of fractional-order sliding mode control for linear motor systems. The last studies the problems of fractional-order sliding mode control for the deployment of space tethered systems.

Specifically, the brief view of fractional-order control strength in modeling and control includes the following:

1. Fractional-order dynamics and control of rigid-flexible coupling space structures,
2. Fractional-order controller of space tethered system.

The main contents of fractional-order sliding mode control for linear motor systems are as follows:

1. Practical tracking control of linear motor via discrete-time fractional-order sliding mode control,
2. Practical tracking control of linear motor by adaptive fractional-order Terminal sliding mode control,
3. Discrete-time fractional-order terminal sliding mode control for the tracking control of linear motor,
4. Fractional-order sliding mode contouring error control for multidimensional systems.

The main contents of fractional-order sliding mode control for space tethered systems are as follows:

1. Fractional-order fuzzy sliding mode control for the deployment of tethered satellite system under input saturation,
2. Fractional-order terminal sliding mode control for the deployment of tethered satellite system,
3. Fractional-order sliding mode control for the deployment of tethered spacecraft systems,
4. Fractional-order adaptive sliding mode control for the deployment of space tethered systems with input limitation.

Among these topics, both simulations and experiments are conducted to validate the effectiveness and advantages of the proposed fractional-order control schemes and fractional-order sliding mode control strategies in this monograph.

Harbin, China Guanghui Sun
Harbin, China Chengwei Wu
Harbin, China Xiaolei Li
Xi An, China Zhiqiang Ma
Nanjing, China Shidong Xu
Harbin, China Xiangyu Shao
February 2024

Acknowledgements

There are numerous individuals without whose help this book will not have been completed. Special thanks go to Prof. Zheng-Hong Zhu from York University, Prof. Ligang Wu, Prof. Jianxing Liu, and Prof. Huijun Gao from Harbin Institute of Technology for their valuable suggestions, constructive comments, and support. Thanks also go to my graduates, Baizheng Huan and Zhihao Cheng, for their collaboration. The authors are especially grateful to their families for their encouragement and never-ending support when it was most required. Finally, we would like to thank the editors at Springer for their professional and efficient handling of this project.

The writing of this book was supported in part by the National Natural Science Foundation of China (Nos. U23A20346, 62173107, 62033005, 62320106001, 62303139) and in part by the Key R&D Program of Heilongjiang Province (No. 2022ZX01A18).

Harbin, China	Guanghui Sun
Harbin, China	Chengwei Wu
Harbin, China	Xiaolei Li
Xi An, China	Zhiqiang Ma
Nanjing, China	Shidong Xu
Harbin, China	Xiangyu Shao
February 2024	

Contents

Part III Fractional-Order SMC for the Deployment of Space Tethered System

Notations and Acronyms

$\|\cdot\|$	Euclidean norm
$\|\cdot\|_1$	1-norm
$\|\cdot\|_2$	2-norm
$\|\cdot\|_\infty$	∞-norm
$\in$	Belongs to
$\mathcal{L}_1$	$\mathcal{L}_1 = \{f(t) : \|f(t)\|_1 < \infty\}$
$\mathcal{L}_2$	$\mathcal{L}_2 = \{f(t) : \|f(t)\|_2 < \infty\}$
$\mathcal{L}_\infty$	$\mathcal{L}_\infty = \{f(t) : \|f(t)\|_\infty < \infty\}$
N^+	Set of positive integers
$\mathrm{sgn}(\cdot)$	Sign function
$\mathbb{R}$	Field of real numbers
$\mathbb{R}^n$	Space of n-dimensional real vectors
$\mathbb{R}^{n\times m}$	space of $n \times m$ real matrices
CSMC	Continuous-time sliding mode control
DSMC	Discrete-time sliding mode control
FNTSM	Fast nonsingular terminal sliding mode
FO	Fractional order
LMI	Linear matrix inequality
SMC	Sliding mode control
SMDO	Sliding mode disturbance observer
STS	Space tethered systems
TSS	Tethered satellite system

List of Figures

Chapter 1
Introduction

FO calculus has a long history which can be traced back to 300 years ago [77]. It can not only describe the objects better than integer-order calculus in natural environment [56], but also has an effect of memory which means that it contains both the present and the past information.

To quickly understand the relationship between FO calculus and classical calculus, Let us review the definition process of classical calculus. Considering a continuous function $y = f(x)$, the well-known calculus definition can be defined by

$$f'(t) = \frac{df}{dt} = \lim_{h\to 0} \frac{f(t) - f(t-h)}{h}. \tag{1.1}$$

Applying the rules twice gives the second-order derivative

$$\begin{aligned} f''(t) = \frac{d^2 f}{dt^2} &= \lim_{h\to 0} \frac{f'(t) - f'(t-h)}{h} \\ &= \lim_{h\to 0} \frac{f(t) - 2f(t-h) + f(t-2h)}{h^2}. \end{aligned}$$

Similarly, we can get

$$\begin{aligned} f'''(t) = \frac{d^3 f}{dt^3} &= \lim_{h\to 0} \frac{f''(t) - f''(t-h)}{h} \\ &= \lim_{h\to 0} \frac{f(t) - 3f(t-h) + 3f(t-2h) + f(t-3h)}{h^3}. \end{aligned}$$

Moreover, we can obtain

$$f^n(t) = \frac{d^n f}{dt^n} = \lim_{h\to 0} \frac{1}{h^n} \sum_{r=0}^{n} (-1)^r \binom{n}{r} f(t - rh),$$

G. Sun et al., *Fractional-Order Sliding Mode Control: Methodologies and Applications*, Studies in Systems, Decision and Control 532,
https://doi.org/10.1007/978-3-031-60847-6_1

where

$$\binom{n}{r} = \frac{n(n-1)(n-2)...(n-r+1)}{r!}.$$

In the last equation, the parameters n and r are integer number. Assuming that the step size is small and the order n is generalized to any rational fractional number. Then, a less rigorous definition for FO differentiation can be stated as follows.

$$f_h^n(t) = \frac{1}{h^p} \sum_{i=0}^{n} (-1)^i \binom{p}{i} f(t - ih).$$

Of course, the above process is unscientific and non-causality, but useful for us to understand the definition process. What's more, we can use it in engineering application because the discrete form is more suitable for modern digital controller. The equation is famously called Grüunwald-Letnikov FO definition (abbreviate as G-L definition). A more rigorous and mathematical process can be found in [77].

The above section is just a first view of FO calculus. In fact the concept of FO calculus has been widely implemented in controller design [58, 111], which attracts attention from researchers in academia and industry. In the past decades, researchers applied FO calculus to SMC schemes and conducted them in industrial fields [100], such as the hydraulic manipulators [91], permanent magnet synchronous motors [106] and so on, which have proved that these control schemes are more effective than their integer-order counterparts. In what follows, preliminaries of FO calculus are provided, and its applications in the linear motor and space tethered system are discussed.

1.1 The Definition of Fraction Order Calculus

The history of FO calculus, which is deemed as the generalization of traditional integer-order calculus, can go back to 1690s. Many great mathematician have contributed to the field, such as Euler, Lagrange, Riemann and so on. Liouville expanded the functions in series of exponentials and defined the nth-order derivative. Riemann proposed a different definition that involved a definite integral and was applicable to power series with non-integer exponents. Grüunwald definite generalized derivative, as stated above, from the starting point of the integer derivative. Several centuries have been past. FO calculus has been slowly developed in the field of mathematics society. The main reason for limiting in engineering applications is the lack of physical meaning. Until now, There are three kinds of common used definitions of FO derivative, which are Riemann-Liouville, Grünwald-Letnikor, and Caputo definitions. More details about the FO calculus can be found in [41, 71, 77]. Here, both the Riemann-Liouville definition and Caputo definition used in this monograph are given.

- The Riemann-Liouville (R-L) definition is addressed as

$$ {}_{t_0}\mathrm{D}_t^{\beta} f(t) = \frac{1}{\Gamma(n-\beta)} \frac{\mathrm{d}^n}{\mathrm{d}t^n} \int_{t_0}^{t} \frac{f(\varepsilon)}{(t-\varepsilon)^{\beta-n+1}} \mathrm{d}\varepsilon, \tag{1.2} $$

where $n-1<\beta<n$, n is an integer, $f(t)$ is a continuous function with respect to t, and $\Gamma(\cdot)$ is a Gamma function.

The Laplace transform of the FO derivative based on Riemann-Liouville definition is expressed as

$$ \mathscr{L}\left\{ {}_{t_0}\mathrm{D}_t^{\beta} f(t) \right\} = s^{\beta} F(s) - \sum_{k=0}^{n-1} s^k \, {}_{t_0}\mathrm{D}_t^{\beta-k-1} f(t)|_{t=t_0}, $$

where $n-1<\beta<n$ and s is the Laplace operator. The Oustaloup's discretization algorithm shown in [75] is used to approximate s^{β} in numerical simulations. For notational simplicity, in what follows, ${}_0\mathrm{D}_t^{\beta}$ is simplified as D^{β}.

- The Caputo definition of the FO derivative is as follows.

$$ D_t^{\alpha} f(t) = \frac{1}{\Gamma(n-\alpha)} \int_0^t \frac{f^{(n)}(\tau)}{(t-\tau)^{\alpha-n+1}} d\tau, n-1<\alpha<n, \tag{1.3} $$

where a is a real number, n is a given integer number and $\Gamma(\cdot)$ is the Gamma function defined as $\Gamma(x) = \int_0^{\infty} t^{x-1} e^{-t} dt$.

It is noted that the FO derivative of a function is related to not only the current state $f(t_n)$ but also the previous states $f(t_i), i<n$. It indicates the FO derivative possesses histroical memory capacity. The stability of FO calculus is very important to its application in dynamics and control. Matignon pioneered in this area [67] by checking the condition of FO system stability by the poles of FO system in the complex plane of its pseudo-state space. Subsequently, many stability conditions have been derived [19, 73], for instance, the LMI condition for linear time-delay systems [19, 73], Mittag-Leffler stability for nonlinear FO systems [50] and so on.

1.2 Realization of Fractional-Order Calculus in Engineering

In order to obtain a FO calculus in engineering application, there are two kinds of methods which are called the discrete numerical method and approximation method in Laplace domain. A natural approach for computing can be got by using the G-L definition which has the discrete form. Using the commonly used integer-order transfer function to approximate the fraction order operators in a given frequency range is the main feature for the approximation method in Laplace domain. Some results have been summarized in Vinagre's work [89].

In this book, considering the discussion background mainly focusing on the control science and engineering, the FO differential equation can be written as

$$D_t^\alpha y(t) = f(t, y(t)). \tag{1.4}$$

For the common sense linear control system, the above equation can be also rewritten as

$$\begin{aligned} a_n D^{\alpha_n} y(t) + a_{n-1} D^{\alpha_{n-1}} y(t) + \cdots + a_0 D^{\alpha_0} y(t) = \\ b_m D^{\beta_m} f(t) + b_{m-1} D^{\beta_{m-1}} f(t) + \cdots + b_0 D^{\beta_0} f(t), \end{aligned} \tag{1.5}$$

or by a continuous transfer function of the form

$$G(s) = \frac{b_m s^{\beta_m} + b_{m-1} s^{\beta_{m-1}} + \cdots + b_0 s^{\beta_0}}{a_n s^{\alpha_n} + a_{n-1} s^{\alpha_{n-1}} + \cdots + a_0 s^{\alpha_0}}. \tag{1.6}$$

It's clear that for the above equation, the first and most important thing is how to solve the equation. Therefore, in this section, we will introduce two kinds of realization examples to compute the FO calculus in our applications.

1.2.1 Discrete Numerical Methods

- **G-L definition based discrete method**

Based on the Grünwald-Letnikov FO definition, for a given FO operator $D^\alpha f(t)$, we can use the following approximation

$$D^\alpha f(t) \approx \Delta_h^\alpha f(t), \tag{1.7}$$

$$\Delta_h^\alpha f(t)|_{t=kh} = h^{-\alpha} \sum_{j=0}^{k} (-1)^j \binom{\alpha}{j} f(kh - jh). \tag{1.8}$$

As we can see in the former equation, as the step $t = kh \to \infty$, more and more items will be computed to solve the equation. In other words, as the time/step grows, more and more memories are needed to store every step value (from $t = 0 \to kh$) to get current solution. It is unpractical to require infinite memories in our application no matter engineering or numerical simulation. To solve this problem, the short memory principle or finite term truncation method is the natural way for computing. Before introducing the principle, define the coefficient parameter as follows

$$w_j^\alpha = (-1)^j \binom{\alpha}{j}. \tag{1.9}$$

It's clearly that the difference equation (1.8) can be viewed as a weighted sum of the values at all time $f(t - jh)$, $j = 0, 1, 2, ...$ with the coefficient equation (1.9). We can observe that the values of the w_j^α near $t = 0$ (or any other time step $t = t_0$ as initial step) have little influence for the current $t = kh \to \infty$. What's more, the initial value will influence the current value more and more small as the current time $t \to \infty$. Thus, The short memory principle or finite term truncation method allows us to approximate the numerical solution by using the information of the "recent past". By using the time interval $[t - L, t]$ instead of the long history $[0, t]$ to get the given length of memory, a window function liked method is easily proposed to compute the FO derivatives. Some results for the error of the approximation can be found in Monje's work [71].

- **Generalized Adams-Bashforth-Moulton method**

Another effective discretization method can be viewed as generalized method from the classical integer-order Adams-Bashforth-Moulton method [21]. Considering the FO differential (1.4) in the time interval $0 \leqslant t \leqslant T$ with initial value $y^{(k)}(0) = y_0^{(k)}$ [20, 22, 77], the equivalent Volterra integral is:

$$y(t) = \sum_{k=0}^{m-1} y_0^{(k)} \frac{t^k}{k!} + \frac{1}{\Gamma(\alpha)} \int_0^t (t - \tau)^{\alpha-1} f(\tau, y(\tau)) d\tau. \tag{1.10}$$

Define $h = T/N$, $t_n = nh$, $n = 0, 1, \cdots N \in Z^+$, the following Adams-Bashforth predictive formula is firstly used.

$$y_h^p(t_{n+1}) = \sum_{k=0}^{m-1} \frac{t_{n+1}^k}{k!} y_0^{(k)} + \frac{1}{\Gamma(\alpha)} \sum_{j=0}^{n} b_{j,n+1} f(t_j, y_h(t_j)), \tag{1.11}$$

where the parameter $b_{j,n+1} = h^\alpha((n + 1 - j)^\alpha - (n - j)^\alpha)/\alpha$, and then the Adams-Moulton correction formula is introduced,

$$\begin{aligned} y_h(t_{n+1}) = &\sum_{k=0}^{m-1} \frac{t_{n+1}^k}{k!} y_0^{(k)} + \frac{h^\alpha}{\Gamma(\alpha+2)} f(t_{n+1}, y_h^p(t_{n+1})) \\ &+ \frac{h^\alpha}{\Gamma(\alpha+2)} \sum_{j=0}^{n} a_{j,n+1} f(t_j, y_h(t_j)), \end{aligned} \tag{1.12}$$

where the parameter $a_{j,n+1}$ can be stated as

$$a_{j,n+1} = \begin{cases} n^{\alpha+1} - (n-\alpha)(n+1)^{\alpha} & , j = 0 \\ (n-j+2)^{\alpha+1} + (n-j)^{\alpha+1} - 2(n-j+1)^{\alpha+1} & , 1 \leqslant j \leqslant n \\ 1 & , j = n+1 \end{cases} \quad (1.13)$$

And the error is

$$\max_{j=0,1,\ldots,N} \left| y(t_j) - y_h(t_j) \right| = O(h^p), \quad (1.14)$$

where $p = \min(2, 1+\alpha)$ [21]. We can find that the generalized Adams-Bashforth-Moulton method has the same disadvantage in computing, which means that infinite memories are needed as the time/step grow. Thus, the same finite term truncation method to get "recent past" effect is also important.

1.2.2 Approximations Method in Laplace Domain

The above discrete methods are very useful in numerical computing. However, many of the engineering problems can be indeed well expressed by transfer functions. Many engineers in the control field are familiar with the transfer function. Considering the FO transfer function (1.6), many researchers have proposed approximation methods in Laplace domain based on the classical transfer function. The main feature is using the classical integer-order transfer function to approximate FO operator s^{α} for a given frequency range. This kind of approximation method can be roughly divided into two categories. One is the approximation method based on continued fraction expansions (CFE) [16] and interpolation techniques, such as Carlson method [9] and Matsuda's algorithm [68]. The other is the approximate method based on Curve fitting or Identification techniques, such as Oustaloup's and Chareff's algorithms [11, 75].

- **Continued fraction expansions and interpolation method**

The CFE is a very famous method for evaluation of functions, which frequently converges much more rapidly than power series expansions, and converges in a much larger domain in the complex plane. The result of such approximation for an irrational function $G(s)$, can be expressed in the form,

$$\begin{aligned} G(s) &\approx a_0(s) + \cfrac{b_1(s)}{a_1(s) + \cfrac{b_2(s)}{a_2(s) + \cfrac{b_3(s)}{a_3(s)+\cdots}}} \\ &= a_0(s) + \frac{b_1(s)}{a_1(s)+}\,\frac{b_2(s)}{a_2(s)+}\,\frac{b_3(s)}{a_3(s)+}\cdots, \end{aligned} \quad (1.15)$$

where $a_i(s)$ and $b_i(s)$ are rational functions of the variable s, or are constant. The application of the method yields a rational function $\hat{G}(s)$, which is an approximation of the irrational function $G(s)$. Based on this principal, three brief methods are presented

(1) **Piecewise linear approximation**: In Laplace domain, FO operator $s^{-\alpha}, 0 < \alpha < 1$ can be approximated by following rational function

$$G_h(s) = \frac{1}{(1+sT)^{\alpha}}, \tag{1.16}$$

$$G_l(s) = \left(1 + \frac{1}{s}\right)^{\alpha}, \tag{1.17}$$

where $G_h(s)$ is the approximation for high frequencies ($\omega \gg 1$), and $G_l(s)$ is the approximation for low frequencies ($\omega \ll 1$).

(2) **Carlson's method** [9]: This method is an iterative approximation algorithm. Derive regular Newton iterative process to approximate the α-th root. The starting point of the method is the statement of the following relationships:

$$(H(s))^{1/\alpha} - G(s) = 0; H(s) = G(s)^{\alpha}. \tag{1.18}$$

Defining $\alpha = 1/q, m = q/2$, in each iteration, starting from the initial value $H_0(s) = 1$, an approximated rational function is obtained in the form

$$H_i(s) = H_{i-1}(s)\frac{(q-m)(H_{i-1}(s))^2 + (q+m)G(s)}{(q+m)(H_{i-1}(s))^2 + (q-m)G(s)}. \tag{1.19}$$

(3) **Matsuda's method** [68]: This method is based on the approximation of an irrational function by a rational one, obtained by CFE and fitting the original function in a set of logarithmically spaced points. Assuming that the selected points are $s_k, k = 0, 1, 2$, the approximation takes on the form:

$$H(s) = a_0 + \frac{s - s_0}{a_1 +} \frac{s - s_1}{a_2 +} \frac{s - s_2}{a_3 +} \cdots, \tag{1.20}$$

where $a_i = \upsilon_i(s_i), \upsilon_0(s) = H(s), \upsilon_{i+1}(s) = \frac{s - s_i}{\upsilon_i(s) - a_i}$.

- **Curve fitting based method**

The curve fitting and identification method is based on the response of fractional integral operator in the frequency domain, and the curve fitting method is adopted to ensure that the following cost function is minimum in the sense of least square criterion:

$$J = \int W(\omega)\left|G(\omega) - \hat{G}(\omega)\right|^2 d\omega, \tag{1.21}$$

where $W(\omega)$ is a weighting function, $G(\omega)$ is the original frequency response, and $\hat{G}(\omega)$ is the frequency response of the approximated rational function. There are two important methods in applications.

(1) **Oustaloup's method** [75]: This method is a proposed by CRONE research group in France. Its highlight is to approximate fractional integral operators in the frequency band of interest through the following approximation formula $H(s) = s^{\mu}$.

$$H(s) \approx C \prod_{k=-N}^{N} \frac{1 + s/\omega_k}{1 + s/\omega'_k}, \tag{1.22}$$

where

$$\begin{aligned}
&\omega'_0 = \alpha^{-0.5}\omega_u, \omega_0 = \alpha^{0.5}\omega_u, \omega_u = \sqrt{\omega_h \omega_l}, \\
&\frac{\omega'_{k+1}}{\omega'_k} = \frac{\omega_{k+1}}{\omega_k} = \alpha\eta > 1, \frac{\omega'_{k+1}}{\omega_k} = \eta > 0, \frac{\omega_k}{\omega'_k} = \alpha > 0, \\
&N = \frac{\log(\omega_N/\omega_0)}{\log(\alpha\eta)}, \mu = \frac{\log\alpha}{\log(\alpha\eta)},
\end{aligned} \tag{1.23}$$

where ω_u is the unit gain frequency and the central frequency of a band of frequencies geometrically distributed around it. ω_b, ω_h are the high and low transitional frequencies, respectively.

(2) **Chareff's method** [11]: This method is very similar with Oustaloup's method. The difference is using the following approximation function instead of fractional operator s^{α}.

$$H(s) = \frac{1}{(1 + s/P_T)^{\alpha}}. \tag{1.24}$$

In the frequency domain, above $H(s)$ can be instead by a quotient of polynomials in s in a factorized form:

$$\hat{H}(s) = \frac{\prod_{i=0}^{n-1}(1 + \frac{s}{z_i})}{\prod_{i=0}^{n}(1 + \frac{s}{p_i})}, \tag{1.25}$$

where the coefficients are computed for obtaining a maximum deviation from the original magnitude response in the frequency domain of y dB. Given the break frequency P_T and frequency range$[\omega_{min}, \omega_{max}]$, define

$$a = 10^{y/10(1-\alpha)}, b = 10^{y/10\alpha}. \tag{1.26}$$

We can obtain $p_0 = P_T\sqrt{b}$,

$$\begin{aligned} p_i &= p_0(ab)^i, \\ z_i &= a\,p_0(ab)^i. \end{aligned} \tag{1.27}$$

The number of poles and zeros is related to the desired bandwidth and the error criteria used by the expression:

$$n = 1 + \left[log(\frac{\omega_{\max}}{p_0})/log(ab) \right]. \tag{1.28}$$

Per the above discussion, we can find that there exist no problems in computing the FO calculus as the rapid development of the computer science. However, an effective approximation method as well as the finite truncation is necessity for the application. The compromise scheme can be selected according to the accuracy and short memory item.

1.2.3 Some Stability Criteria for Fractional-Order System

As a basic problem in control theory, the stability theory of differential equations studies the ability of the system to return to equilibrium after a "long enough" time when the system is subjected to the initial disturbance acting for a long time in the working process. The stability for FO system has some differences with the classical integer-order Lyapunov stability. For example, the stability region is transformed to the Riemann surface not the conventional complex plane. Some great works have been well studied in [50, 51, 56, 73] and others, and some good reviews have been well conclude [80]. Considering the main work of the book, which is focusing on the FO sliding mode. This section will briefly represent some work of the FO stability criteria, the stability definition and the stability condition can be found in [71]. To be more clear and compare with integer-order counterpart, we begin the introduction with the following stability criteria.

Theorem 1.1 ([94]) *A commensurate transfer function $F(s) = T(s^{\alpha})/R(s^{\alpha})$ with the order $0 < \alpha < 1$, where T and R are coprime polynomial, is Bounded-Input Bounded-Output stable if and only if for every $p \in \mathbb{C}$ such that $R(p) = 0$:*

$$|\arg(p)| > \frac{\alpha\pi}{2}. \tag{1.29}$$

Theorem 1.1 states that the stable region of a FO system is generally larger than its integer-order counterpart, see Fig. 1.1. The latter can be viewed as a special case with the order $\alpha = 1$.

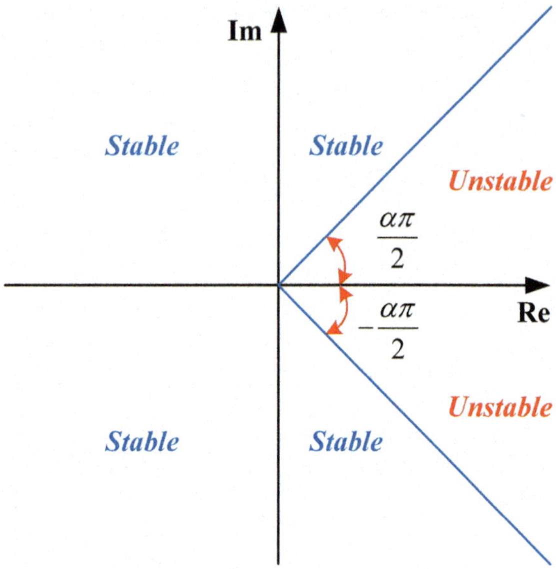

Fig. 1.1 The typical stable and unstable regions of a FO system

Theorem 1.2 ([19]) *Give a non-commensurate FO system:*

$$\begin{aligned} D^{\alpha_1}x_1 &= c_{11}x_1 + c_{12}x_2 + \cdots + c_{1n}x_n, \\ D^{\alpha_2}x_2 &= c_{21}x_1 + c_{22}x_2 + \cdots + c_{2n}x_n, \\ &\vdots \\ D^{\alpha_n}x_n &= c_{n1}x_1 + c_{n2}x_2 + \cdots + c_{nn}x_n, \end{aligned} \tag{1.30}$$

with all fractional orders $\alpha_i \in (0, 1] i = 0, 1, \ldots, n$. Let M be the lowest common multiplier of the denominators $ui, i = 0, 1, \ldots, n$, where the fractional order $\alpha_i = vi/ui, i = 0, 1, \ldots, n$ and $v_i, u_i \in Z^+$. Set $\gamma = 1/M$, then the zero solution of (1.30) is global asymptotically stable if all the roots p_i of (1.31) satisfies $|\arg(p_i)| > \gamma\pi/2$

$$\begin{vmatrix} p^{M\alpha_1} - c_{11} & -c_{12} & \cdots & -c_{1n} \\ -c_{21} & p^{M\alpha_2} - c_{22} & \cdots & -c_{2n} \\ \vdots & \vdots & \cdots & \vdots \\ -c_{n1} & -c_{n2} & \cdots & p^{M\alpha_n} - c_{nn} \end{vmatrix} = 0. \tag{1.31}$$

Theorem 1.3 ([50]) ***Mittag-Leffler Stability***: *Consider the Caputo fractional nonautonomous system*

$$D_t^{\alpha}x(t) = f(t, x), \tag{1.32}$$

where $\alpha \in (0, 1]$, if $x = 0$ is the equilibrium point of the system and then the solution is said to be Mittag-Leffler stable if

$$\|x(t)\| \leqslant \{m[x(t_0)E_\alpha(-\lambda(t-t_0)^\alpha)]\}^b, \tag{1.33}$$

where t_0 is the initial time,$\lambda \geq 0, b > 0, m(0) = 0, m(x) \geq 0$, and $m(x)$ is locally Lipschitz with Lipschitz constant m_0.

Theorem 1.4 ([51]) ***Lyapunov direct method:*** *Let $x = 0$ be an equilibrium point for the system (1.32) and $\mathbb{D} \subset \mathbb{R}^n$ be a domain containing the origin. Let $V(t, x(t))$: $[0, \infty) \times \mathbb{D} \to \mathbb{R}$ be a continuously differentiable function and locally lipschitz with respect to x such that*

$$\begin{aligned} \alpha_1\|x\|^a &\leqslant V(t, x(t)) \leqslant \alpha_2\|x\|^{ab}, \\ D_t^\beta V(t, x(t)) &\leqslant -\alpha_3\|x\|^{ab}, \end{aligned} \tag{1.34}$$

where $t \geqslant 0, x \in \mathbb{D}, \beta \in (0, 1), \alpha_1, \alpha_2, \alpha_3$,$a$ and b are arbitrary positive constants. Then $x = 0$ is Mittag-Leffler stable. If the assumptions hold globally on $\mathbb{R}^n$, then $x = 0$ is globally Mittag-Leffler stable.

1.3 Brief Introduction of Sliding Mode Control Theory

1.3.1 The Historical Development of Sliding Mode Control

Sliding mode variable structure control is a control method proposed by former Soviet Union scholars Emeleyanov, Utkin and Itkin in the early 1960s [88]. Its essence is a kind of special nonlinear control. The fundamental difference from conventional control lies in the discontinuity of control. That is, a switching characteristic that causes the control system structure to change with time. This control property can force the state of the system to be confined to a certain submanifold, namely the "sliding mode" motion. This sliding mode can be designed and is completely independent of parameter perturbation and external disturbance of the system. In this way, the system in sliding mode motion has good robustness. But the sliding mode variable structure control has a serious shortcoming, namely chattering.

Because chattering is easy to arouse the unmodeled characteristics of the system, the control performance of the system is affected, and the practical application of sliding mode variable structure control is still a difficult thing. Thus, many researchers pay much attention in reducing the chattering and finding nonlinear sliding mode surface with a better performance.

In recent years, sliding mode variable structure control has been widely used and developed in engineering related complex nonlinear control systems such as robot control and motor servo control because of its strong robustness.

1.3.2 Fundamentation of Sliding Mode Control

Prior to discussing the SMC, an example is given to show what SMC is. Consider the following linear time-invariant system.

$$\begin{aligned} \dot{x}_1 &= x_2, \\ \dot{x}_2 &= a x_2 + u, \end{aligned} \tag{1.35}$$

where x_1 and x_2 are the system states, u is the control input, and the scalar $a > 0$.

Assume that the controller u is designed as $u = -\varphi x_1$, where $\varphi = \beta$ ($\beta > 0$). Figure 1.2 show the state trajectories of the system (1.35) under the controller $u = -\beta x_1$ and $u = \beta x_1$, respectively. Obviously, neither of these two controllers can guarantee that the system (1.35) is stable. If φ is designed as

$$\varphi = \begin{cases} \beta, & x_1 s > 0, \\ -\beta, & x_1 s < 0, \end{cases}$$

where $s = c x_1 + x_2$, and $0 < c < -\frac{a}{2} + \sqrt{\frac{a^2}{4} + \beta}$. Then, the stability of the system (1.35) is preserved, and the state trajectories are depicted in Fig. 1.3.

As seen from Fig. 1.3, the phase trajectories on both sides of line $s = 0$ can be driven to the line $s = 0$. When these trajectories reach such a line, they will converge to the origin along with the line $s = 0$. The process where the trajectories move to the origin along with the line $s = 0$ is called sliding mode, and the line $s = 0$ is the switching line or named as sliding mode surface.

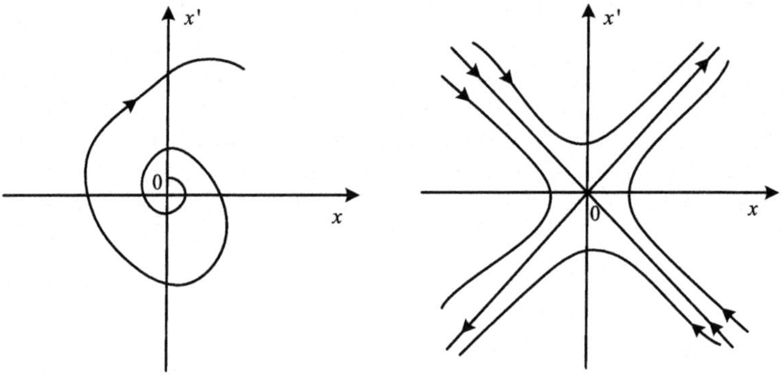

Fig. 1.2 The phase trajectories of the system (1.35). Left: when $\varphi = \beta$. Right: when $\varphi = -\beta$ with parameter $\beta > 0$

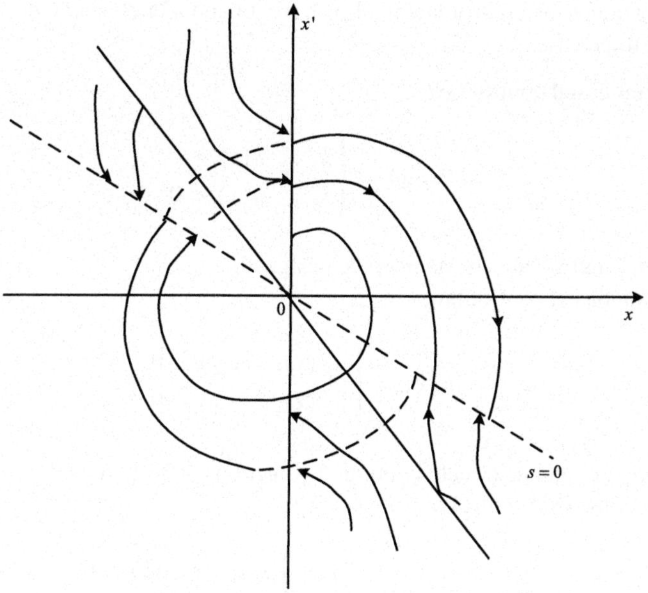

Fig. 1.3 The phase trajectories of the system (1.35) under the sliding mode controller

For system (1.35), the sliding mode equation can be described as

$$\dot{x}_1 + cx_1 = 0,$$

and the solution of the above equation is $x_1(t) = x_1(0)e^{-ct}$.

It is clear that the order of the sliding mode equation is less than that of the system (1.35), and only the parameter c is involved. In this case, the sliding mode dynamics can be robust to uncertain parameters and system disturbances. According to the aforementioned description, the definition of SMC is given as below.

Considering the following system

$$\dot{x} = f(x, u, t), \tag{1.36}$$

where $x \in \mathbb{R}^{n_x}$ is the system state, $u \in \mathbb{R}^{n_u}$ is the control input, and $t \in \mathbb{R}$.

Defining a switching function $s(x)$ with $s \in \mathbb{R}^{n_s}$, if there exists a control law in (1.37) such that the trajectories which are not on the line $s_i(x) = 0$ can reach the sliding mode surface $s_i(x) = 0$ in finite time, and move along with such a surface, the control law is called SMC law, and the system with the SMC law is called a SMC system.

$$u_i(x) = \begin{cases} u_i^+(x), & s_i(x) > 0, \\ u_i^-(x), & s_i(x) < 0, \end{cases} \tag{1.37}$$

To implement the control law in (1.37), $u_i^+(x)$ and $u_i^+(x)$ can be designed in the following three cases.

(1) Constant based control law

$$u_i(x) = \begin{cases} k_i^+, & s_i(x) > 0, \\ k_i^-, & s_i(x) < 0, \end{cases}$$

where k_i^+ and k_i^- are constants to be designed.

(2) Function based control law

$$u_i(x) = \begin{cases} u_i^+(x), & s_i(x) > 0, \\ u_i^-(x), & s_i(x) < 0, \end{cases}$$

where $u_i^+(x)$ and $u_i^-(x)$ are continuous functions to be designed.

(3) Proportion based control law

$$u_j(x) = \psi_{ij}x_i, \ \psi_{ij} = \begin{cases} \alpha_{ij}, & x_i s_j(x) > 0, \\ \beta_{ij}, & x_i s_j(x) < 0. \end{cases}$$

The aim of designing a SMC is to drive the trajectory to the sliding mode surface in a finite time and move on it. Therefore, the existence of the sliding mode, the reachability, and the stability of the sliding motion need to be guaranteed. To guarantee the existence of sliding mode, $\lim_{t\to 0^+} \dot{s}(x) \le 0$ or $\lim_{t\to 0^-} \dot{s}(x) \ge 0$ must hold when the trajectory is around the sliding mode surface $s(x) = 0$. Equivalently, such a condition can be rewritten as $\lim_{t\to 0} \dot{s}(x) < 0$. When the initial state of the system is not around the sliding mode surface $s(x) = 0$, the system states need to approach to the sliding mode surface $s(x) = 0$, that is, satisfy the reachability condition $s(x)\dot{s}(x) < 0$. As a more strict condition, $s(x)\dot{s}(x) < -\varepsilon\|s(x)\|$ can be also used to guarantee the reachability in a finite time, where $\varepsilon > 0$. Moreover, the necessary condition for the stability of the motion along with the sliding surface $s(x) = 0$ is $\lim_{t\to 0} \dot{s}(x) < 0$ and $s(0) = 0$.

The design process of SMC can be summarized as the following two manifolds.

Step 1 One should first construct a sliding mode surface function, which preserves the favorable performance for example stability, and robustness.

Step 2 A sliding mode controller is designed to guarantee the trajectories of the system states can be driven to reach the predefined sliding surface in a finite time and always remain there, under which the performance and stability of the control system can be achieved.

As to Step 1, two different categories of sliding mode surface functions are often designed. One is the linear sliding mode surface function, and the other is the integral sliding mode surface function, which can provides more favorable properties for complicated control systems. When it comes to Step 2, either the equivalent control based design or the reaching law method can be used to design a sliding mode controller. In what follows, a simple example is provided to show the details.

Consider the following nonlinear system

$$\dot{x} = f(x) + Bu, \tag{1.38}$$

where x is the system state, u is the control input to be designed. The matrix B should be full column rank.

A linear sliding mode surface function for the system (1.38) can be designed as

$$s(x) = Gx,$$

where G is a to be designed matrix, which satisfies the that GB is nonsingular.

An integral sliding mode surface function for the system (1.38) can be designed as

$$s(x) = Gx - \int_0^t (GBu + Gf(x))d\tau.$$

Alternatively, an integral sliding mode surface function removing the reaching phase for the system (1.38) can be designed as

$$s(x) = Gx - Gx_0 - \int_0^t (GBu + Gf(x))d\tau,$$

where x_0 is the initial state of the system.

Next, the equivalent control based design process under the linear sliding mode surface function is described.

For the system (1.38), the expression of the sliding mode controller is

$$u = u_{eq} + u_c,$$

where u_{eq} is the equivalent control, and u_c means the discontinuous item.

When $s(x) = \dot{s}(x) = 0$, the equivalent control u_{eq} can be calculated as

$$u_{eq} = -(GB)^{-1}Gf(x),$$

based on which the sliding mode dynamics can be obtained as

$$\dot{x} = f(x) - B(GB)^{-1}Gf(x),$$

To analyze the stability of the sliding mode dynamics, a Lyapunov candidate can be designed, for example, $V(x) = x^T Px$ with P being a positive definite matrix.

Furthermore, the matrix G can be designed as $B^T P$. As to the discontinuous item u_c, it can be designed by analyzing the reachability condition.

Consider the following Lyapunov function

$$V_s = \frac{1}{2} s^T(x)(GB)^{-1} s(x).$$

Then, one can obtain

$$\dot{V}_s = s^T(x)(GB)^{-1}(Gf(x) - GB(GB)^{-1}Gf(x) + GBu_c).$$

If the reachability condition is satisfied, $\dot{V}_s < 0$ holds, which implies that the discontinuous item can be designed as $u_c = -\alpha sign(s(x))$ with $\alpha > 0$.

The reaching law based design method is often utilized to design a sliding mode controller. A common expression of the reaching law is described as below.

$$\dot{s}(x) = -\epsilon sign(s(x)) - h(s(x)),$$

where $\epsilon > 0$, $h(0) = 0$, $s(x)h(s(x)) > 0$ $(s(x)) \neq 0$.

When $h(s(x))$ takes different values, three types of reaching laws can be obtained.

Constant rate reaching law When $h(s(x)) = 0$, the constant rate reaching law is $\dot{s}(x) = -\epsilon sign(s(x))$ with $\epsilon > 0$.
Here, ϵ means the constant reaching rate. When $s > 0$, $s = -t\epsilon$. $s = t\epsilon$ if $s < 0$. Then, the reaching time from the initial state to the sliding mode surface is $t^\star = \frac{s_0}{\epsilon}$ with s_0 being the initial value of the sliding mode surface function. As seen from the reaching time $t^\star$, the smaller ϵ is, the bigger $t^\star$ is, which means that the controller needs more time to driven the system states onto the sliding mode surface. If one sets ϵ big enough, the states can reach the sliding mode surface quickly. However, this can aggravate the chattering.

Exponential rate reaching law When $h(s(x)) = ks(x)$ with $k > 0$, the exponential rate reaching law is $\dot{s}(x) = -\epsilon sign(s(x)) - ks(x)$.
For the exponential rate reaching law, $s = \frac{\epsilon}{k} + (s_0 - \frac{\epsilon}{k})e^{-kt}$ if $s > 0$ holds. When $s < 0$, $s = \frac{-\epsilon}{k} + (s_0 + \frac{\epsilon}{k})e^{-kt}$. Then, the finite reaching time is calculated as $t^\star = \frac{1}{k}(\ln(s_0 + \frac{\epsilon}{k}) - \ln\frac{\epsilon}{k})$. It is obvious that decreasing ϵ and increasing k can reduce the reaching time.

Power rate reaching law When $h(s(x)) = k\|s(x)\|^\alpha sign(s(x)) - \epsilon sign(s(x))$ with $k > 0$ and $0 < \alpha < 1$, the power rate reaching law is $\dot{s}(x) = -k\|s(x)\|^\alpha sign(s(x))$.
Particularly, setting $\alpha = 0.5$ yields

$$\dot{s} = -k\sqrt{\|s\|} sign(s).$$

Taking the integral operation of $\dot{s}$, one can obtain $s^{1-\alpha} = -(1-\alpha)kt + s_0^{1-\alpha}$. Then, the finite reaching time is $t^\star = \frac{s_0^{1-\alpha}}{(1-\alpha)k}$.

Once one choose a reaching law, the sliding mode controller can be calculated by compare the reaching law and the derivative of the designed sliding mode surface function. Taking the exponential rate reaching law as an example, the process of designing a sliding mode controller for the system (1.38) is as below.

Considering the linear sliding mode surface function $s(x) = Gx$, the derivative of which is

$$\dot{s}(x) = Gf(x) + GBu.$$

Then, comparing $\dot{s}(x)$ and the exponential rate reaching law yields

$$u = -(GB)^{-1}(Gf(x) + \epsilon sign(s(x)) + ks(x)).$$

As shown above, the chattering resulting from the switching function $sign(s(x))$ can lead to a poor system performance. Researchers have dedicated themselves to solving such a problem, and tremendous results can be found in the literature. Readers can refer to the references in this book.

The SMC is robust to system uncertainties including external disturbances and parameter uncertainties, which is an important property of SMC. Uncertainties are not considered in the system (1.38). Next, an example is given to show the design process of SMC for uncertain systems. Here, a linear system with uncertainties is given as below.

$$\dot{x} = Ax + Bu + Df, \tag{1.39}$$

where x is the system state, u is the control input to be designed, and Df is the lumped disturbance. A, B and D are known matrices. f is an unknown nonlinear function but the bound is known.

Before showing the SMC designing process for uncertain systems, a condition is first provided to guarantee that the SMC is robust to the uncertainties. Consider the linear sliding mode surface function $s = Gx$. Letting $\dot{s} = 0$ yields

$$GAx + GBu + GDf = 0.$$

Since GB is nonsingular, the equivalent control can be calculated as

$$u_{eq} = -(GB^{-1}G(Ax + Df). \tag{1.40}$$

Substituting (1.40) into the system (1.39), the sliding mode equation is described as

$$\begin{aligned} &\dot{x} = \left[I - B(GB)^{-1}G\right]Ax + \left[I - B(GB)^{-1}G\right]Df, \\ &s = Gx = 0. \end{aligned}$$

Obviously, if $\left[I - B(GB)^{-1}G\right] Df = 0$ holds, the lumped disturbance f cannot affect the stability of the sliding mode equation. $\left[I - B(GB)^{-1}G\right] Df = 0$ is also the necessary and sufficient condition to guarantee that the system disturbance does not affect the sliding mode equation. Next, a theorem is given, and detailed proof is provided.

Theorem 1.5 *The necessary and sufficient condition to guarantee that the disturbance f does not affect the sliding mode equation of system (1.39) is*

$$rank[B,\ D] = rank[B]. \tag{1.41}$$

Proof The necessity is first proved. $\left[I - B(GB)^{-1}G\right] Df = 0$ implies

$$Df = B(GB)^{-1}GDf.$$

The necessary condition to support $Df = B(GB)^{-1}GDf$ is $Df = Bm$, namely, the arbitrary linear combination of the columns in D equals to the linear combination of columns in B, which implies $rank[B,\ D] = rank[B]$. The necessity is proved.

Now consider the sufficiency. if $Df = Bm$ holds, the following equality can be obtained

$$B(GB)^{-1}GDf = B(GB)^{-1}GBm = Bm = Df,$$

which indicates that the disturbance f does not affect the sliding mode equation. The sufficiency is proved.

Remark 1.6 The condition $rank[B,\ D] = rank[B]$ can be also rewritten as $D = B\tilde{D}$ with $\tilde{D}$ being a compatible matrix. In related literature, the condition $rank[B,\ D] = rank[B]$ $(D = B\tilde{D})$ is also named as matching condition. When system disturbances satisfy the matching condition, the sliding mode controller is robust to such disturbances.

In the system (1.39) Df is regarded as the lumped disturbance including the external disturbance and system parameter uncertainties. It can be found that researchers provided effective sliding mode control methods for systems, where only parameter uncertainties are addressed, that is, the uncertain system is described as below.

$$\dot{x} = (A + \Delta A)x + Bu,$$

where ΔA means the system parameter uncertainties.

For such a system, the necessary and sufficient condition to satisfy the matching condition is $rank[B,\ \Delta A] = rank[B]$. The proof can be readily obtained by referring to that in Theorem 1.5, and it is omitted here.

When system uncertainties are considered, the sliding mode controller can be designed based on the the design process aforementioned. By using the matching condition, the stability of the sliding mode equation can be analyzed. Then either the equivalent control based or the reaching law based method can be chosen to design a robust sliding mode controller to guarantee that the system trajectory can be driven onto the sliding mode surface in a finite time.

1.4 Applications of Fractional-Order Sliding Mode Control

The favorite properties of both FO calculus and SMC are discussed in the above sections. To fully take advantages of FO calculus and SMC, FOSMC is proposed to design control schemes for systems. FOSMC preserves the same properties of classical SMC, that is, the existence of the sliding mode, the reachability, and the stability of the sliding motion need to be guaranteed. More importantly, FOSMC can lead to some additional characteristics, which can improve the system performance. The existence of the sliding mode, and the reachability of FOSMC are as the same as those of classical SMC. As to the stability of the sliding motion of FOSMC, a Lyapunov candidate with FO characteristics can be constructed to analyze the stability. Similarly, the design of FOSMC consists of two step. One is to design a sliding mode surface function, and the other is to construct the FOSMC law.

As described in Sect. 1.3.2, both equivalent-based control and reaching law based methods can be chosen to design a sliding mode controller. Here, the design process of reaching law based SMC control is presented. For the system in (1.38), a FO sliding mode surface is designed as below.

$$s = x + \Gamma D^{\alpha-1}\left(sig(x)^{\gamma}\right), \tag{1.42}$$

where Γ is a diagonal matrix with all diagonal elements being positive, $D^{\alpha-1}$ is the $(\alpha-1)th$ order fractional caculus with $0 < \alpha < 1, \gamma > 1$. In fact $D^(\alpha - 1)$ is the fractional integral considering the uniformity of the equation, and the function $sig(x)^{\gamma}$ is defined as

$$sig(x)^{\gamma} = \|x\|^{\gamma} sign(x).$$

In the above equation, the sign function $sign(x) = \frac{x}{\|x\|}$ when $x \neq 0$. Otherwise, $sign(x) = 0$.

Differentiating s in (1.42) yields

$$\dot{s} = \dot{x} + \Gamma D^{1} D^{\alpha-1}\left(sig(x)^{\gamma}\right).$$

According to the definition of Caputo type fractional order derivative in (1.3), one can obtain

$$\dot{s} = \dot{x} + \Gamma^{C} D^{\alpha}\left(sig(x)^{\gamma}\right), \tag{1.43}$$

where ${}^{C}D^{\alpha}$ is the αth Caputo type fractional order derivative.

To design a reaching law based sliding mode controller, a reaching law needs to be designed. A reaching law candidate function is as bellow.

$$\dot{s}_{law} = h(s),$$

where $h(s)$ is a function of the sliding mode surface designed in (1.42).

There exist some candidate expressions for the reaching law, for example the exponential reaching law designed in [72]. The reader can refer to existing results for more details. Here, only the design process of reaching law based SMC is presented.

Substituting the system (1.38) into (1.43) yields

$$\dot{s} = f(x) + Bu + \Gamma^C D^{\alpha} \left(sig(x)^{\gamma} \right),$$

Comparing the expressions of $\dot{s}$ and $\dot{s}_{law}$, the reaching law based sliding mode controller can be designed as

$$u = B^{+} \left(h(s) - f(x) - \Gamma^C D^{\alpha} \left(sig(x)^{\gamma} \right) \right).$$

For the stability of the system (1.38) under the designed sliding mode controller, it can be analyzed based on a common flow. First, a Lyapunov candidate function is designed as $V = \frac{1}{2} s^T s$. Second, differentiating V and substituting the system (1.38) under the designed sliding mode controller into $\dot{V}$ can yield an equation. Then, sufficient conditions can be obtained to guarantee that $\dot{V} < 0$, which implies that the system (1.38) under the designed sliding mode controller is stable.

Some basic preliminaries of FOSMC based on reaching law are discussed. As to the detailed design process of reaching law based FOSMC, it can refer to results in the literature. For the equivalent control based FOSMC, readers can refer to the following chapters/references of this monograph. In what follows, the applications of FOSMC in linear motor systems and space tethered systems are introduced, respectively.

1.4.1 Fractional-Order Sliding Mode Control of Linear Motor System

Linear motor has received extensive attention in recent years [110]. Compared with traditional rotary motors, there is no mechanical transmission such as the belts, leadscrews and worm gear drives. Therefore, high precision, fast response and zero backlash property can be obtained in linear motor system. Moreover, the absence of mechanical wear results in excellent reliability and long lifetime. Nowadays, linear motor has been widely used in train propulsion [37], aircraft launch system [74], industrial gantry [57] and so on.

Many researchers endeavors have been focused on improving the positioning accuracy of the linear motor system during the recent years, such as optimize the mechanism and electronic structures, develop new magnetic materials, use proper winding methods, create a moving magnetic field, to name just a few. Zhang [108] proposed a new ring winding structure for linear motors, resulting in the reduction of the thrust ripple and improvement of the precision further. More recently, a novel structure of stair-shaped auxiliary teeth applied on the linear motors is proposed to reduce the end edge detent force [42]. To achieve the desired positioning accuracy of linear motor systems, suitable controller development efforts are needed. There are many approaches to develop the control method for this system. For example, an adaptive fuzzy-PID controller was designed in the speed loop of the linear motor system [86]. $\mathcal{H}_\infty$ optimal control was conducted to maintain the robustness of the system [84], and a nonsingular kind of sliding mode control (SMC) strategy was adopted [110]. The linear motor-based motion is subjected to the adverse effect of system uncertainties, such as the friction nonlinearity, nominal model errors and uncertain load, and hence the robust control strategy and the disturbance observer are employed to improve precision, i.e., to compensate for the tracking error caused by nonlinear friction. The authors in [52] modeled the nonlinear static friction using Hsieh-Pan model, the parameters of which are evaluated by particle swarm optimization (PSO), and a high positioning precision is achieved by implementing the variable structure controller. Recently, many different control strategies have been proposed for high-precision and fast-response linear motor-based motion, i.e., GA-PID controller [53], desired compensation adaptive robust controller [57], hysteretic relay feedback controller [13], and iterative-learning controller [7]. Among these control methods, SMC has many characteristics, such as insensitivity to external disturbances, high accuracy and simplicity of the design procedure, making it an effective method to deal with the nonlinear behavior of the linear motor systems [82]. Because of these merits, SMC has been widely used in aerospace [30], power electronics [54] and other fields.

Generally speaking, SMC consists of continuous sliding mode control (CSMC) and discrete sliding mode control (DSMC). Researchers firstly paid much attention to CSMC since its clear physical meaning [97]. They find that CSMC is an effective approach to keep robustness when the modeling uncertainties and disturbances exist [26]. However, as more and more modern control systems are controlled by digital processors, researchers find that CSMC algorithms are not proper to be applied in discrete controllers directly, which result in chattering effect or even instability [59]. To solve these problems, the concept of DSMC is proposed and widely studied [85]. Similar to the procedure of classical CSMC, the discrete sliding surface with desired feature should be firstly designed for DSMC. By introducing the proper reaching law [29] or equivalent control [85], the system can be driven to the designed sliding surface monotonically. However, in practical systems where disturbances and uncertainties exist, the sliding variable cannot maintain on the sliding surface at last. Instead, it moves around the sliding surface in a certain boundary which is determined by the scale of uncertainties and disturbances, which is defined as quasi-sliding mode [36, 96]. In practical engineering applications, DSMC has been used extensively because of the promising characteristics of disturbance insensitiveness, fast response and easy implementation. For example, Abidi applied the discrete-time terminal sliding mode (DTSM) controller to Piezomotor-driven linear motion stage [1]. Ma et al. improved the quasi sliding mode domain of DSM controller and applied it to real piezoelectric actuator device [59]. These DSMC schemes are proved to be effective on their corresponding platform. However, few DSMC schemes are implemented on a linear motor platform, due to complexity of the application environment. For example, the existence of unmodeled friction causes great disturbances and the uncertainties such as changing payloads also degrade the control performance [83, 107]. Therefore, a new DSMC scheme is needed for the linear motor platform to overcome these difficulties.

Additionally, one has to resolve the fundamental contradiction between the continuous smooth input and the robustness to the lumped uncertainty when designing a practical sliding-mode-based approach to industrial applications. This is because after determining the desired sliding surface by utilizing the Lyapunov indirect method, a proper control law has to be chosen to drive system states onto the surface and meanwhile has to eliminate the bounded unmodeled lumped uncertainty [104]. In the conventional control law design, a compensation for the uncertainty is to choose the robustness term properly, which takes the form of the sign function with the gain covering the boundary of the uncertainty. However, for the linear motor-based motion control, as the undesired side effect of the compensation, the chattering accompanies to make the actual torque switch in high frequency, and this chattering may cause unmodeled high frequency dynamics to affect the control performance. To suppress chattering, researchers have proposed some strategies, i.e., observer-based compensation, state-dependent gain method, and hysteresis loop method [48]. Besides the methods mentioned above, the boundary layer is a popular technique to trade off between precision enhancement and uncertainty rejection by replacing the sign function with the saturation function due to it can ensure system states reaching the prescript vicinity of the sliding surface [34]. Admittedly, removal of the robustness term leads to a continuous SMC with precision loss. To deal with this issue, a rigorous continuous fast nonsingular terminal sliding mode (FNTSM) controller got rid of sign function to achieve the high-precision and fast-response performance [110].

As mentioned above, the concept of FO calculus has been widely implemented in controller design [58, 111] to get a better performance then the integer counterpart. To suppress the steady-state error, introducing an integer-order integral term into the sliding surface is a good approach, which has been studied sufficiently. FO integral is more flexible than the integer-order one, and provides the controller design with another degree of freedom, namely, the order. Motivated by these benefits, integrating the FO integral in the sliding surface and constructing an adaptive algorithm can effectively reduce loss of control precision. The FO SMC is applied to the industrial applications, such as lighting control [102], rigid robotic manipulator [91], remote electrification systems [18], active suspension actuator [23], linear voice coil actuator [14] and the rotation motor [106]. In order to gain high precision, FO terminal sliding mode (FOTSM) controllers were investigated for robotic manipulators subjected to uncertainties [90, 91]. The FOSMC owns some distinctive features such as global memory, evidently physical significance and slight chattering [77, 106], and hence has been applied widely to the industrial applications, such as lighting control [102] and speed tuning of motor [100, 103, 106]. Typically, these FOSMC strategies are designed based on continuous time for controlled objects, and are directly tested by using digital computer system, which leads to precision loss due to ignoring the effect of sampling interval [49]. Moreover, there exists the non-physical initial condition in experiments if the FO operator is defined by Riemann-Liouville definition [77]. Therefore, Caputo's FO calculus is presented to solve this issue mathematically, but in engineering applications Caputo's definition can only be implemented in the approximation methods based on Laplace transform, which can introduce the extra approximating error into control system [8]. All these flaws can exacerbate the uncertainties existing in the linear motor-based motion control systems.

1.4.2 Fractional-Order Sliding Mode Control of Space Tethered System

The system of satellites connected by space tethers is a kind of artificial objects in space and was proposed by Guiseppe Colombo and Mario Grossi in 1960s. Since then, more than twenty orbital and suborbital experiments have been implemented by many institutes, such as TSS-1R [47], SEDS, MAST, YES2 [94] and so on. The tethered satellite system (TSS), which consists of several satellites linked by tethers is applied widely in scientific tasks, like energy collection with the tether (wire) shuttling in the magnetic field of the Earth [47], deorbit of the subsatellite [40], space tug [92] and space elevator [45]. The TSS is expected to be competent for works like rescuing stranded astronauts, capturing target capsules [33], satellite formation flights [3, 31], establishing the artificial gravity field, providing the microgravity experiment environment [17] and so on. Besides those theoretical researches, plenty of practical experiments, aiming to verify and evaluate the existing control methods, have been implemented by National Aeronautics and Space Administration (NASA) and European Space Agency (ESA) during the last decades [81, 94].

In practical applications, space tether is likely to suffer from serious vibration during deployment process, which may result in the degradation of system performance or even the collapse of system stability in some worse scenarios [10]. Obviously, it is a fundamental and challenging task to implement quick and stable tether deployment [46], due to the coupled and underactuated dynamics. Given the flexibility of space tether [94], the deployment process is prone to suffer from undesirable vibration which may result in tether slackness and deployment deviation, or even stability collapse in some extreme cases.

Over the years, a number of control methods have been developed for the deployment of TSS [43, 69, 79, 112, 113]. According to the type of control action, these control techniques can be classified into two categories: the control methods considering both platform thrusters and tension and the control methods considering tension only. Due to the high-energy consumptions and installation complexities of thrusters, regulating tensile force only becomes a simple, energy-efficient, and effective control approach which has obtained an accelerated tempo with a vast body of research results in recent times. For example, Pradeep proposed a simple linear tension control method for tether deployment of an orbital spacecraft, which was based on linearizing the equations of motion and determined without resorting to trial and error [78]. Fujii and Ishijima presented a mission function controller suitable for both deployment and retrieval cases, without the assumptions of linearity and simple open-loop control [28]. Bainum and Kumar developed an optimal control scheme to achieve a smooth deployment performance of the shuttle-tethered-subsatellite system, whose weighting matrices had piecewise adaptive adjustment law [5]. Kojima and Sugimoto studied a delayed feedback control strategy for electrodynamic tethered system in elliptic and inclined orbits, which was to stabilize the in-plane/out-of-plane coupling librational motion to a periodic motion [44]. By considering tensile force only, an optimal control scheme was proposed to implement the tether deployment of YES-2 (Young Engineers Satellite 2) in [95]. Iki et al. implemented the ground experiment of an electrodynamic tether deployment from a spool-type reel, and the thruster was used to guarantee the deployment of kilometers-length space tether [35]. Aslanov et al. presented a thruster method driving the space tethered tug, and analyzed the dragging dynamics [4]. Huang et al. investigated the attitude and deploying dynamics of the tethered space robot from the standpoint of thruster consumption [32]. Jung et al. adopted thruster scheme to achieve the ideal deployment rate during the deployment of three-body tethered satellites [38]. Liu et al. equipped the thruster system, which consisted of four pairs of thrusters, in the subsatellite dedicating for regulation of the attitude and deployment dynamics [55]. Zakrzhevskii divided the deployment process into two stages, the first stage of which ejected the tether with thruster to position the tether at desired beginning point of the second stage, and the second stage continued to adjust the tether to the stable vertical configuration by using tension generated by friction [105]. Mantellato et al. reported the thrust-aided librating deployment of tape tethers, and the force acting on the tape tether included the tension and thruster force [65]. This hybrid strategy has several advantages, such as simplifying the input model greatly, non-environmental dependence (magnetic field dependence), easy assembly and physical implementation, thus is broadly utilized in reality compared with other control strategies. A novel analytical tension control law fully considering input constraints was applied to deploy the space tether of TSS in [93]. Although effective results are achieved, by using the control efforts aforementioned it is hard to guarantee the control performances and stability of the closed-loop system in the presence of external disturbances and modeling errors, which means these controllers are not robust.

As a popular robust control strategy, SMC has been successfully employed in a variety of engineering applications including tether deployment for its robustness to external perturbations [2, 6, 60, 61]. For example, an adaptive sliding mode control (ASMC) methodology was reported in [12] to realize the deployment of TSS. However, since the presented ASMC was established on a linearized model, the discrepancies between the linearized and actual models may hinder the application of the control method in practice. To overcome this problem, in literature [87] a hierarchical SMC approach was developed to deploy tether based on the nonlinear model of TSS directly. Moreover, the authors of [70] further enhanced the hierarchical SMC approach designed in [87] by tackling tension saturation. Feng et al. developed a nonsingular terminal sliding mode (NTSM) control law in [27], it could guarantee finite-time stability and high robustness to disturbances. For the systems with mismatched disturbances, Yang et al. presented a continuous NTSM control approach to regulate any initial states to reach the desired setpoint within a finite time, which possessed nominal performance recovery and chattering alleviation [99]. To accelerate the deployment process, Zhao et al. utilized super-twisting technique to reform the sliding surface for desired performance [109]. To present the high accurate deployment dynamics, Ma et al. designed a nonlinear sliding surface to generate a rigorous nonlinear reduced system, and well analysed the stability of the motion on the surface. They used an adaptive method to produce the constraint tension control strategy, and solved the underactuated dynamics by reconstructing reduced system with virtual disturbance consisting of actuated dynamics [62]. The results in [62] showed that the adverse effect caused by constrained input may propagate to the reduced system and degenerate the steady-state performance.

On the other hand, FO control, which is the generalization of integer-order control, has emerged as an attractive control strategy [15, 77]. The existence of adjustable FO enables it to achieve the optimum dynamic responses [25, 56]. As mentioned previously, SMC possesses robustness property and can directly deal with nonlinear dynamics, therefore, for the purpose of combining the merits of FO control and SMC, FOSMC is proposed and investigated for various topics in recent years [24, 39, 101]. In fact, FOSMC is an improvement of traditional integer-order SMC. The FO can be regarded as an extra adjustable parameter for controller design. By tuning it appropriately, designers have the opportunity to obtain satisfactory transient and steady-state responses, in the meantime, the advantages of traditional integer-order SMC can still be retained. Hence, the merits of FOSMC are that it can achieve better control performances (such as faster and more smooth dynamic responses) than corresponding integer-order SMC. As for tether deployment, the first attempt was made in [39] to utilize FOSMC to carry out tether deployment with the existence of external disturbances and uncertainties. It is a good work which has demonstrated that FOSMC can attain better tether deployment than classical PD and SMC. Xu et al. presented a FO fuzzy SMC strategy to realize the deployment of tethered satellite system, whose adaptive fuzzy algorithm was to regulate the coupled coefficient of a FO and a constrained integer-order sliding manifolds [98]. The authors of [63] proposed a quaternary FOSMC with the fuzzy logic system, neural network, and adaptive law for the teleoperated cyber-physical system with model uncertainty, external disturbance, and actuator fault. A variable structure control with neural network and optimized FO selection policy has been provided for the sensorless telerobotic system with uncertain time delay, model uncertainty, fractional calculus numerical approximation bias, and external disturbance in [64].

It is noted that FOSMC exhibited its strong ability in suppressing disturbance, reducing settling time and maximum overshoot. However, in the existing researches, there are three main problems remaining to be solved. The first one is input limitation. Tension limitation and thruster saturation are inevitable when the designed control schemes applied in reality, which could cause undesired control results or even concussive system states if no special measures are taken to dispose of these constraints [66]. For example, during TSS-1R mission the tether broke because of the poor consideration of tension limitation [76]. Another problem is that most of the previous controllers are designed based on the linear dynamic model of TSS, which may lead to poor control performance or even control system failure owing to the local linearizing the dynamic model in the neighborhood of the equilibrium point. Finally, most current controllers were designed based on systems in circular orbits while in reality elliptical orbits are more common. As a result, the difference between elliptical orbits and circular orbits could pose a negative effect on the control performance when it is put into practical applications. Hence, a controller to deal with nonlinear dynamic model that is established on elliptical orbits can achieve better performance in real world.

1.5 Publication Contribution

Some state-of-the-art results concerning on FO control, FOSMC and their application in a linear motor and space tethered systems are given in this monograph, the main contents of which includes four parts. Firstly, This chapter introduces the preliminaries of FO calculus and FOSMC and briefly review the development history of FO and FOSMC in application to linear motor and space tethered systems. Secondly, Chaps. 2, 3 discuss how to modeling the rigid-flexible space structures and how to design FO control schemes for space tether system to show the power of fractional order calculus in engineering application. Chapter 2 derives the FO dynamics in modeling the rigid-flexible space structures, and proposes a FO lead-lag compensator design method. Some FO tension control laws are developed for the fast and stable deployment/retrieval of space tethers in Chap. 3. Thirdly, the problems of FOSMC for a linear motor are investigated in Chaps. 4, 5, 6 and 7. In Chap. 4, a practical FOSMC scheme for a linear motor is provided. Chapter 5 focuses on how to design an adaptive FO terminal SMC method for a linear motor. A discrete FO terminal SMC scheme is designed for a linear motor in Chaps. 6, and 7 presents a simplified Newton-based contouring error estimation and a discrete-time FO sliding mode contouring error control strategy, which is applied in a linear motor platform. Fourthly, the issues of FOSMC for space tethered systems are respectively studied in Chaps. 8, 9, 10 and 11. Specifically, Chaps. 8–10 provides FOSMC schemes for the deployment of space tethered systems, and the deployment of space tethered systems under input constraints are discussed in Chap. 11. Finally, Chap. 12 draws conclusions of this monograph and gives the future work.

Compared with existing results, the main contributions of this book can be summarized as follows:

1. FO dynamics of rigid-flexible coupling space structures and space tethered systems are derived, based on which FO control schemes are provided. Such control schemes not only can bridge the gap between FO calculus and its applications in engineering, but also can improve the system performance.
2. Novel FOSMC schemes are proposed for a linear motor under parameter uncertainties, friction, and input limitations. In the design process, different SMC approaches such as adaptive SMC, and terminal SMC are introduced to design FOSMC schemes, under which, the performance of the linear motor is improved effectively.
3. The problem of FO & FOSMC tension control of space tethered systems in the nonlinear form is investigated. Compared with existing results, the space tethered system is described as a nonlinear model rather than a linear one. By combining FO calculus and SMC approaches, effective FOSMC strategies are proposed. The higher robustness, smaller overshoots and shorter settling time can be simultaneously guaranteed.

This publication is a timely reflection of the FO control and their applications. It is a collection of latest research results and therefore serves as a useful textbook for senior and/or graduate students who are interested in knowing recent advances in FO control, FOSMC, and their practical applications in the engineering. It can also be used as a practical research reference for engineers dealing with deploying space tethered systems.

Generally, this is an advanced publication aimed at 3rd/4th-year undergraduates, postgraduates and academic researchers. Prerequisite knowledge includes FO calculus, linear algebra, SMC, adaptive technique, and space systems. Expected readers include (1) control engineers working on designing linear motors and space tethered systems; (2) system engineers working on FO control of dynamical systems; (3) postgraduate students majoring in control engineering.

1.6 Publication Outline

The general layout of presentation of this monograph includes three parts except for the first introduction section. One is to investigate FO control of rigid-flexible coupling space structure, and space tethered system, respectively. Another is to introduce the results of FOSMC of a linear motor, and the last is to present FOSMC schemes for space tethered systems. The main contents of this monograph are shown in Fig. 1.4.

This chapter introduces preliminaries of FO calculus, FO control results, advances of FOSMC and their applications, as well as the research problems including FO control as well as FOSMC design for linearmotor system and space tethered system. The outline of this monograph is given in the end of this chapter.

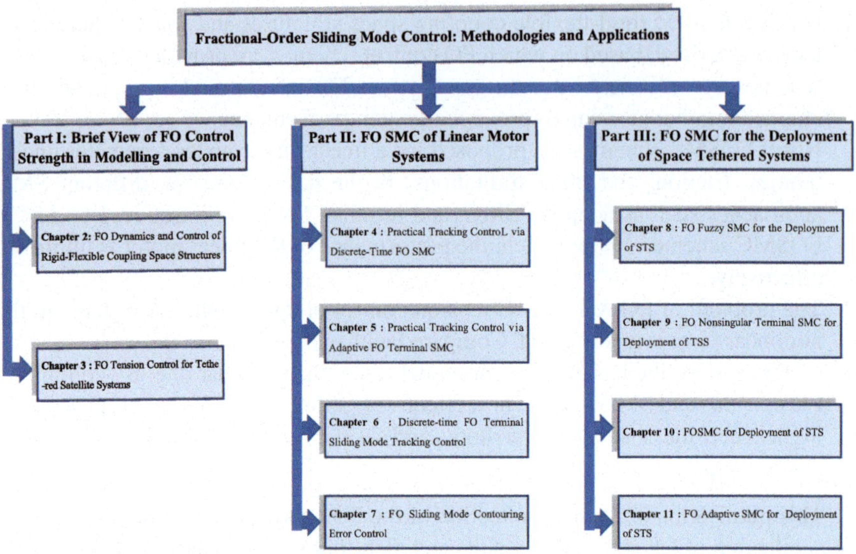

Fig. 1.4 Main contents of this publication

Part I focuses on the FO control for the rigid-flexible coupling space structure and space tethered systems to show the power of fractional order controller in control society. It begins with Chap. 2 and consists of 3 chapters as follows.

Chapter 2 proposes a two-parameter FO model of viscoelastic material and derived the FO dynamics of rigid-flexible coupling structures accordingly to simplify the modeling and controller design for residual vibration suppression. Based on the FO dynamics, a new FO lead-lag compensator design method is proposed. Simulation results demonstrate the effectiveness and robustness of the FO dynamics and control for the rigid-flexible coupling structures.

Chapter 3 provides two new FO tension control laws for the fast and stable deployment and retrieve problem of space tethered system. It is shown that the stable region of classic integer-order tension control is the subset of the stable region of the FO tension control. The newly proposed FO control law is discretized and implemented by the frequency approximation method and Tustin operator. Unlike the existing integer-order control laws, which are based on the feedback of current state and memoryless, the FO control law has the memory of previous states and thus controls the tether deployment/retrieval more smoothly while maintaining the speed. The effectiveness and advantage of the new FO tension control law is demonstrated numerically by comparing with its integer-order counterpart. The results show that the new control law not only deployment/retrieval the subsatellite without winding the main satellite, but also provides a better control performance with smaller libration angles.

Part II focuses on the FOSMC design of a linear motor. It begins with Chap. 4 and consists of 4 chapters as follows.

Chapter 4 proposes a novel discrete-time FO sliding mode controller to guarantee the desired tracking performance of the linear motor control system. A discrete-time linear motor model is firstly presented by using Euler's discretization method. By introducing a FO difference of the tracking error, a new FO sliding surface is proposed to obtain a better performance than integer-order counterparts due to the memory effect introduced by FO calculus. Then, a novel equivalent controller is introduced to drive the system to the newly proposed sliding surface in finite sampling steps. Meanwhile, a theoretical tracking error of the linear motor system is analyzed and the practical stability of the sliding surface is validated by numerical simulations. Finally, the effectiveness of the proposed control strategy is verified by a group of tracking experiments on the linear motor platform.

Chapter 5 presents a practical adaptive FO terminal SMC strategy for tracking control of the linear motor. Compared with conventional fast nonsingular SMC, the proposed approach, with a FO integral sliding surface and the adaptive switching input, can obtain higher convergence precision, even though the motion control system suffers from system uncertainties. The adaptive term is designed to guarantee finite-time high-precision convergence of the sliding mode variable, and meanwhile to degenerate the effect of uncertainties by selecting the proper adaptive gain. Moreover, continuous input due to cancel the sign term ensures that the motion control system is chattering-free. Finally, to further improve precision, we introduce the super-twisting sliding mode disturbance observer for reducing unknown bounded disturbance, i.e., the quantization noise caused by velocity estimation. Experimental results indicate that the control system with the proposed controller is easily implemented, and has higher tracking precision and considerable robustness to uncertainties compared with the existing controllers.

Chapter 6 presents a practical discrete-time FO terminal SMC strategy for high-precision tracking tasks based on a linear motor. In particular, the practical parametric uncertainties involving sliding friction, uncertain payload, and disturbance in tracking tasks are considered in this chapter. Combining Grünwald-Letnikov FO definition and terminal sliding mode technique, the proposed method synthesizes a novel discrete-time FO terminal SMC law to drive the sliding mode dynamics into the stable region in finite steps theoretically, even though the system is suffering from uncertainties and disturbances, and the motion on the surface can guarantee higher tracking precision than the conventional discrete-time terminal sliding surface by selecting suitable controller parameters. The theoretical analyses give out the guideline of parameter selection, and experiments are carried out on the linear-motor-based test platform to demonstrate that the proposed controller is easily implemented and can achieve high-precision tracking, fast response, and considerable robustness to uncertainties.

Chapter 7 presents a simplified Newton-based contouring error estimation method and a discrete-time FO sliding-mode contouring error control strategy to obtain an excellent contouring motion performance for multidimensional systems. Specifically, the simplified Newton-based contouring error estimation is modified from the Newton extremum seeking algorithm. Compared with existing contouring error estimation, the simplified Newton-based contouring error estimation method keeps the high precision; meanwhile, it requires less computing resources and avoids the possible singularity as no derivatives are demanded during the calculation. Moreover, the form of simplified Newton-based contouring error estimation is rather simple in a multidimensional system as the designed cost function in the simplified Newton-based contouring error estimation takes the vectors into consideration innovatively. To improve the performance further, a discrete-time FO sliding-mode contouring error control is proposed when the contouring error control strategy is designed. A novel fractional-order sliding surface is addressed, and the conclusion on its stability in a linear matrix inequality form is given out. Furthermore, the dynamics of the entire system are also analyzed in this chapter. At last, groups of comparative experiments are implemented on a 2-degree-of-freedom linear-motor table, whose results validate the efficiency of the proposed strategy.

Part III is concerned with the FOSMC design of space tethered systems. It begins with Chap. 8 and consists of 4 chapters as follows.

Chapter 8 develops a novel FO fuzzy sliding mode control strategy to realize the deployment of the tethered satellite system with input saturation. The considered TSS is modeled as an underactuated system. By decoupling the underactuated system into two subsystems, a FO and a constrained integer-order sliding surfaces are designed for the actuated and underactuated subsystems, respectively. Then, a new hybrid sliding manifold is obtained by coupling the two subsliding surfaces. Adaptive fuzzy algorithm is used to regulate the coupling coefficient in the newly proposed hybrid sliding manifold in order to procure satisfactory performance. Meanwhile, the saturation nonlinearity of control input is also considered. The asymptotic stability of the closed-loop system is demonstrated theoretically. With the existence of FO operator, the presented controller can perform faster and more smooth tether deployment when compared with conventional ones. Finally, the effectiveness and superiority of the proposed control approach are validated by illustrative simulations.

Chapter 9 provides a novel FO nonsingular terminal SMC scheme, which guarantees the desired deployment performance of space tethered satellite (STS) system. The presented control scheme is to stabilize the underactuated deployment mission with only tension regulation. Based on the underactuated model, a practicable integer-order nonsingular terminal sliding mode controller is designed to realize the deployment of STS system, with the finite-time stability. Then, to improve the deployment performance, the FO calculus is introduced to raise the FO nonsingular terminal sliding mode controller, whose MittagLeffler-based uniform ultimate boundedness is proved to ensure the finite-time convergence and steady-state performance. In the FO nonsingular terminal sliding mode controller, a FO disturbance observer and a FO adaptive compensation law are proposed, to eliminate the adverse impacts of external disturbances and input saturation respectively.

Moreover, compared to the reported FOSMC strategy of STS system, a better deployment performance can be obtained by the integer-order and FO nonsingular terminal sliding mode controllers in this work. Finally, the effectiveness of the proposed controllers is verified by groups of deployment simulations of STS system.

Chapter 10 proposes a FO integral SMC with the order $0 < \nu < 1$ to stabilize the deployment of tethered spacecraft system with only tension regulation. The work in this chapter is partially based on integer-order nonlinear sliding mode controller and improves its performance with FO calculus. The proposed scheme makes use of integral sliding surface to obtain smaller convergence regions of state errors, and the fractional derivative is synthesized to enhance the flexibility of controller design by fining parameters for better dynamic and steady-state performance. Fractional-order observers help to eliminate external disturbances while the adaptive law is presented to remove the adverse effect in stability analyses, and FO uniform ultimate boundedness is proved to guarantee the existence of the proposed sliding surface. According to theoretical analyses, the fractional order will indeed affect the dynamic and steady-state performance of control system, and the proposed method will be verified in numerical simulations compared with the nonlinear sliding mode counterpart.

Chapter 11 develops a novel FOSMC strategy for the deployment of space tethered system with consideration of constrained input. The simplified nonlinear dynamic model of space tethered system in elliptical orbits is first modeled by using the Euler-Lagrange mechanical equation. Considering the flexibility of tether, the compression or any component of shear forces for the tether is assumed to be input limitation. By introducing FO operator and saturation function into the sliding surface, a new adaptive FO sliding mode control strategy is introduced based on the proposed nonlinear dynamic model. Compared with classical sliding mode methods, a faster deployment time without overshoot and chattering-reduced performance can be achieved. Finally, numerical simulations are illustrated to validate the effectiveness of our methods.

Chapter 12 draws conclusions on the book, and points out some possible research directions related to the work done in this book.

References

1. Abidi, K., Xu, J.X., She, J.H.: A discrete-time terminal sliding-mode control approach applied to a motion control problem. IEEE Trans. Ind. Electron. **56**(9), 3619–3627 (2008)
2. Alali, M.A.E., Shtessel, Y.B., Barbot, J.P.: Grid-connected shunt active LCL control via continuous sliding modes. IEEE/ASME Trans. Mechatron. **24**(2), 729–740 (2019)
3. Alary, D., Andreev, K., Boyko, P., Ivanova, E., Pritykin, D., Sidorenko, V., Tourneur, C., Yarotsky, D.: Dynamics of multi-tethered pyramidal satellite formation. Acta Astronautica **117**, 222–230 (2015)
4. Aslanov, V.S., Yudintsev, V.V.: Dynamics of large debris connected to space tug by a tether. J. Guidance Control Dyn. **36**(6), 1654–1660 (2013)

5. Bainum, P.M., Kumar, V.: Optimal control of the shuttle-tethered-subsatellite system. Acta Astronautica **7**(12), 1333–1348 (1980)
6. Basin, M.V., Yu, P., Shtessel, Y.B.: Hypersonic missile adaptive sliding mode control using finite- and fixed-time observers. IEEE Trans. Ind. Electron. **65**(1), 930–941 (2017)
7. Butcher, M., Karimi, A.: Linear parameter-varying iterative learning control with application to a linear motor system. IEEE/ASME Trans. Mechatron. **15**(3), 412–420 (2009)
8. Caputo, M.: Linear models of dissipation whose Q is almost frequency independent. Ann. Geophys. **19**(4), 383–393 (1966)
9. Carlson, G., Halijak, C.: Approximation of fractional capacitors $(1/s)^{(1/n)}$ by a regular newton process. IEEE Trans. Circuit Theory **11**(2), 210–213 (1964)
10. Carroll, J.A.: Guidebook for Analysis of Tether Applications. Technical report (1985)
11. Charef, A., Sun, H., Tsao, Y., Onaral, B.: Fractal system as represented by singularity function. IEEE Trans. Autom. Control **37**(9), 1465–1470 (1992)
12. Chen, S., Li, A., Wang, C., Liu, C.: Adaptive sliding mode control for deployment of electro-dynamic tether via limited tension and current. Acta Astronautica **177**, 842–852 (2020)
13. Chen, S.L., Tan, K.K., Huang, S., Teo, C.S.: Modeling and compensation of ripples and friction in permanent-magnet linear motor using a hysteretic relay. IEEE/ASME Trans. Mechatron. **15**(4), 586–594 (2009)
14. Chen, S.Y., Lee, C.Y.: Digital signal processor based intelligent fractional-order sliding-mode control for a linear voice coil actuator. IET Control Theory Appl. **11**(8), 1282–1292 (2017)
15. Chen, Y., Petras, I., Xue, D.: Fractional order control-a tutorial. In: 2009 American Control Conference, pp. 1397–1411. IEEE (2009b)
16. Chen, Y.Q., Moore, K.L.: Discretization schemes for fractional-order differentiators and integrators. IEEE Trans. Circuits Syst. I: Fundam. Theory Appl. **49**(3), 363–367 (2002)
17. Chung, S.J., Slotine, J.J.E., Miller, D.W.: Nonlinear model reduction and decentralized control of tethered formation flight. J. Guidance Control Dyn. **30**(2), 390–400 (2007)
18. Delghavi, M.B., Shoja-Majidabad, S., Yazdani, A.: Fractional-order sliding-mode control of islanded distributed energy resource systems. IEEE Trans. Sustain. Energy **7**(4), 1482–1491 (2016)
19. Deng, W., Li, C., Lü, J.: Stability analysis of linear fractional differential system with multiple time delays. Nonlinear Dyn. **48**(4), 409–416 (2007)
20. Diethelm, K., Ford, N.J.: Analysis of fractional differential equations. J. Math. Anal. Appl. **265**(2), 229–248 (2002)
21. Diethelm, K., Ford, N.J., Freed, A.D.: A predictor-corrector approach for the numerical solution of fractional differential equations. Nonlinear Dyn. **29**(1), 3–22 (2002)
22. Diethelm, K., Ford, N.J., Freed, A.D.: Detailed error analysis for a fractional adams method. Numer. Algorithms **36**, 31–52 (2004)
23. Dong, X., Zhao, D., Yang, B., Han, C.: Fractional-order control of active suspension actuator based on parallel adaptive clonal selection algorithm. J. Mech. Sci. Technol. **30**(6), 2769–2781 (2016)
24. Efe, M.Ö.: Fractional fuzzy adaptive sliding-mode control of a 2-DOF direct-drive robot arm. IEEE Trans. Syst. Man Cybern. Part B (Cybernetics) **38**(6), 1561–1570 (2008)
25. Efe, M.Ö.: Fractional order systems in industrial automation-a survey. IEEE Trans. Ind. Inf. **7**(4), 582–591 (2011)
26. Erbatur, K., Kaynak, M.O., Sabanovic, A.: A study on robustness property of sliding-mode controllers: A novel design and experimental investigations. IEEE Trans. Ind. Electron. **46**(5), 1012–1018 (1999)
27. Feng, Y., Yu, X., Man, Z.: Non-singular terminal sliding mode control of rigid manipulators. Automatica **38**(12), 2159–2167 (2002)
28. Fujii, H.A., Ishijima, S.: Mission function control for deployment and retrieval of a subsatellite. J. Guidance Control Dyn. **12**(2), 243–247 (1989)
29. Gao, W., Hung, J.C.: Variable structure control of nonlinear systems: A new approach. IEEE Trans. Ind. Electron. **40**(1), 45–55 (1993)

30. Giri, D.K., Sinha, M.: Finite-time continuous sliding mode magneto-coulombic satellite attitude control. IEEE Trans. Aerosp. Electron. Syst. **52**(5), 2397–2412 (2016)
31. Hallaj, M.A.A., Assadian, N.: Tethered satellite system control using electromagnetic forces and reaction wheels. Acta Astronautica **117**, 390–401 (2015)
32. Huang, P., Hu, Z., Meng, Z.: Coupling dynamics modelling and optimal coordinated control of tethered space robot. Aerosp. Sci. Technol. **41**, 36–46 (2015)
33. Huang, P., Wang, D., Meng, Z., Zhang, F., Guo, J.: Adaptive postcapture backstepping control for tumbling tethered space robot-target combination. J. Guidance Control Dyn. **39**(1), 150–156 (2016)
34. Huang, Y.S., Sung, C.C.: Implementation of sliding mode controller for linear synchronous motors based on direct thrust control theory. IET Control Theory Appl. **4**(3), 326–338 (2010)
35. Iki, K., Kawamoto, S., Morino, Y.: Experiments and numerical simulations of an electrodynamic tether deployment from a spool-type reel using thrusters. Acta Astronautica **94**(1), 318–327 (2014)
36. Janardhanan, S., Bandyopadhyay, B.: Multirate output feedback based robust quasi-sliding mode control of discrete-time systems. IEEE Trans. Autom. Control **52**(3), 499–503 (2007)
37. Jo, J.M., Lee, J.H., Han, Y.J., Lee, C.Y., Lee, K.S.: Development of propulsion inverter control system for high-speed maglev based on long stator linear synchronous motor. Energies **10**(2), 170 (2017)
38. Jung, W., Mazzoleni, A.P., Chung, J.: Nonlinear dynamic analysis of a three-body tethered satellite system with deployment/retrieval. Nonlinear Dyn. **82**(3), 1127–1144 (2015)
39. Kang, J., Zhu, Z.H., Wang, W., Li, A., Wang, C.: Fractional order sliding mode control for tethered satellite deployment with disturbances. Adv. Space Res. **59**(1), 263–273 (2017)
40. Khan, S.B., Sanmartin, J.R.: Analysis of tape tether survival in LEO against orbital debris. Adv. Space Res. **53**(9), 1370–1376 (2014)
41. Kilbas, A.A., Srivastava, H.M., Trujillo, J.J.: Theory and Applications of Fractional Differential Equations, vol. 204. Elsevier (2006)
42. Kim, S.J., Park, E.J., Jung, S.Y., Kim, Y.J.: Optimum design of end edge in discontinuous armature permanent magnet linear motors for automation transportation systems. Int. J. Precis. Eng. Manuf. **18**(3), 317–323 (2017)
43. Kojima, H., Aoki, H.: Stabilisation of in-plane periodic motion of electrodynamic tether system by combining tether length control and current control. Int. J. Space Sci. Eng. **3**(4), 318–341 (2015)
44. Kojima, H., Sugimoto, T.: Stability analysis of in-plane and out-of-plane periodic motions of electrodynamic tether system in inclined elliptic orbit. Acta Astronautica **65**(3–4), 477–488 (2009)
45. Kojima, H., Fukatsu, K., Trivailo, P.M.: Mission-function control of tethered satellite/climber system. Acta Astronautica **106**, 24–32 (2015)
46. Kumar, K.: Review on dynamics and control of nonelectrodynamic tethered satellite systems. J. Spacecraft Rockets **43**(4), 705–720 (2006)
47. Lanoix, E.L., Misra, A.K., Modi, V.J., Tyc, G.: Effect of electromagnetic forces on the orbital dynamics of tethered satellites. J. Guidance Control Dyn. **28**(6), 1309–1315 (2005)
48. Lee, H., Utkin, V.I.: Chattering suppression methods in sliding mode control systems. Ann. Rev. Control **31**(2), 179–188 (2007)
49. Li, S., Du, H., Yu, X.: Discrete-time terminal sliding mode control systems based on Euler's discretization. IEEE Trans. Autom. Control **59**(2), 546–552 (2013)
50. Li, Y., Chen, Y., Podlubny, I.: Mittag-leffler stability of fractional order nonlinear dynamic systems. Automatica **45**(8), 1965–1969 (2009)
51. Li, Y., Chen, Y., Podlubny, I.: Stability of fractional-order nonlinear dynamic systems: Lyapunov direct method and generalized Mittag-Leffler stability. Comput. Math. Appl. **59**(5), 1810–1821 (2010)
52. Lin, C.J., Yau, H.T., Tian, Y.C.: Identification and compensation of nonlinear friction characteristics and precision control for a linear motor stage. IEEE/ASME Trans. Mechatron. **18**(4), 1385–1396 (2012)

53. Lin, C.L., Jan, H.Y., Shieh, N.C.: GA-based multiobjective PID control for a linear brushless DC motor. IEEE/ASME Trans. Mechatron. **8**(1), 56–65 (2003)
54. Liu, J., Vazquez, S., Wu, L., Marquez, A., Gao, H., Franquelo, L.G.: Extended state observer-based sliding-mode control for three-phase power converters. IEEE Trans. Ind. Electron. **64**(1), 22–31 (2016)
55. Liu, Y., Zhou, J.: Attitude dynamics and thrust control for short tethered sub-satellite in deployment. Proc. Inst. Mech. Eng. Part G: J. Aerosp. Eng. **229**(8), 1407–1422 (2015)
56. Lu, J.G., Chen, Y.Q.: Robust stability and stabilization of fractional-order interval systems with the fractional order α: The $0<\alpha<1$ case. IEEE Trans. Automa. Control **55**(1), 152–158 (2009)
57. Lu, L., Chen, Z., Yao, B., Wang, Q.: Desired compensation adaptive robust control of a linear-motor-driven precision industrial gantry with improved cogging force compensation. IEEE/ASME Trans. Mechatron. **13**(6), 617–624 (2008)
58. Luo, Y., Chen, Y.: Stabilizing and robust fractional order PI controller synthesis for first order plus time delay systems. Automatica **48**(9), 2159–2167 (2012)
59. Ma, H., Wu, J., Xiong, Z.: Discrete-time sliding-mode control with improved quasi-sliding-mode domain. IEEE Trans. Ind. Electron. **63**(10), 6292–6304 (2016)
60. Ma, Z., Huang, P.: Discrete-time sliding mode control for deployment of tethered space robot with only length and angle measurement. IEEE Trans. Aerosp. Electron. Syst. **56**(1), 585–596 (2020)
61. Ma, Z., Huang, P.: Nonlinear analysis of discrete-time sliding mode prediction deployment of tethered space robot. IEEE Trans. Ind. Electron. **68**(6), 5166–5175 (2021)
62. Ma, Z., Sun, G., Cheng, Z., Li, Z.: Pure-tension non-linear sliding mode control for deployment of tethered satellite system. Proc. Inst. Mech. Eng. Part G: J. Aerosp. Eng. **232**(13), 2541–2551 (2018)
63. Ma, Z., Liu, Z., Huang, P.: Fractional-order control for uncertain teleoperated cyber-physical system with actuator fault. IEEE/ASME Trans. Mechatron. **26**(5), 2472–2482 (2021)
64. Ma, Z., Liu, Z., Huang, P., Kuang, Z.: Adaptive fractional-order sliding mode control for admittance-based telerobotic system with optimized order and force estimation. IEEE Trans. Ind. Electron. **69**(5), 5165–5174 (2021)
65. Mantellato, R., Valmorbida, A., Lorenzini, E.: Thrust-aided librating deployment of tape tethers. J. Spacecraft Rockets **52**(5), 1395–1406 (2015)
66. Martinez-Sanchez, M., Pollard, J.E.: Spacecraft electric propulsion-an overview. J. Propul. Power **14**(5), 688–699 (1998)
67. Matignon, D.: Stability results for fractional differential equations with applications to control processing. Comput. Eng. Syst. Appl. Citeseer **2**, 963–968 (1996)
68. Matsuda, K., Fujii, H.: H (infinity) optimized wave-absorbing control-analytical and experimental results. J. Guidance Control Dyn. **16**(6), 1146–1153 (1993)
69. Meng, Z., Huang, P., Guo, J.: Approach modeling and control of an autonomous maneuverable space net. IEEE Trans. Aerosp. Electron. Syst. **53**(6), 2651–2661 (2017)
70. Meng, Z., Wang, B., Huang, P.: A space tethered towing method using tension and platform thrusts. Adv. Space Res. **59**(2), 656–669 (2017)
71. Monje, C.A., Chen, Y., Vinagre, B.M., Xue, D., Feliu-Batlle, V.: Fractional-Order Systems and Controls: Fundamentals and Applications. Springer Science & Business Media (2010)
72. Mozayan, S.M., Saad, M., Vahedi, H., Fortin-Blanchette, H., Soltani, M.: Sliding mode control of PMSG wind turbine based on enhanced exponential reaching law. IEEE Trans. Ind. Electron. **63**(10), 6148–6159 (2016)
73. Moze, M., Sabatier, J., Oustaloup, A.: Lmi tools for stability analysis of fractional systems. Int. Des. Eng. Tech. Conf. Comput. Inf. Eng. Conf. **47438**, 1611–1619 (2005)
74. Mu, S., Chai, J., Sun, X., Wang, S.: A variable pole pitch linear induction motor for electromagnetic aircraft launch system. IEEE Trans. Plasma Sci. **43**(5), 1346–1351 (2015)
75. Oustaloup, A., Levron, F., Mathieu, B., Nanot, F.M.: Frequency-band complex noninteger differentiator: characterization and synthesis. IEEE Trans. Circuits Syst. I: Fundam. Theory Appl. **47**(1), 25–39 (2000)

76. Papadopoulos, K., Chang, C., Drobot, A.: Ion reflection by the TSS-1R satellite. Geophys. Res. Lett. **25**(5), 745–748 (1998)
77. Podlubny, I.: Fractional Differential Equations. Academic, New York (1999)
78. Pradeep, S.: A new tension control law for deployment of tethered satellites. Mech. Res. Commun. **24**(3), 247–254 (1997)
79. Qi, R., Misra, A.K.: Dynamics of double-pyramid satellite formations interconnected by tethers and coulomb forces. J. Guidance Control Dyn. **39**(6), 1265–1277 (2016)
80. Rivero, M., Rogosin, S.V., Tenreiro Machado, J.A., Trujillo, J.J.: Stability of fractional order systems. Mathematical Problems in Engineering (2013)
81. Robitaille, T.P., Whitney, B.A., Indebetouw, R., Wood, K., Denzmore, P.: Interpreting spectral energy distributions from young stellar objects. i. a grid of 200,000 YSO model SEDs. Astrophys. J. Suppl. Ser. **167**(2), 256 (2006)
82. Shi, P., Xia, Y., Liu, G., Rees, D.: On designing of sliding-mode control for stochastic jump systems. IEEE Trans. Autom. Control **51**(1), 97–103 (2006)
83. Shi, P., Liu, M., Zhang, L.: Fault-tolerant sliding-mode-observer synthesis of Markovian jump systems using quantized measurements. IEEE Trans. Ind. Electron. **62**(9), 5910–5918 (2015)
84. Shieh, N., Tung, P., Lin, C.: Robust output tracking control of a linear brushless DC motor with time-varying disturbances. IEE Proc.-Electr. Power Appli. **149**(1), 39–45 (2002)
85. Su, W.C., Drakunov, S.V., Ozguner, U.: An O (T^2) boundary layer in sliding mode for sampled-data systems. IEEE Trans. Autom. Control **45**(3), 482–485 (2000)
86. Su, X., Yang, X., Shi, P., Wu, L.: Fuzzy control of nonlinear electromagnetic suspension systems. Mechatronics **24**(4), 328–335 (2014)
87. Sun, L., Zhao, G., Huang, H.: Stability and control of tethered satellite with chemical propulsion in orbital plane. Nonlinear Dyn. **74**(4), 1113–1131 (2013)
88. Utkin, V.: Variable structure systems with sliding modes. IEEE Trans. Autom. control **22**(2), 212–222 (1977)
89. Vinagre, B., Podlubny, I., Hernandez, A., Feliu, V.: Some approximations of fractional order operators used in control theory and applications. Fractional Calculus Appl. Anal. **3**(3), 231–248 (2000)
90. Wang, Y., Gu, L., Xu, Y., Cao, X.: Practical tracking control of robot manipulators with continuous fractional-order nonsingular terminal sliding mode. IEEE Trans. Ind. Electron. **63**(10), 6194–6204 (2016)
91. Wang, Y., Luo, G., Gu, L., Li, X.: Fractional-order nonsingular terminal sliding mode control of hydraulic manipulators using time delay estimation. J. Vib. Control **22**(19), 3998–4011 (2016)
92. Wen, H., Zhu, Z.H., Jin, D., Hu, H.: Constrained tension control of a tethered space-tug system with only length measurement. Acta Astronautica **119**, 110–117 (2016)
93. Wen, H., Zhu, Z.H., Jin, D., Hu, H.: Space tether deployment control with explicit tension constraint and saturation function. J. Guidance Control Dyn. **39**(4), 916–921 (2016)
94. Williams, P.: A review of space tether technology. Recent Patents Space Technol. **2**(1), 22–36 (2012)
95. Williams, P., Hyslop, A., Stelzer, M., Kruijff, M.: YES2 optimal trajectories in presence of eccentricity and aerodynamic drag. Acta Astronautica **64**(7–8), 745–769 (2009)
96. Wu, L., Gao, H., Wang, C.: Quasi sliding mode control of differential linear repetitive processes with unknown input disturbance. IEEE Trans. Ind. Electron. **58**(7), 3059–3068 (2010)
97. Wu, L., Zheng, W.X., Gao, H.: Dissipativity-based sliding mode control of switched stochastic systems. IEEE Trans. Autom. Control **58**(3), 785–791 (2012)
98. Xu, S., Sun, G., Ma, Z., Li, X.: Fractional-order fuzzy sliding mode control for the deployment of tethered satellite system under input saturation. IEEE Trans. Aerosp. Electron. Syst. **55**(2), 747–756 (2018)
99. Yang, J., Li, S., Su, J., Yu, X.: Continuous nonsingular terminal sliding mode control for systems with mismatched disturbances. Automatica **49**(7), 2287–2291 (2013)
100. Yin, C., Chen, Y., Zhong, S.M.: Fractional-order sliding mode based extremum seeking control of a class of nonlinear systems. Automatica **50**(12), 3173–3181 (2014)

101. Yin, C., Cheng, Y., Chen, Y., Stark, B., Zhong, S.: Adaptive fractional-order switching-type control method design for 3D fractional-order nonlinear systems. Nonlinear Dyn. **82**(1), 39–52 (2015)
102. Yin, C., Stark, B., Chen, Y., Sm, Zhong, Lau, E.: Fractional-order adaptive minimum energy cognitive lighting control strategy for the hybrid lighting system. Energy Build. **87**, 176–184 (2015)
103. Yin, C., Huang, X., Chen, Y., Dadras, S., Sm, Zhong, Cheng, Y.: Fractional-order exponential switching technique to enhance sliding mode control. Appl. Math. Model. **44**, 705–726 (2017)
104. Young, K.D., Utkin, V.I., Ozguner, U.: A control engineer's guide to sliding mode control. IEEE Trans. Control Syst. Technol. **7**(3), 328–342 (1999)
105. Zakrzhevskii, A.: Method of deployment of a space tethered system aligned to the local vertical. J. Astronaut. Sci. **63**, 221–236 (2016)
106. Zhang, B., Pi, Y., Luo, Y.: Fractional order sliding-mode control based on parameters auto-tuning for velocity control of permanent magnet synchronous motor. ISA Trans. **51**(5), 649–656 (2012)
107. Zhang, J., Shi, P., Lin, W.: Extended sliding mode observer based control for Markovian jump linear systems with disturbances. Automatica **70**, 140–147 (2016)
108. Zhang, Z., Zhou, H., Duan, J., Kou, B.: Design and analysis of a new ring winding structure for permanent magnet linear synchronous motors. IEEE Trans. Plasma Sci. **44**(12), 3311–3321 (2016)
109. Zhao, Y., Huang, P., Zhang, F.: Dynamic modeling and super-twisting sliding mode control for tethered space robot. Acta Astronaut. **143**, 310–321 (2018)
110. Zheng, J., Wang, H., Man, Z., Jin, J., Fu, M.: Robust motion control of a linear motor positioner using fast nonsingular terminal sliding mode. IEEE/ASME Trans. Mechatron. **20**(4), 1743–1752 (2014)
111. Zheng, W., Luo, Y., Wang, X., Pi, Y., Chen, Y.: Fractional order $PI^{\lambda}D^{\mu}$ controller design for satisfying time and frequency domain specifications simultaneously. ISA Trans. **70** (2017)
112. Zhong, R., Wang, Y.: Research on the tether assisted observation of an asteroid. Acta Astronautica **123**, 310–319 (2016)
113. Zhong, R., Zhu, Z.: Long term dynamics and optimal control of nano-satellite deorbit using a short electrodynamic tether. Adv. Space Res. **52**(8), 1530–1544 (2013)

Part I
Brief View of Fractional-Order Control Strength in Modelling and Control

Chapter 2
Fractional-Order Dynamics and Control of Rigid-Flexible Coupling Space Structures

2.1 Introduction

Rigid-flexible coupling structures have been widely used in spacecraft with flexible attachments, robotic manipulators, helicopter blades, and other flexible mechanical/structural systems [8, 9, 11]. Generally, the damping effect generated by the flexible components in these structures is used to improve the overall performance of the systems under dynamic loads by suppressing residual vibration [15]. However, a challenge arises when the precision positioning and motion control of the rigid-flexible coupling structures are attempted, where the residual vibration may occur due to the structural inertia and time delay resulting from the flexible components of structures [13, 14]. The problem is further complicated by viscoelasticity in the flexible components of structures.

It is well known that viscoelasticity is characterized by the combination of elastic springs and viscous fluids, represented by Hooke's elasticity and Newtonian viscosity, respectively. Currently, most existing works modeled viscoelasticity by the Kelvin-Voigt damping model that compromises between complicacy and accuracy [19, 21]. However, with the rapid development of advanced space materials, the Kelvin-Voigt model was found inadequate in describing composite materials, especially for space composite materials [3]. The need for a more refined model has pushed the development of viscoelasticity. In fact, as early as in the 1920s, Nutting [18] found that the viscous deformation of solids, which is always interpreted as the stress history under strain, could be best described by a Fo power law decay. Nutting's work attracted little attention due to the mathematical complexity of FO calculus and the associated extensive computational effort until the pioneering work on the FO Maxwell fluid by Caputo in 1960s [5] who made the FO calculus applicable for real engineering problems. Since then, it has been proved that the FO calculus can describe the dynamic response of viscoelastic flexible systems more accurately than its classical integer-order counterpart due to its unique time memory capacity [19]. Nowadays, FO calculus has been proven a powerful method in describing the

G. Sun et al., *Fractional-Order Sliding Mode Control: Methodologies and Applications*,
Studies in Systems, Decision and Control 532,
https://doi.org/10.1007/978-3-031-60847-6_2

dynamics of various materials, especially viscoelastic materials for vibration control [20]. More details of the related works can be found in [10, 19, 20]. However, there are no clear criteria for the selection of viscous models. The actual model is usually optimized by trial and error for different materials [16].

The control of the rigid-flexible coupling structures is addressed extensively by many control strategies and modeling schemes in literature [12, 17, 24]. However, the dynamic models of these works were generally derived from the Newtonian or Lagrangian mechanics with the Kelvin-Voigt or Rayleigh damping models. Even though the FO viscous model provides an excellent alternative in describing the high stiffness and lightweight composite materials, the FO control strategy is seldom used to control rigid-flexible coupling structures [2, 22].

In this chapter, we propose a two-parameter FO model of viscoelastic material and derive the FO dynamics of rigid-flexible coupling structures to simplify the modeling and controller design for residual vibration suppression. A new FO lead-lag compensator design method is proposed based on the FO dynamics. Simulation results demonstrate the effectiveness and robustness of the FO dynamics and control for the rigid-flexible coupling structures.

2.2 Fractional-Order Constitutive Law for Viscoelastic Material

As early as the 1920s, Nutting [18] proposed firstly a FO constitutive law for viscoelastic materials. Later, it was presented in terms of the elastic modulus of the material,

$$G(t) = \frac{G_\beta}{\Gamma(1-\beta)} t^{-\beta}, 0 \leq \beta \leq 1, \tag{2.1}$$

where $\Gamma(\cdot)$ is the gamma function, G_β is the nominal shear modulus, t is the time, and β is the FO power representing the viscoelastic characteristics of the material. In fact, Hooke's law and Newtonian fluid model provide two mathematical models for pure elastic solids and pure viscous fluids, respectively. Neither of them exists in the real world. All real materials generally exhibit properties between the pure elastic solids ($G(t) = G_\beta$ at $\beta = 0$ in (2.1)) and the pure viscous fluids ($G(t) = 0$ at $\beta = 1$ in (2.1)) [7].

In a viscoelastic medium, the shear stress depends on the history of the shear strain rate, such that,

$$\tau(t) = \int_0^t G(t-\xi)\dot{\gamma}(\xi)d\xi, \tag{2.2}$$

where γ is the shear strain, the overhead dot denotes the time derivative.

Based on Caputo's definition of the FO derivative, the shear stress can be expressed in the term of the FO derivative by substituting (2.1) into (2.2),

$$\begin{aligned}\tau(t) &= \frac{G_\beta}{\Gamma(1-\beta)}\int_0^t \frac{\dot{\gamma}(\xi)}{(t-\xi)^\beta}d\xi \\ &= G_\beta\left({}_0^cD_t^\beta\right)\gamma(t) = G_\beta\left(D_t^\beta\right)\gamma(t),\end{aligned} \tag{2.3}$$

where the symbol ${}_0^cD_t^\beta$ is the Caputo's FO derivative of time t for a given initial time $t_0 = 0$. In the rest of this work, ${}_0^cD_t^\beta$ is abbreviated as D_t^β for simplicity. By using Fourier transform with zero initial condition, the FO constitutive law in the frequency domain can be expressed as $\tau(j\omega) = G_\beta \cdot (j\omega)^\beta \cdot \gamma(j\omega)$.

It should be noted that the two-parameter constitutive law described in (2.3) is the simplest model in describing viscoelastic behaviors. By properly adjusting the parameters, it is able to model accurately the constitutive behavior of a viscoelastic material in a specified frequency range. Taking the Hubble Telescope Mission 3 as an example, the fundamental bending mode of wings is between 1.2 and 1.6 Hz with 2.3% of critical damping [15]. The corresponding equivalent damping parameter η of the classical Kelvin-Voigt model is about 0.0046 $\sim$ 0.0061. As a comparison, the magnitude of frequency responses of the classical Kelvin-Voigt model with a damping parameter $\eta = 0.005$ and $\tau(j\omega)/[G_\beta\gamma(j\omega)] = 1 + \eta(j\omega)$ is plotted in solid line in Fig. 2.1. Meanwhile, by using curve fitting technology, the FO constitutive law in (2.3) with a fractional order $\beta = 0.002129$ and $\tau(j\omega)/[G_\beta\gamma(j\omega)] = 0.9966(j\omega)^\beta$ is plotted in dash line. It shows clearly that the FO constitutive law fits with the classical Kelvin-Voigt model in the given frequency range (1Hz $\sim$ 2Hz) except for the different curvature, the FO constitutive law instead of Kelvin-Voigt model coincides with the experimental data shown in Ref. [15]. Thus, we may conclude that the two-parameter FO constitutive law provides a simple and effective model for modeling space structures.

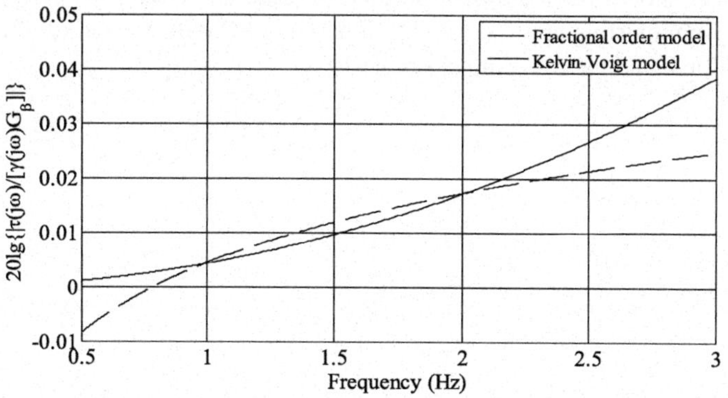

Fig. 2.1 Magnitude of the frequency response of normalized constitutive laws

2.3 Fractional-Order Dynamics of Rigid-Flexible Coupling Structures

2.3.1 Dynamics of Rigid-Flexible Coupling Structures

Many attitude dynamics and control problems of spacecraft with rigid-flexible coupling structures can be idealized to a fundamental problem shown in Fig. 2.2. The spacecraft's main body is simplified to a rigid disc with a moment of inertia of I_1. It is connected to an equivalent flexible shaft of radius R and length l. The materials of the shaft and disc are assumed to be homogeneous and isotropic. The control torque T_1 and react torque T_2 acting on the rigid disc are shown in Fig. 2.2. The rigid disc may experience residual vibration after applying the control torque T_1 due to the flexibility and inertia of the shaft. For the shaft with a circular cross-section, the internal torque $T(t,x)$ can be related to the rotational deformation $\theta(t,x)$ of the shaft by the FO constitutive law in (2.3), such that, $T(t,x) = G_\beta J\left(D_t^\beta\right)(\partial\theta(t,x)/\partial x)$. Thus, the torsional equilibrium equation of the flexible shaft is expressed as

$$\rho J\frac{\partial^2\theta(t,x)}{\partial t^2} = \frac{dT(t,x)}{dx} = \frac{d\left\{G_\beta J\left(D_t^\beta\right)(\partial\theta(t,x)/\partial x)\right\}}{dx}, \tag{2.4}$$

where ρ and J are the material density and polar moment of the shaft, respectively. Assume the rotational displacement $\theta(t,x)$ can be represented by two independent variables x and t, which represent the spatial and time variables, respectively. Then, (2.4) leads to

$$\rho\frac{\partial^2\theta(t,x)}{\partial t^2} = G_\beta\left({}_cD^\beta\right)\left(\frac{\partial^2\theta(t,x)}{\partial x^2}\right). \tag{2.5}$$

(2.5) is the FO wave equation for rigid-flexible coupling structures with viscoelastic materials. It should be noted that there is no stand-alone damping term, which is different from existing approaches [1]. In fact, the FO dynamics in (2.5) covers the damping force and the restoring elastic force.

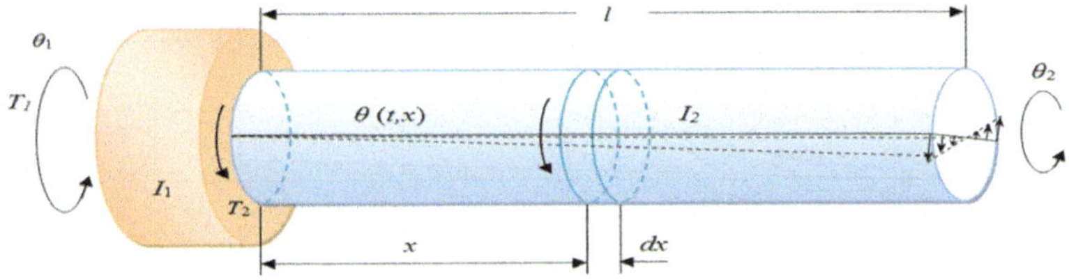

Fig. 2.2 Simplified fundamental model

By applying the Laplace transform to (2.5) in the time domain with zero initial condition and defining a new pseudo-phase velocity of the torsional wave in the shaft as $v = \sqrt{G_\beta/\rho}$ (it becomes the phase velocity in case of pure elastic solid when $\beta = 0$), we derive the following FO modal equation of the shaft

$$\left(\frac{\partial^2\theta(s,x)}{\partial x^2}\right) - \frac{s^{2-\beta}}{v^2}\theta(s,x) = 0. \tag{2.6}$$

The general solution concerning the spatial scale x of (2.6) can be determined by boundary conditions.

2.3.2 Fractional-Order Transfer Function

For the rigid-flexible coupling structure shown in Fig. 2.2, the continuity condition for rotational displacement and torque at the disc/shaft interface and the free-end condition lead to

$$\theta(s,x)|_{x=0} = \theta_1(s), \quad \left.\frac{d\theta(s,x)}{dx}\right|_{x=l} = 0. \tag{2.7}$$

Thus, the general solution can be stated as follows

$$\theta(s,x) = \frac{\theta_1(s)e^{-l\lambda}}{e^{l\lambda}+e^{-l\lambda}}e^{\lambda x} + \frac{\theta_1(s)e^{l\lambda}}{e^{l\lambda}+e^{-l\lambda}}e^{-\lambda x}, \tag{2.8}$$

where $\lambda = s^{1-\beta/2}/v$. The torsional equilibrium equation of the rigid disc can be written as

$$I_1\frac{d^2\theta_1(t)}{dt^2} = T_1 + T_2, \tag{2.9}$$

where $\theta_1(t)$ is the rotational angle of the disc. Accordingly, the torque T_2 is now written as

$$\begin{aligned} T_2(s) &= G_\beta J s^\beta \left.\frac{\partial\theta(s,x)}{\partial x}\right|_{x=0} = -G_\beta J s^\beta \theta_1(s)\lambda\tanh(l\lambda) \\ &= -G_\beta J\theta_1(s)\frac{s^{1+\beta/2}}{v}\tanh\left(\frac{l}{v}s^{1-\beta/2}\right). \end{aligned} \tag{2.10}$$

By applying the Laplace transform to (2.9) and substituting (2.10) into (2.9), we derive the FO transfer function of the disc from $\theta_1(s)$ to $T_1(s)$, such that,

$$\frac{\theta_1(s)}{T_1(s)} = \frac{1}{I_1 s^2 + G_\beta J \frac{s^{1+\beta/2}}{v} \tanh(\frac{l}{v} s^{1-\beta/2})}. \tag{2.11}$$

Consider that $I_2 = \rho l J$ and define the pseudo-fundamental or natural frequency of the flexible shaft as $\omega_{nf} = \sqrt{(G_\beta J/l)/I_2} = v/l$. Then, the transfer function can be further simplified

$$\frac{\theta_1(s)}{T_1(s)} = \frac{1}{I_1 s^2} \times \frac{1}{1 + \frac{I_2}{I_1} \cdot \frac{\tanh(s^{1-\beta/2}/\omega_{nf})}{s^{1-\beta/2}/\omega_{nf}}}, \tag{2.12}$$

where l is the length of the shaft. Note that ω_{nf} will be the true fundamental frequency of the flexible shaft for pure elastic solid at $\beta = 0$.

It is worth noting that the transfer function in (2.12) consists of two parts. The first is the integer-order transfer function of the rigid disc $1/I_1 s^2$. The second is the coupling effect from the flexible shaft, which is FO due to the FO constitutive law. It shows clearly that the residual vibration of the structure depends on the inertia ratio (I_2/I_1) and the pseudo-natural frequency of the shaft.

If there is no flexible shaft ($I_2/I_1 = 0$) or the inertia of the shaft is negligible compared with the disc ($I_2/I_1 << 1$), (2.12) is reduced to the transfer function of the rigid disc itself $1/\left(I_1 s^2\right)$. If the shaft is rigid or very stiff ($s^{1-\beta/2}/\omega_{nf} \approx 0$), (2.12) is reduced to the transfer function of two connected rigid bodies $1/\left[(I_1 + I_2)\, s^2\right]$. In order to illustrate the frequency responses of the FO system more clearly, the ideal Bode plots of the system are shown in Figs. 2.3, 2.4 and 2.5, where $1/s^\nu$ is the straight line with a constant slope -20μ dB/dec, and the phase is $-\nu\pi/2$ at infinity. Three limit cases are discussed in the following

(1) No or small flexible shaft: $I_2/I_1 = 0$ or $<< 1$
The Bode magnitude plot with different inertia ratios of (I_2/I_1) is shown in Fig. 2.3, where the shaft material is assumed as the pure elastic solid ($\beta = 0$). The shaft's natural frequency is assumed as $\omega_{nf} = 5$ Hz. Figure 2.3 shows many vibration modes in the given frequency range due to the flexibility of the shaft. The modal magnitudes and modal frequency spacing will be reduced as the ratio I_2/I_1 decreases, as shown in the inserts in Fig. 2.3. The curve with a smaller ratio I_2/I_1 is closer to the ideal transfer function described by $1/\left(I_1 s^2\right)$ than that of the larger one.

(2) Low frequency response: $s^{1-\beta/2}/\omega_{nf} << 1$
In this case, the frequency response of the disc is approximately independent of the pseudo-natural frequency of shaft in the low-frequency range, as if the shaft is rigid indicated by $1/\left[(I_1 + I_2)\, s^2\right]$. Figure 2.4 shows the Bode magnitude with the assumption of unit inertia ratio ($I_2/I_1 = 1$) and pure elastic solid ($\beta = 0$). The transfer functions for the pseudo-frequencies of $\omega_{nf} = 2$ Hz, $\omega_{nf} = 5$ Hz and rigid shaft are shown in the solid, dashed and dot straight lines, respectively. It shows clearly that the frequency responses of $\omega_{nf} = 2$ and 5 Hz are almost identical to the frequency response of rigid shaft in the low-frequency range of $\omega \in (0, 0.06]$ Hz.

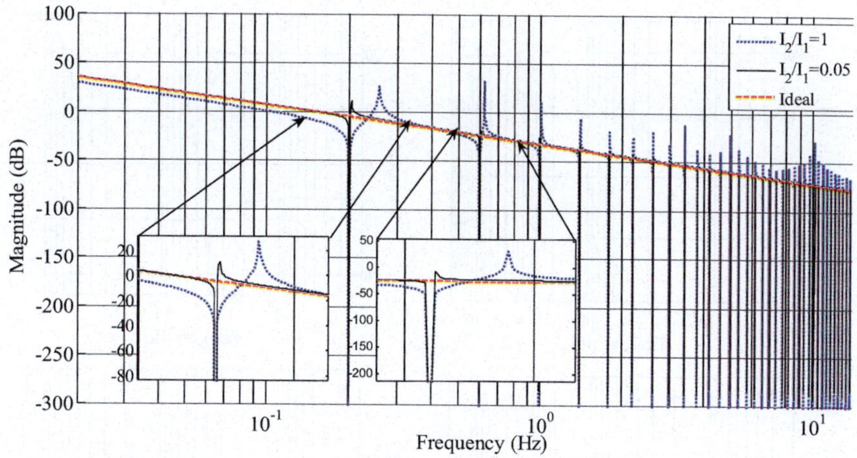

Fig. 2.3 Bode magnitude plot for different ratios of I_2/I_1

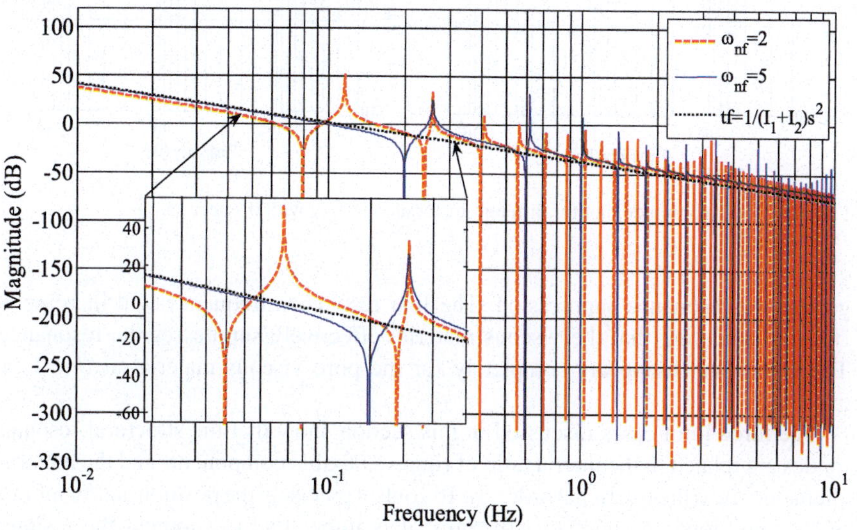

Fig. 2.4 Bode magnitude plot for different natural frequencies ω_{nf}

(3) Viscoelastic shaft: $0 < \beta < 1$
In this case, we focus on the second part of the transfer function in (2.12) with the assumption of unit ratio of I_2/I_1 to understand the impact of the shaft's viscoelasticity on the disc's frequency response. For the given fractional orders $\beta = 0, 0.1, 0.5, 1$ and the pseudo-natural frequency of the shaft $\omega_{nf} = 2\,\text{Hz}$, the ideal Bode plots of $1/\left[1+\omega_{nf}\tanh\left(s^{1-\beta/2}\omega_{nf}\right)s^{1-\beta/2}\right]$ are shown in Fig. 2.5. It clearly shows violent resonances for the pure elastic solid ($\beta = 0$). The resonance will be suppressed due to the viscosity in the shaft once $\beta \neq 0$. The greater the fractional order, the

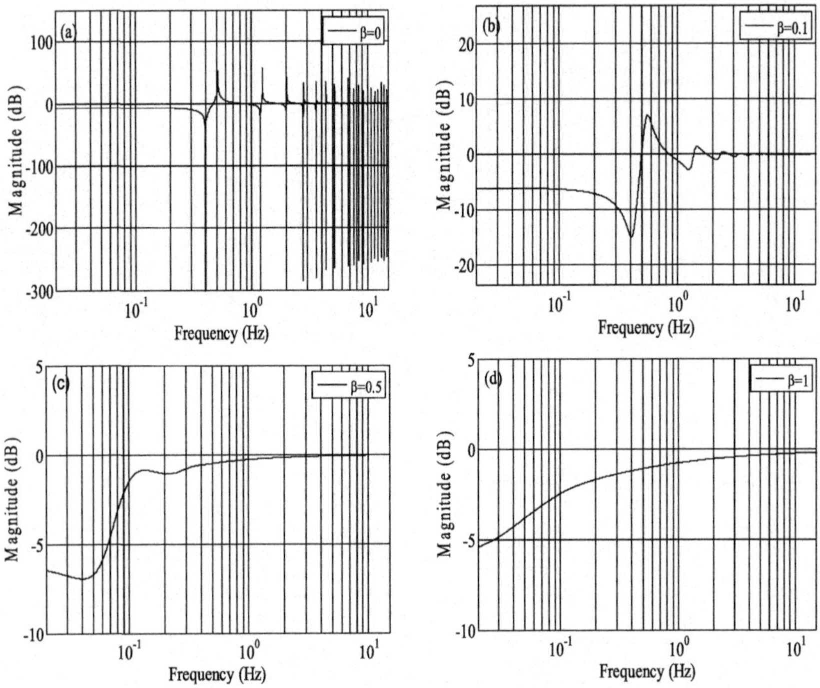

Fig. 2.5 Bode magnitude plot for different material viscosity with $0 < \beta < 1$

greater the resonance suppression. The first resonance frequency will increase as $\beta \to 1$, indicating that the viscous material effectively suppresses the resonance. The resonance disappears completely for the pure viscous material at $\beta = 1$, as expected.

The three limit cases discussed in this section show that the structural resonant modes are related to the inertia ratio of rigid vs flexible components and the material characteristics (the fractional order β). To control precisely the position and/or motion of the rigid-flexible coupling structures, it is imperative to suppress the residual vibrations with a proper controller.

2.4 Fractional-Order Controller Design

There are many control methods to suppress the residual vibration of rigid-flexible coupling systems [12, 22, 24], while only a few approaches have been proposed for the FO system [4]. With the derivation of the FO transfer function in (2.12), the new controller design can take advantage of the explicit coupling form of the transfer function.

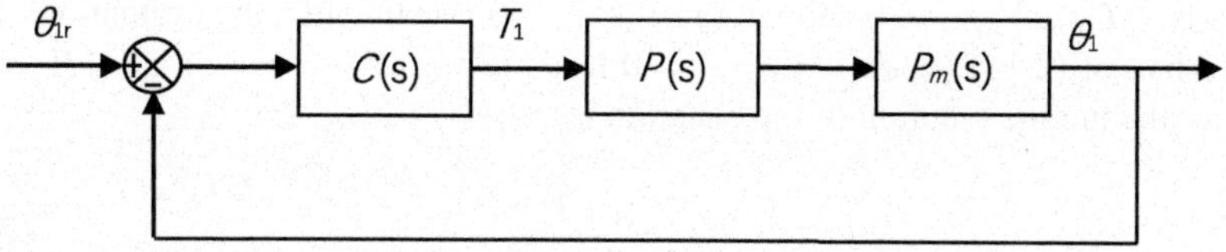

Fig. 2.6 Closed-loop control of the disc-shaft coupling structures

For a typical attitude control problem of flexible spacecraft, the natural frequency of spacecraft structure is usually quite low [15]. The previous section shows that a good performance is hard for an open-loop system given the unstable phase margin and violent resonant effect with $\beta \to 0$. For example, assume the upper bound of the pseudo-natural frequency of the shaft is $\omega_{nf} = 2\,\text{Hz}$, the inertia ratio is $I_2/I_1 = 2$, and the fractional order is $\beta = 0.002$. It will be shown in the following simulations that the bandwidth and the phase margin of the open-loop system are approximately 0.065 Hz and 0, respectively. A closed-loop control for the attitude control is usually adopted to improve the performance while eliminating the resonant effect. The typical closed-loop control of disc-shaft coupling structures is shown in Fig. 2.6, where $C(s)$ is the controller to be designed, $P(s) = 1/\left(I_1 s^2\right)$ is the disc transfer function described in (2.12), $P_m(s)$ is the remaining part of (2.12) that stands for the resonant effect.

The controller should satisfy the following specifications for a typical spacecraft attitude control. First, the designed bandwidth of the closed-loop should be smaller than the resonant frequency, while the curve in the ideal Bode diagram should cross the 0 dB with a slope of –30 dB [23]. Second, both the amplitude and phase stabilization conditions should satisfy the Bode stability criterion. It means that the magnitude of the resonant frequency in the Bode diagram should be as small as possible, at least smaller than the 0 dB line by the controller. At the same time, the phase should be compensated by the designed controller to make the phase margin more significant than 40 at a more comprehensive frequency range.

In this work, a new FO lead-lag compensator design method has been proposed,

$$C(s) = K_c\left(\frac{s + 1/f}{s + 1/(\varepsilon f)}\right)^{\alpha}, 0 < \varepsilon < 1, \tag{2.13}$$

where α is the fractional order of the controller, $1/f$ and $1/(\varepsilon f)$ are the zero and pole of the controller if $\alpha > 0$, respectively. It can be observed that the parameter determines the distance between zero and pole. The compensator is a FO lag compensator if $\alpha > 0$ or a FO lead compensator if $\alpha < 0$ [6]. To evaluate the maximum phase that can be achieved by the controller, the following phase equation is obtained by substituting $s = j\omega$ into (2.13),

$$\angle C(j\omega) = \alpha\left(\angle\left(\frac{1}{f} + j\omega\right) - \angle\left(\frac{1}{\varepsilon f} + j\omega\right)\right). \tag{2.14}$$

Clearly, $\angle C(j\omega)|_{\omega=0} = 0$ and $\angle C(j\omega)|_{\omega=\infty} = 0$. Meanwhile, the evaluation of the derivative of (2.14) $(d/d\omega)\angle C(j\omega) = 0$ leads in $\varepsilon f^2\omega^2 + \varepsilon = \varepsilon^2 f^2\omega^2 + 1$. Since $\varepsilon \neq 1$, the unique solution of the equation is

$$\omega_m = \frac{1}{f\sqrt{\varepsilon}}. \tag{2.15}$$

It is clear that the frequency locates the geometric mean of the corner frequency $1/f$ and $1/(\varepsilon f)$. The result is the same as that of the integer-order case, which is well-known in the control theory. From (2.14) and (2.15), it is concluded that the maximum compensated phase provided by the FO controller is

$$\varphi_m = \alpha\left(\tan^{-1}(\sqrt{\varepsilon}) - \tan^{-1}(1\sqrt{\varepsilon})\right). \tag{2.16}$$

Generally, the FO lead-lag compensator is typically designed based on the knowledge of the plant, the specifications of error constant Kss, the gain crossover frequency ω_{cg} and the associated phase margin ψ_m [6]. However, it is not suitable for the present application since the objective of the present study is to provide a required magnitude and phase margin at a given frequency range to meet the stability condition, not the specification of error. To address the requirement, a new design method has been proposed here. First, for a given gain crossover frequency ω_{cg} or frequency range (ω_L, ω_H), the parameters (λ, ε) can be obtained using any two of ω_L, ω_H and ω_{cg}, based on the fact that the frequency ω_m always coincides with the gain crossover frequency ω_{cg} in order to get the maximum compensated phase. For example, given a frequency range (ω_L, ω_H), setting $1/f = \omega_L$ and $1/(\varepsilon f) = \omega_H$, the expected gain crossover frequency $\omega_{cg} = \omega_m = 1/(f\sqrt{\varepsilon})$. Then, for a specified phase margin ψ_m, the next relationship for the open-loop system is

$$G(j\omega_{cg})K_c\left(\frac{j\omega_{cg} + 1/f}{j\omega_{cg} + 1/(\varepsilon f)}\right)^{\alpha} = e^{i(\pi+\psi_m)}, \tag{2.17}$$

where $G(j\omega_{cg})$ is the phase angle of a general plant model of the controlled system. In the current work, $G(j\omega_{cg}) = P(j\omega_{cg})P_m(j\omega_{cg})$. Finally, the fractional order can be determined by the phase margin relationship

$$\angle G(j\omega_{cg}) + \varphi_m = \pi - \psi_m, \tag{2.18}$$

and the controller gain can be resolved by the magnitude relationship

$$20\log\left[\left|G(j\omega_{cg})\right| K_c \left|\left(\frac{j\omega_{cg} + 1/f}{j\omega_{cg} + 1/(\varepsilon f)}\right)^{\alpha}\right|\right] = 0. \tag{2.19}$$

Based on Eqs. (2.16), (2.18)–(2.19), we conclude that the larger fractional order α will lead to a larger compensated phase and magnitude. For example, given a frequency range (ω_L, ω_H) = (0.02 Hz, 0.2 Hz) and the gain crossover frequency is

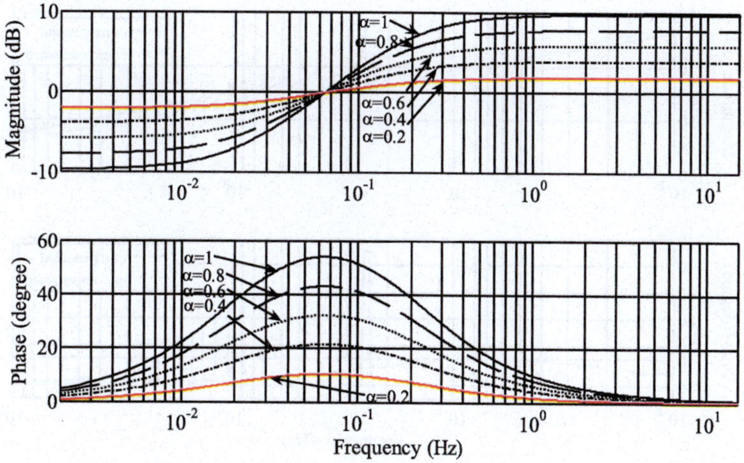

Fig. 2.7 Bode diagram for the FO controller

the same as the open-loop system $\omega_{cg} = 0.065\,\text{Hz}$, a series of Bode diagram of the controller (2.13) is shown in Fig. 2.7.

Figure 2.7 shows that, by keeping the same gain crossover frequency, the compensated phase and magnitude increase as the fractional order increases. The classical integer-order compensator can be viewed as the case with $\alpha = 1$. For the vibration suppression, there is a minimum value for the fractional order α constrained by (2.16) to satisfy the phase margin requirement while keeping the fractional order as small as possible to reduce the extra compensator on magnitude. This is because the vibration modes will also be increased. The fractional order α compromises phase and magnitude. Next, the proposed controller is applied to the rigid-flexible coupling system shown in Fig. 2.2. For the given frequency $\omega_H = 0.2\,\text{Hz}$, the gain crossover frequency is the same as that of the open-loop system. According to (2.15), the parameters $\omega_{cg} = \omega_m = 0.065$ lead to $\omega_L = 0.0212\,\text{Hz}$. Then, for the typical attitude control problem, the required phase margin ψ_m is assumed to $40°$ and the value of fractional order should be greater than 0.74 by combining (2.18) and (2.16) and $\angle G(j\omega_{cg}) = 179.7172°$. Finally, the gain of the controller $K_c = 2.3$ can be achieved by (2.19) and the designed FO controller can be written as

$$C(s) = 2.3\left(\frac{s + 0.1331}{s + 1.256}\right)^{0.74}. \tag{2.20}$$

By introducing the FO controller in (2.20), the ideal Bode diagram of the compensated open-loop transfer function $C(s)P(s)P_m(s)$ and the closed-loop transfer function $C(s)P(s)P_m(s)/\left[1 + C(s)P(s)P_m(s)\right]$ are shown in Figs. 2.8, and 2.9, respectively, with $\omega_{nf} = 2\,\text{Hz}$, $I_2/I_1 = 2$ and the fractional order $\beta = 0.002$.

The bandwidth of the open-loop control system in Fig. 2.8 (dot line) is approximately equal to 0.065Hz, and the phase is $-180°$. According to the classical control

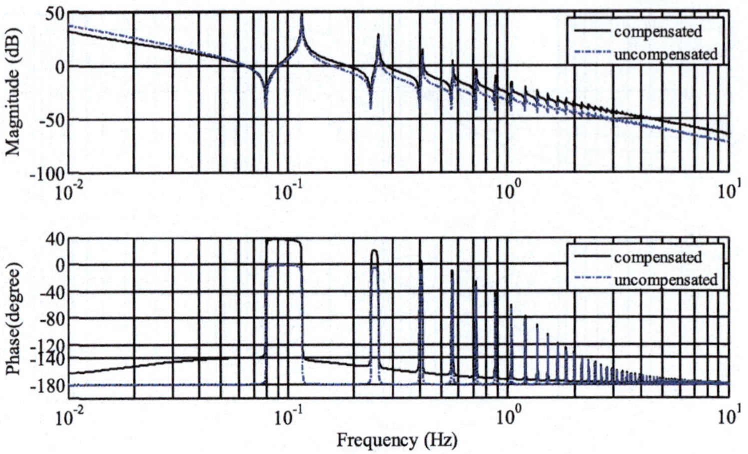

Fig. 2.8 Bode diagram of the compensated open-loop transfer function

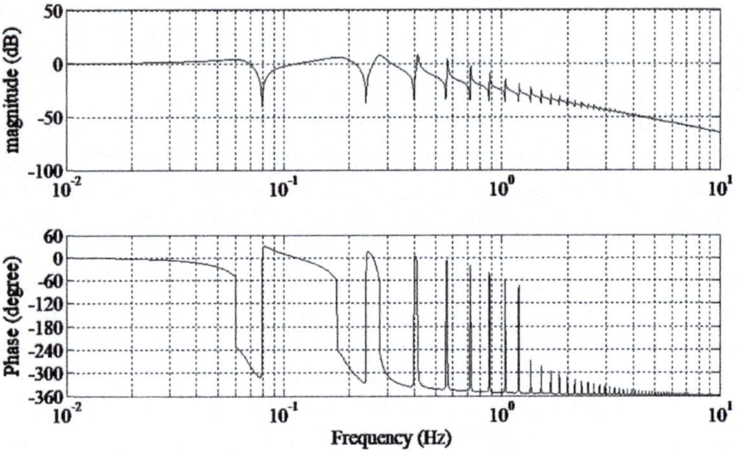

Fig. 2.9 Bode diagram of the closed-loop transfer function

theory, stability and good performance are hard to achieve in the open-loop system due to the unsatisfied magnitude margin and phase margin. Using the proposed controller, the magnitude and phase margin have been compensated to satisfy the stability condition (solid line in Fig. 2.8). The compensated system crosses the 0dB line with a slope more diminutive than the uncompensated open-loop system while keeping the maximum vibration mode almost the same. The phase margin has been compensated to be 40°. The ideal Bode diagram of the closed-loop system in Fig. 2.9 shows that all the magnitudes of resonant modes are below 8dB, indicating that all the resonant modes have been effectively suppressed.

2.5 Conclusion

Based on the FO constitutive law of viscoelastic material, a generic FO dynamics has been derived for general rigid-flexible coupling viscoelastic structures in a uniform framework ranging from purely elastic solids to purely viscous fluids. A new FO transfer function of the coupling structures has been derived by separating the dynamic responses of the rigid and flexible components in a distinguished and coupled form: the transfer function of the rigid component and the FO transfer function of the flexible component. Based on the ideal Bode plot, it has been found that the resonant modes of the coupling structure depend only on the inertia ratio of the rigid component over the flexible component and the natural frequency of the flexible component. Finally, a new FO lead-lag compensator design method is developed to suppress the residual vibration for a given frequency range (ω_L, ω_H) instead of the error constant gain K_{ss} crossover frequency ω_{cg}, and phase margin Φ_m. Numerical results show that the proposed FO controller is adequate and robust in achieving the control requirement.

References

1. Adhikari, S.: Structural Dynamic Analysis with Generalized Damping Models: Analysis. Wiley, New York (2013)
2. Aghababa, M.P.: A fractional-order controller for vibration suppression of uncertain structures. ISA Trans. **52**(6), 881–887 (2013)
3. Amosov, A., Kostin, I., Panasenko, G., Smyshlyaev, V.P.: Homogenization of a thermo-chemo-viscoelastic kelvin-voigt model. J. Math. Phys. **54**(8), 081501 (2013)
4. Bingi, K., Ibrahim, R., Karsiti, M.N., Hassan, S.M., Harindran, V.R., et al.: Fractional-Order Systems and PID Controllers. Springer (2020)
5. Caputo, M.: Linear models of dissipation whose Q is almost frequency independent. Ann. Geophys. **19**(4), 383–393 (1966)
6. Chen, Y.Q., Moore, K.L.: Discretization schemes for fractional-order differentiators and integrators. IEEE Trans. Circuits Syst. I: Fundam. Theory Appl. **49**(3), 363–367 (2002)
7. Di Paola, M., Heuer, R., Pirrotta, A.: Fractional visco-elastic Euler-Bernoulli beam. Int. J. Solids Struct. **50**(22–23), 3505–3510 (2013)
8. Di You, B., Wen, J.M., Zhao, Y.: Nonlinear analysis and vibration suppression control for a rigid-flexible coupling satellite antenna system composed of laminated shell reflector. Acta Astronautica **96**, 269–279 (2014)
9. Dwivedy, S.K., Eberhard, P.: Dynamic analysis of flexible manipulators, a literature review. Mech. Mach. Theory **41**(7), 749–777 (2006)
10. Freed, A., Diethelm, K., Luchko, Y.: Fractional-order viscoelasticity (FOV): Constitutive development using the fractional calculus: First annual report. Technical report, NASA 1.15: 211914 (2002)
11. Hu, Z., Hong, J.: Modeling and analysis of a coupled rigid-flexible system. Appl. Math. Mech. **20**(10), 1167–1174 (1999)
12. Huang, S.T., Davison, E.J., Kwong, R.H.: Decentralized robust servomechanism problem for large flexible space structures under sensor and actuator failures. IEEE Trans. Autom. Control **57**(12), 3219–3224 (2012)
13. Korayem, M.H., Ghariblu, H.: Analysis of wheeled mobile flexible manipulator dynamic motions with maximum load carrying capacities. Robot. Auton. Syst. **48**(2–3), 63–76 (2004)

14. Ma, C., Hori, Y.: Fractional-order control: theory and applications in motion control [past and present]. IEEE Ind. Electron. Mag. **1**(4), 6–16 (2007)
15. Maly, J.R., Pendleton, S.C., Salmanoff, J., Blount, G.J., Mathews, K.: Hubble space telescope solar array damper. In: Smart Structures and Materials 1999: Passive Damping and Isolation, SPIE, vol. 3672, pp. 186–197 (1999)
16. Manabe, S.: A suggestion of fractional-order controller for flexible spacecraft attitude control. Nonlinear Dyn. **29**(1), 251–268 (2002)
17. Mohamed, Z., Tokhi, M.O.: Command shaping techniques for vibration control of a flexible robot manipulator. Mechatronics **14**(1), 69–90 (2004)
18. Nutting, P.: A new general law of deformation. J. Franklin Inst. **191**(5), 679–685 (1921)
19. Podlubny, I.: Fractional Differential Equations. Academic, New York (1999)
20. Pritz, T.: Analysis of four-parameter fractional derivative model of real solid materials. J. Sound Vib. **195**(1), 103–115 (1996)
21. Rutkovskii, V.Y., Sukhanov, V.M., Glumov, V.M.: Stabilization of low-frequency vibrations of a large satellite structure with powered gyro control. Autom. Remote Control **74**(3), 413–425 (2013)
22. Saito, E., Katsura, S.: Vibration control of resonant system by using reflected wave rejection with fractional order low-pass filter. In: 2013 IEEE International Conference on Mechatronics (ICM), pp. 853–858. IEEE (2013)
23. Tustin, A., Allanson, J., Layton, J., Jakeways, R.: The design of systems for automatic control of the position of massive objects. Proc. IEE-Part C: Monographs **105**(1S), 1–57 (1958)
24. Ulrich, S., Sasiadek, J.Z., Barkana, I.: Modeling and direct adaptive control of a flexible-joint manipulator. J. Guidance Control Dyn. **35**(1), 25–39 (2012)

Chapter 3
Fractional-Order Tension Control for Tethered Satellite Systems

3.1 Introduction

The past decades have witnessed 26 tethered satellite missions [34] due to its great potential in space exploration, microgravity, and upper atmospheric studies, electrodynamic propulsion, and deep space observation [6, 15, 28]. A typical tethered satellite mission generally involves tether deployment, state keeping, and tether retrieval, where the tether deployment and retrieval are the most critical stages [3, 6, 32]. The deployment process is generally stable due to the stabilizing effect of the gravity gradient field and can be controlled by an open-loop and linear-exponential control law [10]. However, the existing tension control laws either take a relatively long time to deploy the tether with an overshoot in tether length or employ complex nonlinear feedback control strategies that require precise dynamic models of the space-tethered system. On the other hand, the deployment/retrieval process is inherently unstable [20], especially as the subsatellite is very close to the main satellite. Many control methods have been proposed to deploy/retrieve the tethered satellite in a stable and fast manner in the past decades, such as tension feedback control strategy to retrieve the tether with a windlass mechanism [27, 29], open-loop control strategy to retrieve the tether uniformly or exponentially [10], optimal control [29], delayed feedback control [13], and others [11, 18]. Among them, the tension control is most appealing due to its simplicity, effectiveness, and implementability [30].

The tension control law was first introduced by Rupp [27], where the tether tension was regulated by a predetermined control law to retrieve the subsatellite to the required length while keeping the tether libration as small as possible. Baker et al. [2] applied the tension control law to control the out-of-plane libration through the higher-order modal coupling between the in-plane and out-of-plane libration modes. Pradeep [26] developed a simple tension control law based on the linearization of the equation of in-plane motion of tether in the neighborhood of a system stationary point using the KTC (Kelvin-Talt-Chetayev) theorem for in-plane libration control. Kumar and Pradeep [14] generalized Pradeep's simple tension control law to

G. Sun et al., *Fractional-Order Sliding Mode Control: Methodologies and Applications*, Studies in Systems, Decision and Control 532,
https://doi.org/10.1007/978-3-031-60847-6_3

three-dimensional tethered satellite systems. Modi et al. [21] derived a nonlinear feedback control law that included the out-of-plane tether angular velocity to eliminate the librations. Liaw and Abed [16] proposed a controller based on the Hopf bifurcation theorem. Licata [17] studied a fuzzy logic feedback tension control scheme for the in-plane tether deployment. A detailed review can be found in [31]. Although effective, the existing control schemes cannot simultaneously achieve stable and fast deployment/retrieval of tethered subsatellites. To achieve the balance between stability and efficiency, a memory feedback control is highly desirable to adjust the control performance based on its time history automatically.

FO calculus possesses the historical memory capacity by generalizing the classical integer-order differential and integral calculus to a non-integer order combined differentiation/integration operator, called differintegral. It dates back to the birth of classical integer-order calculus but received little attention due to its inherent mathematic complexity and lack of clear physical meaning. Over the past decades, it has been observed that many physical systems can be described more accurately by FO calculus than classical calculus due to its unique historical memory capacity, for instance, the viscoelastic systems [1], polymeric chemistry systems [12], biomedical systems [19], and signal processing [7], to name a few. FO control theory attracts great interest in dynamics and control since it can provide a better and more robust performance for a closed-loop control system. Over the years, many effective FOC methods have been proposed, such as the CRONE (Crontrôle Robuste d'Ordre Non-Entier) strategy [23], fractional controller [25], and FO lead-lag compensators [22]. Compared with its integer-order counterpart, the FO control has three unique advantages that specifically appeal to space tether deployment/retrieval. First, it is more suitable for modeling flexible structures because of its unique historical memory effect. Second, it provides a practical and clear-cut robust control strategy, especially for distributed parameter systems. Third, it responds to control input faster with smaller overshoots than its integer counterpart.

In this chapter, two FO tension control laws are proposed for the retrieval and deployment of tethered satellites. The nonlinear tether dynamics are linearized in the neighborhood of a given equilibrium point. The linearized dynamic equations become asymptotically stable by introducing the linear tension feedback and FO items. Compared to the existing tension control law, the main advantages of the proposed FO tension control law are as follows

- A better control strategy to overcome the instability experienced by the integer-order tension control laws
- A faster retrieval/deployment time than its integer-order counterparts
- A high efficiency and real-time control law that requires less feedback and is easy to implement in engineer applications.

3.2 Dynamics Behaviors of Tethered Satellite System

3.2.1 Dynamics of Space 3D Tethered Satellite System

A tethered satellite system (TSS) usually consists of a long, thin tether connecting two satellites at its ends. Generally, the TSS can be modeled in five categories [31] depending on the level of detail required in the analysis. To focus our attention on the control strategy, it is assumed that the tether is straight and inextensible, and its mass is negligible compared to the primary satellite. Due to the large ratio of tether length over satellites' dimensions, the satellites are modeled as lumped masses attached to the ends of the tether. Moreover, the center of mass of the TSS is assumed to reside in the main satellite throughout the retrieval process. Based upon the assumptions, the TSS motion can be safely decomposed into the orbital motion of the main satellite around the Earth and the local libration motion of the tether about the main satellite, as shown in Fig. 3.1. It is assumed that the TSS orbits the Earth in a circular orbit, and the subsatellite has no thruster. Then, a local orbital reference frame is defined with the origin at the main satellite. The y-axis of the orbital frame points toward the Earth's center. The x-axis lies in the orbital plane perpendicular to the y-axis and points towards the moving direction of main satellite. The z-axis completes a right-hand coordinate system.

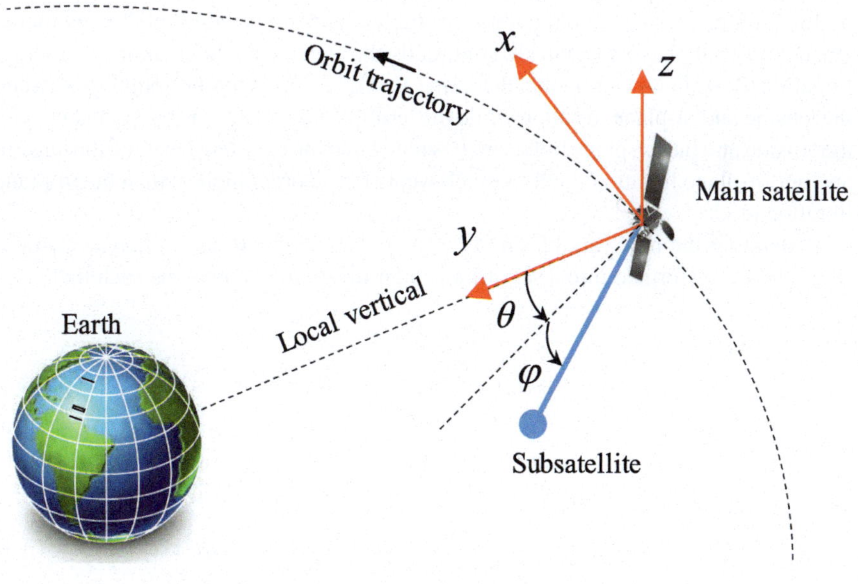

Fig. 3.1 The local orbital reference frame of TSS

Thus, the equations of motion of the TSS without thrust at the subsatellite can be expressed as [11]

$$
\begin{aligned}
&\ddot{l} - l[\dot{\varphi}^2 + \cos^2\varphi(\dot{\theta} + \Omega)^2 - \Omega^2 + 3\Omega^2\cos^2\varphi\cos^2\theta] = -T/m,\\
&\ddot{\theta} + 2[(\dot{l}/l) - \dot{\varphi}\tan\varphi](\dot{\theta} + \Omega) + 3\Omega^2\cos\theta\sin\theta = 0,\\
&\ddot{\varphi} + 2(\dot{l}/l)\dot{\varphi} + \cos\varphi\sin\varphi[(\dot{\theta} + \Omega)^2 + 3\Omega^2\cos^2\theta] = 0.
\end{aligned}
\tag{3.1}
$$

where l is the instant tether length, m is the mass of subsatellite, T is the tension in tether, Ω is the orbit rate of the TSS, θ and ϕ are the in-plane and out-of-plane libration angles, respectively. The dot denotes the derivative with respect to time.

Introduce the following dimensionless variables

$$
\tau = \Omega t,\ \lambda = l/L,\ \hat{T} = T/[m\Omega^2 L]. \tag{3.2}
$$

Equation (3.1) can be rewritten in the dimensionless form

$$
\begin{aligned}
&\lambda - \lambda[\varphi'^2 + \cos^2\varphi(1 + \theta')^2 - 1 + 3\cos^2\theta\cos^2\varphi] = -\hat{T},\\
&\theta + 2[(\lambda'/\lambda) - \varphi'\tan\varphi](1 + \theta') + 3\cos\theta\sin\theta = 0,\\
&\varphi + 2(\lambda'/\lambda)\varphi' + \cos\varphi\sin\varphi[(1 + \theta')^2 + 3\cos^2\theta] = 0,
\end{aligned}
\tag{3.3}
$$

where L is the total length of the tether before retrieval, and the prime denotes the dimensionless time derivative. $\hat{T}$ is the dimensionless tether tension and will be the designed control input in the tension control law. Obviously, this is a single-input-multiple-output system, and the libration angles cannot be controlled directly by the tension. Instead, they are controlled indirectly through the nonlinear modal coupling. Equation (3.3) shows clearly that the term $(\lambda'/\lambda)(1 + \theta')$ is for the coupling between the tension and in-plane libration while the term $(\lambda'/\lambda)\varphi'$ is for the coupling between the tension and out-of-plane libration. It is noted that the coupling between the tension and out-of-plane libration mode is much weaker (one order higher) than the in-plane libration mode.

Redefining the new variables as $x_1 = \lambda$, $x_2 = \lambda'$, $x_3 = \theta$, $x_4 = \theta'$, $x_5 = \varphi$, $x_6 = \varphi'$, (3.3) is transformed into a set of first-order differential equations such that

$$
\begin{aligned}
&{x_1}' = x_2,\\
&{x_2}' = x_1 x_6^2 + x_1(1 + x_4)^2\cos^2 x_5 - x_1 + 3x_1\cos^2 x_5\cos^2 x_3 - \hat{T},\\
&{x_3}' = x_4,\\
&{x_4}' = -2\left[(x_2/x_1) - x_6\tan x_5\right](1 + x_4) - 3\cos x_3\sin x_3,\\
&{x_5}' = x_6,\\
&{x_6}' = -2\,(x_2/x_1)\,x_6 - \cos x_5\sin x_5\left[(1 + x_4)^2 + 3\cos^2 x_3\right].
\end{aligned}
\tag{3.4}
$$

Assume the TSS system is in equilibrium at any point between the completely retrieved and fully deployed states, such that, $x_1 = a$, $x_2 = x_3 = x_4 = x_5 = x_6 = 0$ with $0 < a \leq 1$, where $a = 1$ or 0 means the tethered satellite is fully deployed or completely retrieved. Change the original coordinates to the equilibrium point by defining new coordinates as $y_1 = x_1 - a$, $y_2 = x_2$, $y_3 = x_3$, $y_4 = x_4$, $y_5 = x_5$, $y_6 = x_6$. Then, (3.4) becomes

$$\begin{aligned}
&y_1{}' = y_2, \\
&y_2{}' = (y_1 + a)y_6^2 + (y_1 + a)(1 + y_4)^2\cos^2 y_5 - (y_1 + a) \\
&\qquad + 3(y_1 + a)\cos^2 y_5 \cos^2 y_3 - \hat{T}, \\
&y_3{}' = y_4, \\
&y_4{}' = -2\left[(y_2/y_1 + a) - y_6 \tan y_5\right](1 + y_4) - 3\cos y_3 \sin y_3, \\
&y_5{}' = y_6, \\
&y_6{}' = -2\,(y_2/y_1 + a)\,y_6 - \cos y_5 \sin y_5 \left[(1 + y_4)^2 + 3\cos^2 y_3\right].
\end{aligned} \tag{3.5}$$

Thus, the linearized form of (3.5) in the neighborhood of equilibrium ($y_1 = y_2 = y_3 = y_4 = y_5 = y_6 = 0$) becomes

$$\begin{Bmatrix} y_1{}' \\ y_2{}' \\ y_3{}' \\ y_4{}' \\ y_5{}' \\ y_6{}' \end{Bmatrix} = \begin{bmatrix} 0 & 1 & 0 & 0 & 0 & 0 \\ 3 & 0 & 0 & 2a & 0 & 0 \\ 0 & 0 & 0 & 1 & 0 & 0 \\ 0 & -\frac{2}{a} & -3 & 0 & 0 & 0 \\ 0 & 0 & 0 & 0 & 0 & 1 \\ 0 & 0 & 0 & 0 & -4 & 0 \end{bmatrix} \begin{Bmatrix} y_1 \\ y_2 \\ y_3 \\ y_4 \\ y_5 \\ y_6 \end{Bmatrix} + \begin{Bmatrix} 0 \\ 3a - \hat{T} \\ 0 \\ 0 \\ 0 \\ 0 \end{Bmatrix}, \tag{3.6}$$

where the constant $3a$ is the static tension at the equilibrium.

3.2.2 *Simplified 2D in-Plane Dynamic of Tethered Satellite System*

For satellite control verification, a semi-physical air-breathing satellite platform is the most commonly used equipment. Sometimes the platform is called a zero-gravity test-bench platform (shown in Fig. 3.2). However, the platform has a disadvantage in that it can only verify the controller in the degree perpendicular to the platform surface. Thus, for the tethered system, the in-plane angle dynamic can be viewed as the reduction dynamic model for the last section by setting the out-plane angle $\varphi = 0$ in (3.3). It is much simpler than the 3-D models but is useful for the workbench platform on the ground.

To verify the effectiveness of the controller in the workbench platform, the in-plane dynamic can be easily obtained by

$$\begin{aligned}
&\ddot{\lambda} - \lambda[(1 + \dot{\theta})^2 - 1 + 3\cos^2\theta] = -\hat{T}, \\
&\ddot{\theta} + 2(\dot{\lambda}/\lambda)(1 + \dot{\theta}) + 3\cos\theta \sin\theta = 0,
\end{aligned} \tag{3.7}$$

Fig. 3.2 A kind of air-bearing workbench platform on the ground

with the same definition λ, θ, $\hat{T}$ for (3.3). With the same process, the Eq. (3.5) can be simplified with $a = 1$ as

$$\begin{aligned} y_1{}' &= y_2, \\ y_2{}' &= (1 + y_1)(1 + y_4)^2 - (1 + y_1) + 3(1 + y_1)\cos^2 y_3 - \hat{T}, \\ y_3{}' &= y_4, \\ y_4{}' &= -2\frac{y_2}{y_1+1}(1 + y_4) - 3\cos y_3 \sin y_3. \end{aligned} \tag{3.8}$$

For the deployment problem, the Eq. (3.8) can be linearized in the neighborhood of the equilibrium point by using $a = 1$ such that

$$\begin{Bmatrix} y_1{}' \\ y_2{}' \\ y_3{}' \\ y_4{}' \end{Bmatrix} = \begin{bmatrix} 0 & 1 & 0 & 0 \\ 3 & 0 & 0 & 2 \\ 0 & 0 & 0 & 1 \\ 0 & -2 & -3 & 0 \end{bmatrix} \begin{Bmatrix} y_1 \\ y_2 \\ y_3 \\ y_4 \end{Bmatrix} + \begin{Bmatrix} 0 \\ 3 - \hat{T} \\ 0 \\ 0 \end{Bmatrix}. \tag{3.9}$$

3.3 Fractional-Order Controller for Space Tethered Satellite Retrieval

3.3.1 PID-Liked Simple Tension Control Law for Retrieval

As a benchmark for the effectiveness of the proposed FO tension control law, we first derive a simple tension control law for retrieval as a baseline. Consider the governing equation of the TSS in (3.5). The simplest tension control law in the literature has the following form [14]

$$\hat{T} = k_1 y_1 + k_2 y_2 + k_3 y_3 + k_4 y_4 + k_5 y_5 + k_6 y_6 + k_7, \tag{3.10}$$

where the parameters k_i, $i = 1, 2, \ldots, 7$ are the control gains to be determined by the stability condition and control performance.

Substituting (3.10) into (3.6) leads to

$$\begin{Bmatrix} y_1' \\ y_2' \\ y_3' \\ y_4' \\ y_5' \\ y_6' \end{Bmatrix} = \begin{bmatrix} 0 & 1 & 0 & 0 & 0 & 0 \\ 3-k_1 & -k_2 & -k_3 & 2a-k_4 & -k_5 & -k_6 \\ 0 & 0 & 0 & 1 & 0 & 0 \\ 0 & -\frac{2}{a} & -3 & 0 & 0 & 0 \\ 0 & 0 & 0 & 0 & 0 & 1 \\ 0 & 0 & 0 & 0 & -4 & 0 \end{bmatrix} \begin{Bmatrix} y_1 \\ y_2 \\ y_3 \\ y_4 \\ y_5 \\ y_6 \end{Bmatrix} + \begin{Bmatrix} 0 \\ 3a - k_7 - \hat{T} \\ 0 \\ 0 \\ 0 \\ 0 \end{Bmatrix}. \tag{3.11}$$

For the tethered system in (3.11) to be stable at the equilibrium, there must be $k_7 = 3a$ and all eigenvalues of (3.11) are located in the open left half-plane. The eigenfunction of (3.11) is

$$\begin{aligned} &\begin{vmatrix} p & -1 & 0 & 0 & 0 & 0 \\ k_1-3 & p+k_2 & +k_3 & k_4-2a & k_5 & k_6 \\ 0 & 0 & p & -1 & 0 & 0 \\ 0 & \frac{2}{a} & 3 & p & 0 & 0 \\ 0 & 0 & 0 & 0 & p & -1 \\ 0 & 0 & 0 & 0 & 4 & p \end{vmatrix} \\ &= p^6 + k_2 p^5 + (8 - 2k_4/a + k_1)p^4 + (7k_2 - 2k_3/a)p^3 \\ &+ (7 - 8k_4/a + 7k_1)p^2 + (12k_2 - 8k_3/a)p - 36 + 12k_1 = 0 \end{aligned} \tag{3.12}$$

It is interesting to note that the parameters k_5 and k_6 are not presented in the eigenfunction (3.12). This implies that the stability is independent of the feedback gain k_5 and k_6. In other words, the simple tension control law is completely decoupled from the out-of-plane mode in the linearization process due to the high-order nonlinear coupling nature of the out-of-plane angle φ with the other two variables λ, θ is eliminated. Thus, an out-of-plane libration cannot be controlled effectively alone by the tension feedback. Previous work showed that a thrust was necessary to suppress the libration angle φ to zero quickly in Ref. [18]. It should be noted that this limitation is due to the simple tension control strategy.

The stability condition of (3.11) can be expressed as per Routh-Hurwitz criterion

$$\begin{aligned} &k_2 > 0;\ 1 - \tfrac{2}{a}k_4 + k_1 + \tfrac{2k_3}{ak_2} > 0;\ 3k_1 - 9 > 0, \\ &12k_2 - \tfrac{6}{a}k_2k_4 + \tfrac{4}{a}k_3 - \tfrac{4k_3k_4}{a^2} - \tfrac{2k_3k_1}{a} - \tfrac{4k_3^2}{a^2k_2} > 0. \end{aligned} \tag{3.13}$$

From (3.13), the stable ranges of (k_1, k_2) can be derived as $k_1 > 3$ and $k_2 > 0$, explicitly. The stable range of k_3 can be determined as $k_3 = 0$ by the KTC asymptotical stability theorem. Substituting $k_3 = 0$ into (3.13) leads to

$$k_4 < \frac{a}{2}(1 + k_1). \tag{3.14}$$

Obviously, k_4 will approach zero as the tether is retrieved. For the sake of simplicity in constructing the tension control law, it is assumed $k_4 = 0$ and $k_5 = k_6 = 0$ in the following work. Then, the simple tension control law for the tether retrieval becomes

$$\hat{T} = k_1y_1 + k_2y_2 + 3a, \text{ with } k_1 > 3, k_2 > 0, 0 < a \leq 1. \tag{3.15}$$

The tether can be retrieved accordingly by varying the parameter a from 1 to 0. It should be noted that the simple tension control law cannot be used at $a = 0$ since the linearized system (3.6) becomes singular. Furthermore, the simple tension control law is asymptotically stable at the equilibrium point a due to the application of the KTC theorem and no limitation being imposed on the magnitude of libration angle as the system approaches the equilibrium. As a result, the tether may wind around the main satellite when $a \to 0$, which will be a catastrophic failure. The performance of the control law is illustrated numerically as follows.

Assume that the TSS is running in a circular orbit with a 20 km long tether at an altitude of 220 km. The corresponding orbital rate is 1.1804×10^{-3} rad/s. To circumvent the singularity at $a = 0$, a small value ($a = 0.02$) is used instead in the analysis. The initial condition for retrieval is $y_1 = 1$, $y_2 = 0$, $y_3 = 0.02$, $y_4 = 0$, $y_5 = 0$, $y_6 = 0.02$, where the conditions $y_3 = -0.02$ and $y_5 = 0.02$ are randomly selected to introduce an initial perturbation in libration angles. Furthermore, the control gains (k_1, k_2) are optimized for the tethered satellite to be retrieved within two orbits, with $k_1 = 4$ and $k2 = 2.8$. The system responses are shown in Figs. 3.3, 3.4, 3.5 and 3.6. Figure 3.3 shows that the simple tension control law can retrieve the subsatellite within three orbits. However, the in-plane libration angle shown in Fig. 3.4 increases as the subsatellite approaches the main satellite. After one and a half orbits, the tether length has been reduced to less than 5% of the total length, and the libration angle suddenly increases over 180°, which implies the tether wound around the main satellite or the subsatellite hit the main satellite. This can be best illustrated by the trajectory of the subsatellite in the orbital plane in Fig. 3.5, where

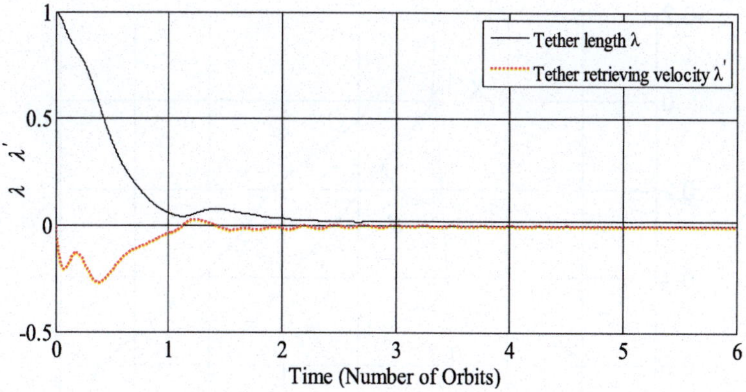

Fig. 3.3 Time history of dimensionless tether length and velocity in retrieval

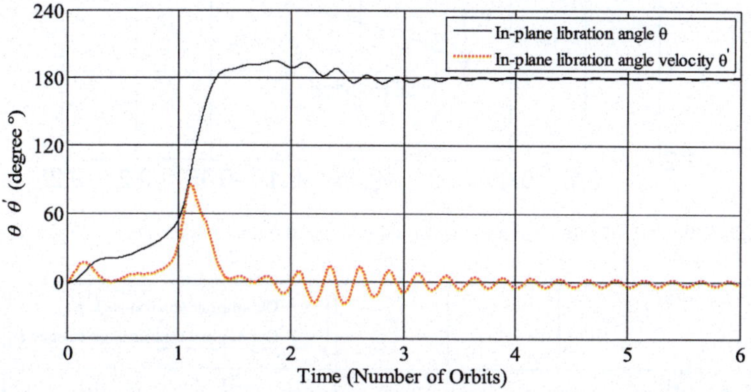

Fig. 3.4 Time history of in-plane libration angle and velocity of retrieval. Note that the unit for the libration velocity is degree per dimensionless time (τ)

the subsatellite starts to wind around the main satellite when the tether length is very short. Moreover, Fig. 3.6 shows violent out-of-plane libration motion for the non-zero initial value $y_5 = 0.02$. Such results are expected, as discussed before.

3.3.2 Improved Fractional-Order Tension Control Law for Retrieval

To retrieve the subsatellite stably and quickly, the existing control strategies commonly employ two-stage control laws: fast retrieval followed by slow retrieval after the tether is less than 1% of its original length [33]. In the current work, a new FO tension control is proposed, which can control the system performance with the infor-

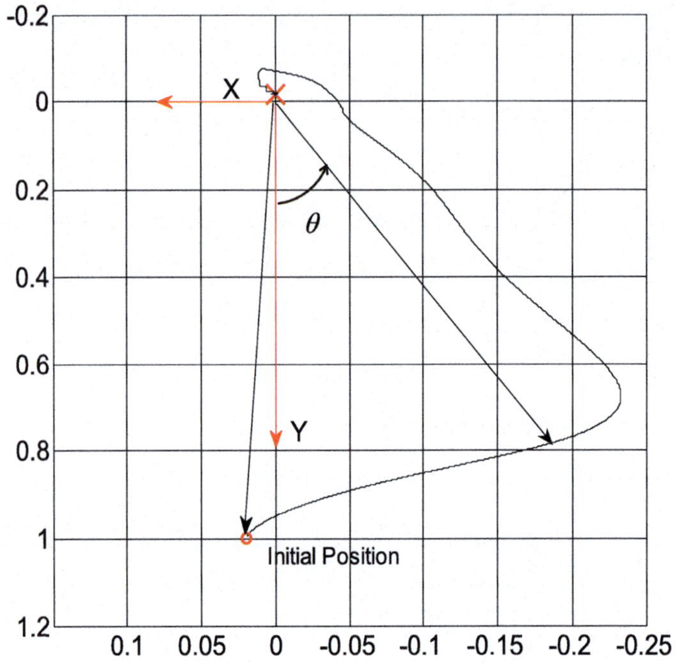

Fig. 3.5 Trajectory of subsatellite in the orbital plane in the retrieval process

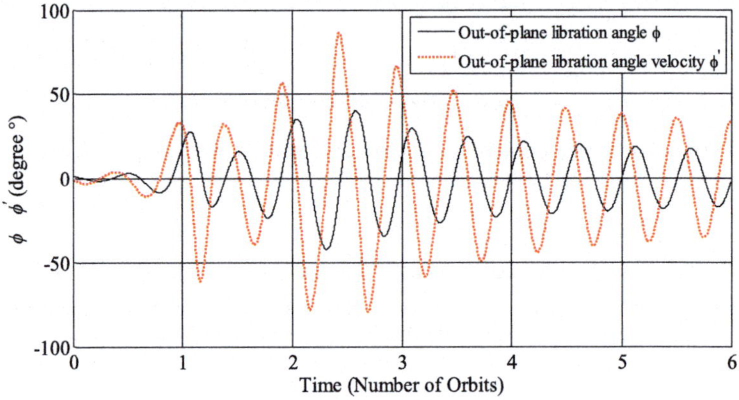

Fig. 3.6 Time history of out-of-plane libration angle and velocity of retrieval. Note that the unit for the libration velocity is degree per dimensionless time (τ)

mation of previous states. Based on the simple tension control law outlined above, the FO tension control law is defined as follows

$$\hat{T} = D^{\alpha} y_2 - y_2{}' + k_1 y_1 + k_2 y_2 + 3a, \tag{3.16}$$

where $D^{\alpha} y_2$ is the Caputo definition of FO differential. Substituting (3.16) into (3.6) leads to a non-commensurate FO system in the neighborhood of the equilibrium point such that

$$\begin{Bmatrix} y_1{}' \\ D^{\alpha} y_2 \\ y_3{}' \\ y_4{}' \\ y_5{}' \\ y_6{}' \end{Bmatrix} = \begin{bmatrix} 0 & 1 & 0 & 0 & 0 & 0 \\ 3-k_1 & -k_2 & 0 & 0.04 & 0 & 0 \\ 0 & 0 & 0 & 1 & 0 & 0 \\ 0 & -100 & -3 & 0 & 0 & 0 \\ 0 & 0 & 0 & 0 & 0 & 1 \\ 0 & 0 & 0 & 0 & -4 & 0 \end{bmatrix} \begin{Bmatrix} y_1 \\ y_2 \\ y_3 \\ y_4 \\ y_5 \\ y_6 \end{Bmatrix}. \tag{3.17}$$

Here, it is assumed that $a = 0.02$ to avoid the singularity at $a = 0$. According to the FO stability result Theorem 1.1 presented in the introduction section, (3.17) can asymptotically approach the equilibrium point for any given FO with properly selected values of (k_1, k_2). Furthermore, Theorem 1.1 indicates that the stable region of a FO system is always larger than its integer-order counterpart. Thus, we choose the same values for (k_1, k_2) as in the integer-order control law in (3.15) to satisfy the stability condition instead of finding the eigenvalues of the FO system in (3.17). By replacing the FO derivative with the FO integral, the FO tension law is written as

$$\begin{aligned} \hat{T} &= D^{\alpha} y_2 - y_2{}' + 4y_1 + 2.8y_2 + 0.06 \\ &= y_2{}'(I^{1-\alpha} y_2 - 1) + 4y_1 + 2.8y_2 + 0.06, \end{aligned} \tag{3.18}$$

and

$$\begin{Bmatrix} y_1{}' \\ D^{\alpha} y_2 \\ y_3{}' \\ y_4{}' \\ y_5{}' \\ y_6{}' \end{Bmatrix} = \begin{bmatrix} 0 & 1 & 0 & 0 & 0 & 0 \\ 3 & 0 & 0 & 0.04 & 0 & 0 \\ 0 & 0 & 0 & 1 & 0 & 0 \\ 0 & -100 & -3 & 0 & 0 & 0 \\ 0 & 0 & 0 & 0 & 0 & 1 \\ 0 & 0 & 0 & 0 & -4 & 0 \end{bmatrix} \begin{Bmatrix} y_1 \\ y_2 \\ y_3 \\ y_4 \\ y_5 \\ y_6 \end{Bmatrix}, \tag{3.19}$$

where $I^{1-\alpha}$ denotes the FO integral operator, which satisfies the equation $D^{1-\alpha}(I^{1-\alpha} y_2) = y_2$.

Next, we will develop a numerical procedure to implement the fraction order controller. Assume the fractional order $\alpha = 0.05$. The FO integral $I^{1-\alpha}$ in (3.19) can be approximated by the Laplace transform and procedures in Chareff's method in Sect. 1.2.2 such that

$$L\left\{I^{0.95}y_2\right\} = \frac{1}{s^{0.95}} = \frac{2.977\times 10^{-7}s^2+0.6159s+78.481}{1.1289\times 10^{-7}s^3+0.3793s^2+78.481s+1}. \quad (3.20)$$

The inclusion of the feedback of derivative y_2' (acceleration) in (3.18) may lead to some unwanted oscillations in the system responses. To avoid possible oscillations, the Laplace transform is modified by the following items,

$$s \leftarrow \frac{s}{T_1 s+1}, \quad (3.21)$$

where T_1 is the parameter to be determined by the control requirement. The current work defines it as $T_1 = 0.2$.

Assume the sampling time step is $Ts = 0.002$, (3.20) is discretized by the Tustin operator [5], $S = \frac{2}{T_s}\left[\frac{1-z^{-1}}{1+z^{-1}}\right]$ such that

$$Z^{-1}\left\{\frac{1}{s^{0.95}}\right\} \approx \frac{0.00151z^3+0.00185z^2-0.000832z-0.001173}{z^3-0.6577z^2-0.999z+0.6567}. \quad (3.22)$$

Hence, the implementation procedure of the FO tension control law in (3.18) is established.

Using the same initial condition employed in the case of the integer-order tension control law, the performance of the newly devised FO control law is shown in Figs. 3.7, 3.8, 3.9, 3.10 and 3.11 in comparison with its integer-order counterpart. The solid lines denote the FO controller, while the dotted lines represent the simple integer-order tension controller. The results show that the FO control provides a more stable and fast retrieval than its integral-order counterpart. The FO control works well

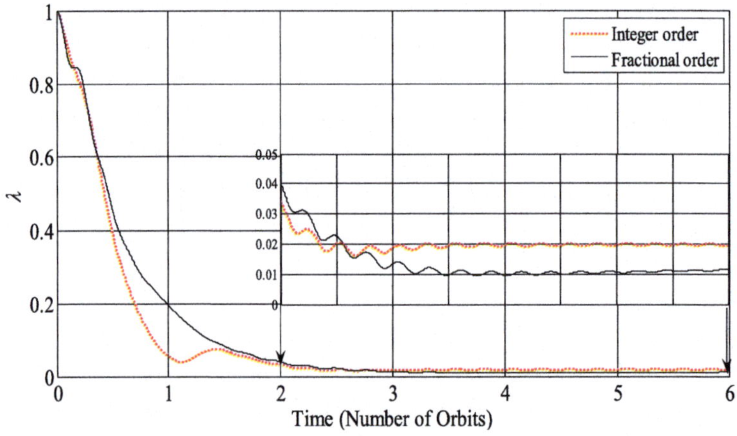

Fig. 3.7 Time history of out-of-plane libration angle and velocity of retrieval. Note that the unit for the libration velocity is degree per dimensionless time (τ)

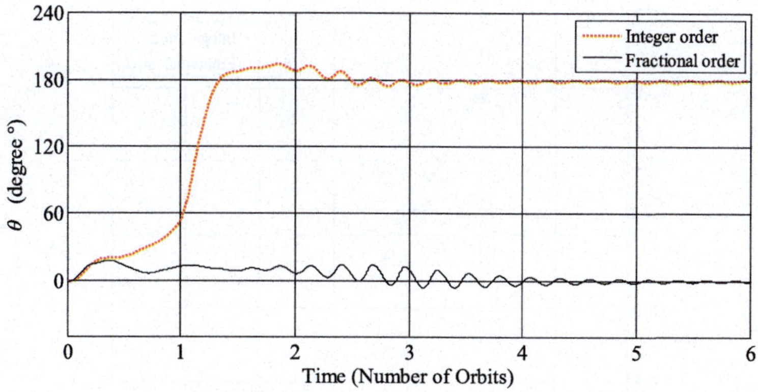

Fig. 3.8 Time history of out-of-plane libration angle and velocity of retrieval. Note that the unit for the libration velocity is degree per dimensionless time (τ)

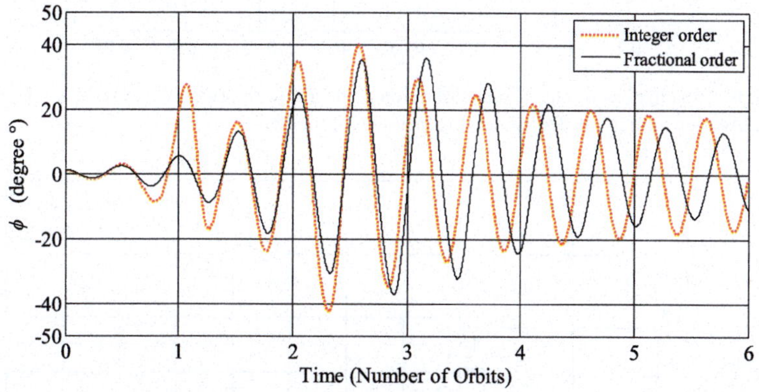

Fig. 3.9 Time history of out-of-plane libration angle and velocity of retrieval. Note that the unit for the libration velocity is degree per dimensionless time (τ)

even near singularity at $a = 0.01$. The tether length reduction is smoother with the FO controller as shown in Fig. 3.7. Figure 3.8 shows the maximum in-plane libration angle is less than 17°. Figure 3.9 illustrates that, since the existence of the nonlinear coupling in the governing equation (3.5), the maximal value of out-of-plane libration (37°) is also smaller than that of the integer-order tension control (42°), and approaches to zero more quickly. Figure 3.10 indicates the subsatellite is retrieved without the tether winding or hitting the main satellite when the tether length is short under the FO control. Finally, Fig. 3.11 shows that the tether tension is also bounded in an acceptable region even though there are some oscillations in the initial retrieval stage compared with the integer-order tension controller.

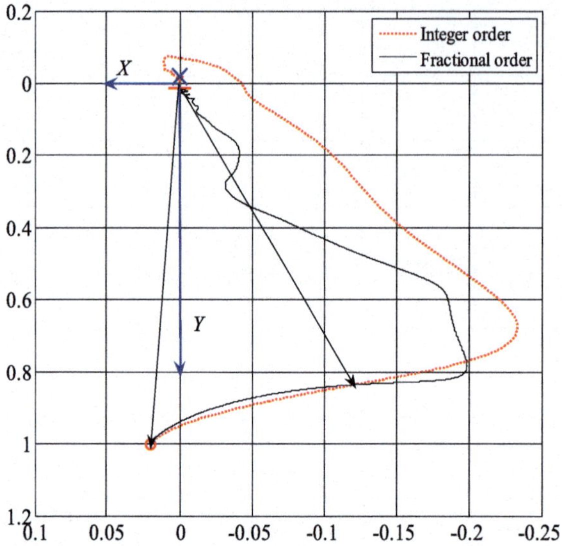

Fig. 3.10 Time history of out-of-plane libration angle and velocity of retrieval. Note that the unit for the libration velocity is degree per dimensionless time (τ)

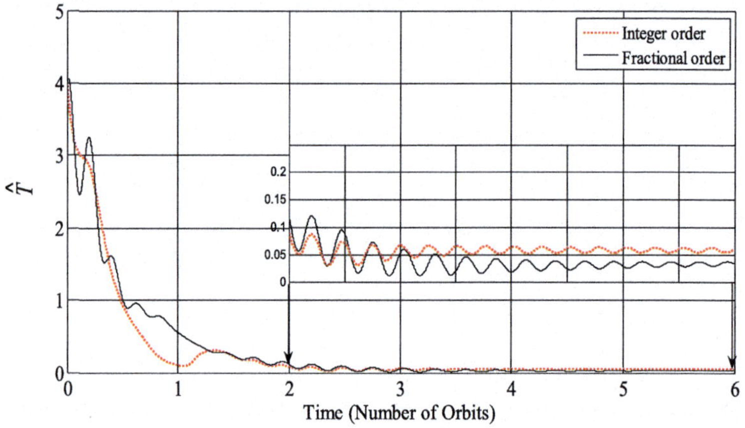

Fig. 3.11 Time history of out-of-plane libration angle and velocity of retrieval. Note that the unit for the libration velocity is degree per dimensionless time (τ)

3.4 In-Plane Controller for Deployment of Tethered Satellite

3.4.1 Simple In-Plane Tension Control Law for Deployment

The same process is in the last section for the tethered satellite retrieval. The PID-liked simple closed-loop linear tension control law has the following form [14, 26] such that

$$\hat{T} = k_1 y_1 + k_2 y_2 + k_3 y_3 + k_4 y_4 + k_5, \tag{3.23}$$

where $k_i, i = 1, 2, 3, 4$ are the control gains determined by the stability condition and control performance, and k_5 represents the tension offset introduced to balance the static tension. Introducing the parameter $k_5 = 3$ and substituting (3.23) into (3.8) lead to

$$\begin{Bmatrix} \dot{y}_1 \\ \dot{y}_2 \\ \dot{y}_3 \\ \dot{y}_4 \end{Bmatrix} = \begin{bmatrix} 0 & 1 & 0 & 0 \\ 3-k_1 & -k_2 & -k_3 & 2-k_4 \\ 0 & 0 & 0 & 1 \\ 0 & -2 & -3 & 0 \end{bmatrix} \begin{Bmatrix} y_1 \\ y_2 \\ y_3 \\ y_4 \end{Bmatrix}, \tag{3.24}$$

with the eigenfunction

$$\begin{aligned} \det & \begin{bmatrix} -\lambda & 1 & 0 & 0 \\ 3-k_1 & -k_2-\lambda & -k_3 & 2-k_4 \\ 0 & 0 & -\lambda & 1 \\ 0 & -2 & -3 & -\lambda \end{bmatrix} \\ &= \lambda^4 + k_2\lambda^3 + (4 - 2k_4 + k_1)\lambda^2 + (3k_2 - 2k_3)\lambda + 3k_1 - 9 \\ &= 0. \end{aligned} \tag{3.25}$$

The stability of the simple tension control law requires that all its eigenvalues must be in the open left half-plane. Based on the Routh-Hurwitz criterion, the stability condition can be expressed as

$$\begin{cases} k_2 > 0, \\ k_2(4 - 2k_4 + k_1) > 3k_2 - 2k_3, \\ k_2(4 - 2k_4 + k_1)(3k_2 - 2k_3) > (3k_2 - 2k_3)^2 + k_2^2(3k_1 - 9), \\ 3k_1 - 9 > 0. \end{cases} \tag{3.26}$$

From (3.26), the stable ranges of k_1 and k_2 can be derived as $k_1 > 3$ and $k_2 > 0$ explicitly, the stable ranges of k_3 and k4 have to be determined by the Kelvin-Tait-Chetaev (KTC) theorem [22, 26] if one accepts the asymptotic stability condition. From the KTC theorem, k_3 must be zero to eliminate the circulatory forces. Substituting $k_3 = 0$ into (3.26) leads to

$$k_4 < \frac{1}{2}(1 + k_1). \tag{3.27}$$

It is noted that the value of k_1 bounds the maximum value of k_4 and cannot be any value without disturbing the stability of the tension control law as concluded in [26]. For the sake of simplicity in constructing the tension control law, the value $k_4 = 0$ is accepted to eliminate the need for the feedback of pitch velocity such as

$$\hat{T} = k_1 y_1 + k_2 y_2 + 3, \tag{3.28}$$

where the constant 3 represents the static tether tension. To show the power of the fractional order controller, the performance of the simple tension control law will be shown in the following subsection as a benchmark, and the undesired overshoot of the tether length is observed, which cannot be solved without extending the deployment time.

3.4.2 *Improved In-Plane Fractional-Order Tension Control Law for Deployment*

Based on (3.9), the new FO tension control law is defined as

$$\hat{T} = D^{\alpha} y_2 - \dot{y}_2 + k_1 y_1 + k_2 y_2 + 3, \tag{3.29}$$

where the *Caputo* definition of the fractional derivative [24] is computed as

$$D_t^{\alpha} y_2(t) = \frac{1}{\Gamma(n-\alpha)} \int_0^t \frac{y_2^{(n)}(\tau)}{(t-\tau)^{\alpha-n+1}} d\tau, n < \alpha < n+1. \tag{3.30}$$

It shows the fractional derivative of y_2 depends on its time history due to the integration of the state variable from its initial condition to the current time.

Substituting (3.29) into (3.9) leads to

$$\begin{Bmatrix} \dot{y}_1 \\ D^{\alpha} y_2 \\ \dot{y}_3 \\ \dot{y}_4 \end{Bmatrix} = \begin{bmatrix} 0 & 1 & 0 & 0 \\ 3-k_1 & -k_2 & 0 & 2 \\ 0 & 0 & 0 & 1 \\ 0 & -2 & -3 & 0 \end{bmatrix} \begin{Bmatrix} y_1 \\ y_2 \\ y_3 \\ y_4 \end{Bmatrix}. \tag{3.31}$$

Equation (3.31) is a non-commensurate FO system. Assume the FO as $\alpha = N/M$, where N and M are integers and satisfy $N < M$. Then, the eigenfunction of the FO system has the following form,

$$\det\begin{bmatrix} \lambda^M & -1 & 0 & 0 \\ -(3-k_1) & \lambda^N+k_2 & 0 & -2 \\ 0 & 0 & \lambda^M & -1 \\ 0 & 2 & 3 & \lambda^M \end{bmatrix} = \lambda^{3M+N} + k_2\lambda^{3M} + 3\lambda^{M+N} \\ + (1+k_1)\lambda^{2M} + 3k_2\lambda^M - 3(3-k_1) = 0. \tag{3.32}$$

The non-commensurate stability theorem [8] states that the (3.31) is Lyapunov globally and asymptotically stable if all roots of (3.32) satisfy the condition $|\arg(\lambda_i)| > \pi/2/M$. Generally, there is no analytic solution of (3.32) if $3M + N > 4$ and a numerical approach is used. Considering the fact that the stable region ($|\arg(\lambda_i)| > \pi/2/M$) of a FO system is generally greater than its integer-order counterpart ($|\arg(\lambda_i)| > \pi/2$), it is a good starting point by adopting the same stability condition for k_1, k_2 of its integer-order counterpart even though it is conservative. Assume the fractional order is $\alpha = 1/5$ with $N = 1$ and $M = 5$, then (3.32) becomes

$$\lambda^{16}+k_2\lambda^{15} + (1+k_1)\lambda^{10} + 3\lambda^6+3k_2\lambda^5 - 3(3-k_1) = 0. \tag{3.33}$$

Based on the stability condition for k_1 and k_2 of the integer-order system outlined in Sect. 3.2, Fig. 3.12 shows the roots for $k_1 \in (3, 6)$ and $k_2 \in (0, 4)$ with a step of $\Delta k_1 = \Delta k_2 = 0.1$, respectively. Obviously, all roots are located within the region of $|\arg(\lambda_i)| > \pi/10$. Thus, the FO system is stable for the given ranges of k_1, k_2. This demonstrates that the stability condition for an integer-order control law is a subset of its FO counterpart.

In real engineering applications, two kinds of methods called direct and indirect discretization methods have been proposed for the realization of discrete FO calculus

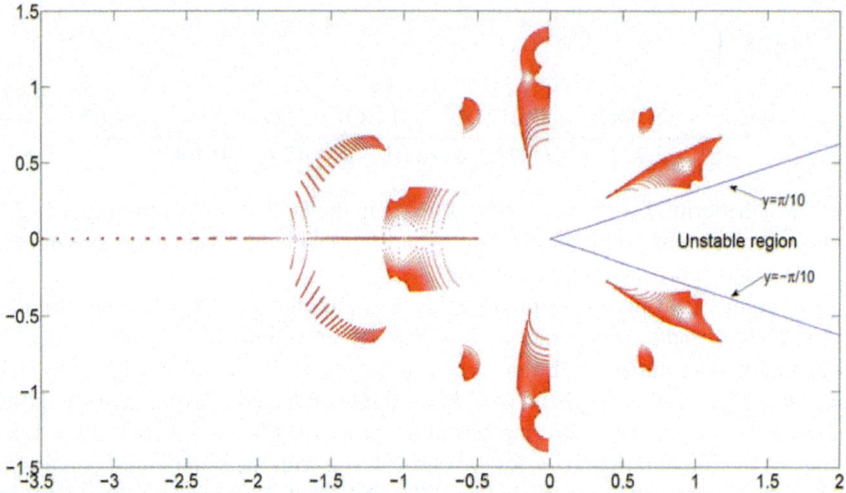

Fig. 3.12 Root distribution of the FO control law in s-plane

[5, 9]. In this chapter, the indirect method has implemented the FO tension control law, which follows two steps. The first step is to achieve an approximation model by the division of two polynomials using Charef's method in Sect. 1.2.2 [4],

$$\{I^{\alpha} y_2\} = \frac{y_2(s)}{s^{\alpha}} \approx \frac{1}{(1+s/p_T)^{\alpha}} = \frac{P(s)}{Q(s)}, \tag{3.34}$$

where I^{α} denotes the FO integral operator satisfying $D^{\alpha}(I^{\alpha} y_2) = y_2$, $\mathfrak{L}$ stands for Laplace transform, and $P(s)$ and $Q(s)$ are the polynomials of Laplace operator s. For instance, the FO integrator $1/s^{0.3}$ can be approximated by the following transfer function when the desired bandwidth $\Omega_{\max} = 2000$, corner frequency $p_T = 0.01$ and error criteria $y = 2dB$,

$$\begin{aligned}\mathfrak{L}\left\{I^{0.3} y_2\right\} &= \frac{1}{s^{0.3}} \\ &\approx \frac{1.96 \times 10^{-4} s^4 + 0.3338 s^3 + 20.41 s^2 + 46.38 s + 3.788}{1.63 \times 10^{-7} s^5 + 2.78 \times 10^{-3} s^4 + 1.704 s^3 + 38.77 s^2 + 32.85 s + 1}.\end{aligned} \tag{3.35}$$

The term with small coefficient, $1.63 \times 10^{-7} s^5$ in (3.35) can be safely neglected in engineering application. The next step is replacing the Laplace operator with the Tustin operator,

$$s = \frac{2}{T_s}\left[\frac{1 - z^{-1}}{1 + z^{-1}}\right], \tag{3.36}$$

where T_s is the sample time step. For instance, the discrete transform for $1/s^{0.3}$ can be expressed as follows at $T_s = 0.001$

$$\begin{aligned}Z^{-1}&\left\{\frac{1}{s^{0.3}}\right\} \\ &\approx \frac{0.0926 z^5 - 0.1887 z^4 + 0.01218 z^3 + 0.1801 z^2 - 0.1048 z + 0.00865}{z^5 - 2.727 z^4 + 1.792 z^3 + 1.001 z^2 - 1.474 z + 0.4067}.\end{aligned} \tag{3.37}$$

Thus, the term $D^{\alpha} y_2 - \dot{y}_2 = \dot{y}_2(I^{\alpha} y_2 - 1)$ in the FO tension control input (3.29) can be easily evaluated by substituting $\dot{y}_2$ and $I^{\alpha} y_2$ with the Tustin operator with the similar discrete process outlined above.

For comparison purposes, the same system parameters (220 km orbit altitude, 1.1804×10^{-3}rad/s orbital rate and 100km tether length) in [26] are used as a benchmark. The initial conditions are assumed as $y_1 = l = -0.99$, $y_2 = \dot{l} = 0.5$, $y_3 = \theta = 0$, $y_4 = \dot{\theta} = 0$, where $y_1 = l = -0.99$ is selected to avoid the singularity of (3.8) at $y_1 = -1$. The performances of the proposed FO tension control law for a series of fractional orders $\alpha = 1, 0.9, 0.7, 0.4, 0.2$ with the same initial conditions are evaluated using the same gains $k_1 = 4$ and $k_2 = 2$. Note that $\alpha = 1$ means the classic integer-order tension control law.

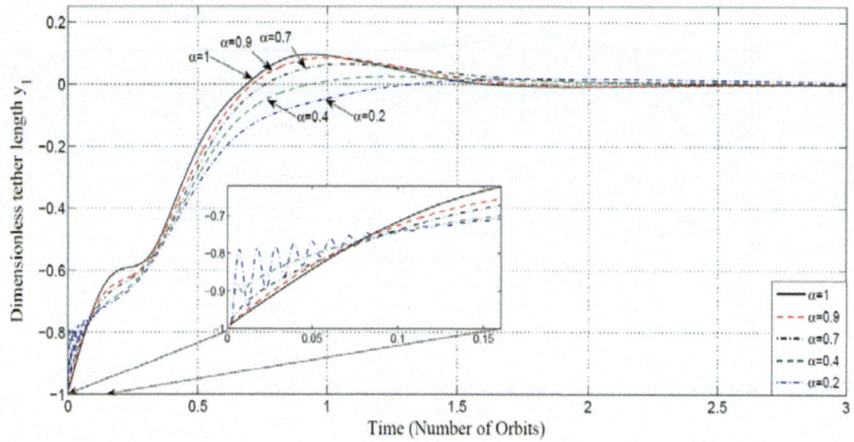

Fig. 3.13 Time history of tether length with different fractional orders

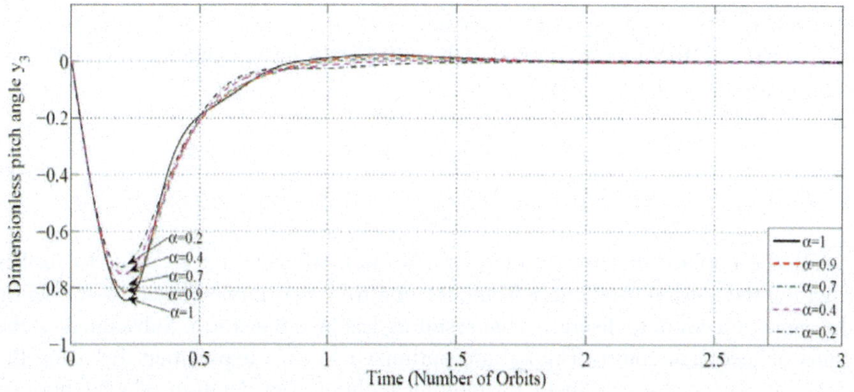

Fig. 3.14 Time history of pitch angle with different fractional orders

Figure 3.13 shows the time histories of tether length, and the overshoot of the tether length y_1 has been reduced significantly by the FO control. The smaller the fractional order, the less the overshoot and the faster the system achieves stability. The integer control law achieves the stable state in two orbits with 10% overshoot in tether length. In comparison, the FO control law with $\alpha = 0.2$ achieves the stable state in 1.5 orbits almost without overshoot. The same improvements are observed in the pitch angle, see Fig. 3.14, which is indirectly achieved through the nonlinear mode coupling. However, the small fractional order α causes unwanted oscillations of the tether length at the beginning of deployment; see the inset in Fig. 3.13 with order $\alpha = 0.2$. This is caused by the time delay of the FO differential $D^{\alpha}y_2$ with respect to the acceleration of tether length $\dot{y}_2$ in the tension control law. Furthermore, the difference of $(D^{\alpha}y_2 - \dot{y}_2)$ is also related to the initial value of y2, where the more

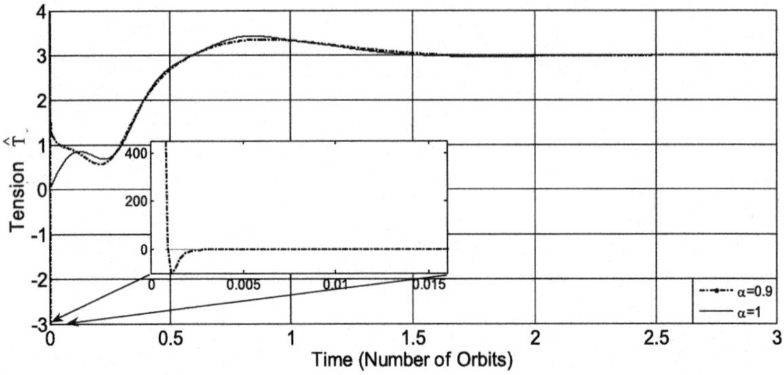

Fig. 3.15 Tension of different fractional orders with the same initial condition

significant initial deployment velocity will cause a higher acceleration. Figure 3.15 shows that the tension variation arises from the initial difference of $((D^{\alpha}y_2 - \dot{y}_2))$, resulting in an unrealistic negative tension.

Since the oscillations are limited to the initial state, an exponential weight parameter is introduced in (3.29), such that,

$$\begin{aligned}\hat{T} &= (1 - e^{-2t})(D^{\alpha}y_2 - \dot{y}_2) + k_1y_1 + k_2y_2 + k_5 \\ &= (1 - e^{-2t})\dot{y}_2(I^{1-\alpha}y_2 - 1) + k_1y_1 + k_2y_2 + k_5.\end{aligned} \tag{3.38}$$

Apply the Laplace transform to (3.38) and the modification $s \leftarrow \frac{s}{T_1s+1}$ to further filter out the tension oscillation. Then, the control law is implemented by following the same procedure to discretize the resulting Laplace transform. Subsequently, the gains of feedback control (k_1, k_2) are optimized for fast deployment by using the classical pole-zero assignment technology without consideration of overshoot of the tether length. Once the gains are defined, the fraction order α is optimized for smaller overshoot. For the same orbital parameters outlined above, assume $T_1 = 0.1$, and with an initial condition, the time histories of the tether length, pitch angle, and tension input are shown in Figs. 3.16, 3.17 and 3.18, respectively.

For the same orbital parameters outlined above, assume $k_1 \approx 5.4$, $k_2 \approx 3.5$, $T_1 = 0.1$ and $\alpha = 0.1$ with an initial condition $(y_1, y_2, y_3, y_4) = (-0.99, 0.7, 0, 0)$, the time histories of the tether length, pitch angle and tension input are shown on Figs. 3.16, 3.17 and 3.18, respectively.

It is observed from Fig. 3.16 that the subsatellite achieved the equilibrium state in 0.667 orbits without overshoot, which is much less than the two orbits by the classic integer-order tension control law reported in [26]. Furthermore, Fig. 3.16 also shows that the revised weighted FO tension control law has successfully suppressed the violent variation of tether length. This is evident in Fig. 3.17 that the tension control input is kept in an acceptable range of variation and is smooth. Finally, it is noted in Fig. 3.18 that little improvement in the magnitude of pitch angle response is

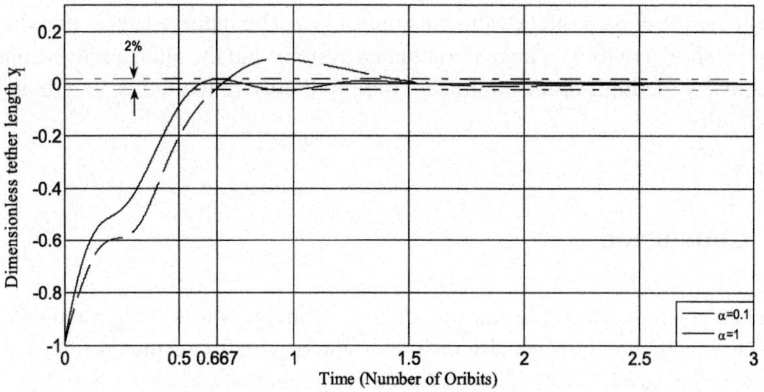

Fig. 3.16 Deployment dynamics of tether length with $k_1 = 5.4$, $k_2 = 3.5$ and $\alpha = 0.1$

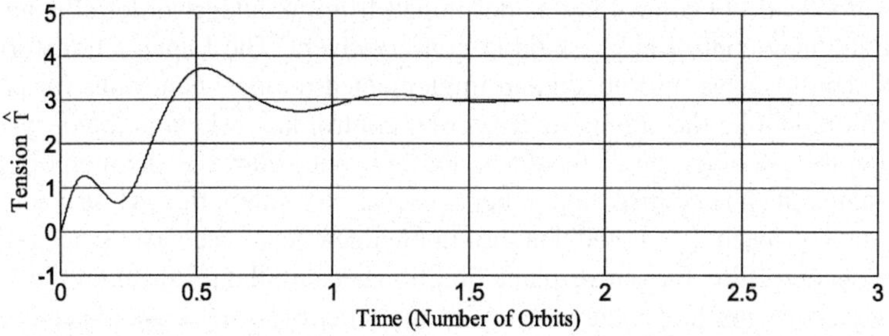

Fig. 3.17 Tension control input with $k_1 = 5.4$, $k_2 = 3.5$ and $\alpha = 0.1$

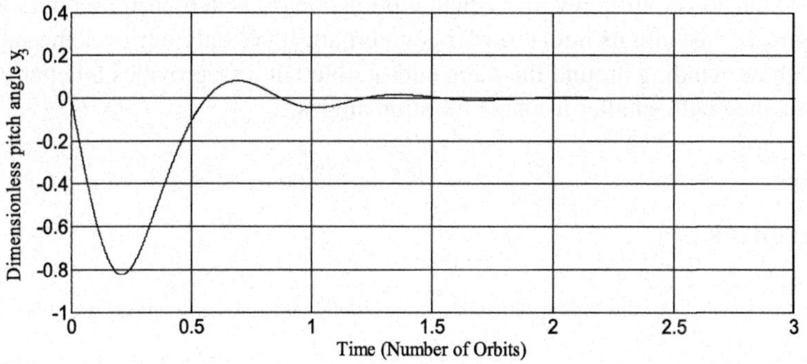

Fig. 3.18 Deployment dynamics of pitch angle with $k_1 = 5.4$, $k_2 = 3.5$ and $\alpha = 0.1$

gained over the classic integer tension control law. The main reason is that the tether dynamics shown in (3.3) is an underactuated system, and the pitch angle is controlled indirectly by the high-order nonlinear coupling between the tether length and pitch angle modes.

3.5 Conclusion

This chapter investigated the stability control problem of retrieving a tethered satellite system, which is intrinsically unstable. The originality of the current work is to control the retrieval of a tethered satellite stably and quickly by a new FO tension control law. The governing equation of tethered satellite systems and classic linear feedback tension control law were reviewed and examined as a benchmark test. The newly proposed FO control law is developed from its integer-order counterpart to ensure the new control law's stability conservatively. The Laplace transform, and Tustin operator have discretized and implemented the new controller for practical application. Unlike the classic integer-order control law, which is memoryless and based on the previous state's feedback, the FO control law has the memory of previous states. It thus controls the tether retrieval more smoothly while maintaining the retrieving speed. It is noted that the simple linear tension control strategy is ineffective in controlling the out-of-plane libration because the feedback of the states of that mode does not affect the stability of the tethered system. As expected, based on the same control strategy, the linear FO tension control law is not practical in controlling the out-of-plane libration either. The effectiveness and advantage of the new FO tension control law for in-plan motion control is demonstrated numerically by comparisons with its integer-order counterpart. It not only retrieves the subsatellite without winding around the main subsatellite but also provides a better control performance with smaller in-plane libration angles.

References

1. Atanackovic, T.M.: On a distributed derivative model of a viscoelastic body. Comptes Rendus Mecanique **331**(10), 687–692 (2003)
2. Baker, W.P., Dunkin, J.A., Galaboff, Z.J., Johnston, K.D., Kissel, R.R., Rheinfurth, M.H., Siebel, M.P.L.: Tether Subsatellite Study. NASA TMX-73314 (1976)
3. Cai, Z., Li, X., Wu, Z.: Deployment and retrieval of a rotating triangular tethered satellite formation near libration points. Acta Astronautica **98**, 37–49 (2014)
4. Charef, A., Sun, H., Tsao, Y., Onaral, B.: Fractal system as represented by singularity function. IEEE Trans. Autom. Control **37**(9), 1465–1470 (1992)
5. Chen, Y.Q., Moore, K.L.: Discretization schemes for fractional-order differentiators and integrators. IEEE Trans. Circuits Syst. I: Fundam. Theory Appl. **49**(3), 363–367 (2002)

6. Cosmo, M.L., Lorenzini, E.C.: Tethers in space handbook. Technical report, Citeseer (1997)
7. Das, S., Pan, I.: Fractional Order Signal Processing: Introductory Concepts and Applications. Springer Science & Business Media (2011)
8. Deng, W., Li, C., Lü, J.: Stability analysis of linear fractional differential system with multiple time delays. Nonlinear Dyn. **48**(4), 409–416 (2007)
9. Diethelm, K., Ford, N.J., Freed, A.D.: A predictor-corrector approach for the numerical solution of fractional differential equations. Nonlinear Dyn. **29**(1), 3–22 (2002)
10. Djebli, A., Pascal, M., El Bakkali, L.: Laws of deployment/retrieval in tether connected satellites systems. Acta Astronautica **45**(2), 61–73 (1999)
11. Fujii, H.A., Ishijima, S.: Mission function control for deployment and retrieval of a subsatellite. J. Guidance Control Dyn. **12**(2), 243–247 (1989)
12. Gorenflo, R., Mainardi, F.: Fractional Calculus. Springer (1997)
13. Kojima, H., Sugimoto, T.: Stability analysis of in-plane and out-of-plane periodic motions of electrodynamic tether system in inclined elliptic orbit. Acta Astronautica **65**(3–4), 477–488 (2009)
14. Kumar, K., Pradeep, S.: Strategies for three dimensional deployment of tethered satellites. Mech. Res. Commun. **25**(5), 543–550 (1998)
15. Kumar, K., Yasaka, T., Sasaki, T.: Orbit transfer of service vehicle/payload through tether retrieval. Acta Astronautica **54**(9), 687–698 (2004)
16. Liaw, D.C., Abed, E.H.: Stabilization of tethered satellites during station keeping. IEEE Trans. Autom. Control **35**(11), 1186–1196 (1990)
17. Licata, R.: Tethered system deployment controls by feedback fuzzy logic. Acta Astronautica **40**(9), 619–634 (1997)
18. Liu, Y., Zhou, J., Chen, H.: Variable structure control for tethered satellite fast deployment and retrieval. In: Future Control and Automation, pp. 157–164. Springer (2012)
19. Magin, R.L.: Fractional calculus in bioengineering: a tool to model complex dynamics. In: Proceedings of the 13th International Carpathian Control Conference (ICCC), pp. 464–469. IEEE (2012)
20. Misra, A.K.: Dynamics and control of tethered satellite systems. Acta Astronautica **63**(11–12), 1169–1177 (2008)
21. Modi, V., Chang-Fu, G., Misra, A., et al.: On the control of the space shuttle based tethered systems. Acta Astronautica **9**(6–7), 437–443 (1982)
22. Monje, C.A., Chen, Y., Vinagre, B.M., Xue, D., Feliu-Batlle, V.: Fractional-Order Systems and Controls: Fundamentals and Applications. Springer Science & Business Media (2010)
23. Oustaloup, A., Sabatier, J., Lanusse, P., Malti, R., Melchior, P., Moreau, X., Moze, M.: An overview of the CRONE approach in system analysis, modeling and identification, observation and control. IFAC Proc. Vol. **41**(2), 14254–14265 (2008)
24. Podlubny, I.: Fractional Differential Equations. Academic, New York (1999)
25. Podlubny, I.: Fractional-order systems and PID controllers. IEEE Trans. Autom. Control **44**(8) (1999b)
26. Pradeep, S.: A new tension control law for deployment of tethered satellites. Mech. Res. Commun. **24**(3), 247–254 (1997)
27. Rupp, C.C.: A tether tension control law for tethered subsatellites deployed along local vertical. Technical report, NASA-TM-X-64963 (1975)
28. Sanmartín, J.R., et al.: Propellantless deorbiting of space debris by bare electrodynamic tethers. In: 62nd International Astronautical Congress 2011, vol. 3, pp. 2239–2248 (2011)
29. Steindl, A., Steiner, W., Troger, H.: Optimal control of retrieval of a tethered subsatellite. In: IUTAM Symposium on Chaotic Dynamics and Control of Systems and Processes in Mechanics, pp. 441–450. Springer (2005)
30. Tomlin, D., Mowery, D., Bodley, C.: Tethered satellite system control system design. In: 3rd Tethers in Space/Toward Flight International Conference, p. 1568 (1989)
31. Williams, P.: A review of space tether technology. Recent Patents Space Technol. **2**(1), 22–36 (2012)

32. Yu, B., Jin, D.: Deployment and retrieval of tethered satellite system under J2 perturbation and heating effect. Acta Astronautica **67**(7–8), 845–853 (2010)
33. Zhong, R., Zhu, Z.: Dynamic analysis of deployment and retrieval of tethered satellites using a hybrid hinged-rod tether model. Int. J. Aerosp. Lightweight Struct. **1**(2), 239–259 (2011)
34. Zhu, Z., Zhong, R.: Deorbiting dynamics of electrodynamic tether. Int. J. Aerosp. Lightweight Struct. (IJALS) **1**(1) (2011)

Part II
Fractional-Order SMC of Linear Motor Systems

Chapter 4
Practical Tracking Control via Discrete-Time Fractional-Order SMC

4.1 Introduction

In the past decades, researchers applied FO calculus to CSMC schemes and conducted them in industrial fields [13], such as the hydraulic manipulators [12], permanent magnet synchronous motors [14] and so on, which have proved that these control schemes are more effective than their integer-order counterparts. However, only a few works focusing on applying the FO calculus to DSM schemes can be found in the literature. Therefore, such an issue is investigated in this chapter. Inspired by the merits of FO calculus, a new kind of controller, that is, discrete-time FO sliding mode (DFSM) controller is proposed and implemented on the linear motor platform in this chapter. In the procedure to design the controller, a new FO sliding surface in discrete form is introduced for the first time, which is universal when other controllers are designed. Since no similar sliding surface or controller has been designed before, the characteristics of the sliding surface and the dynamics of the sliding state are studied and proved strictly in this chapter. Moreover, the experiments prove not only that the DFSM controller is effective from the precision and robustness point of view, but also that the chattering effect is alleviated when DFSM is conducted.

4.2 Plant Modeling of X-Y Servo Platform

To facilitate the discussion, this section first introduces the experimental platform for the linear motor positioning system (production of Artbus company) which is shown in Fig. 4.1. The 2-DOF positioning platform consists of a moving stage and three linear motors, which has a 700 mm travel range in X-direction and 650 mm in Y-direction. In this chapter, as an example, the positioning of the moving stage in X-direction is considered, and the positioning precision is guaranteed by a

G. Sun et al., *Fractional-Order Sliding Mode Control: Methodologies and Applications*, Studies in Systems, Decision and Control 532,
https://doi.org/10.1007/978-3-031-60847-6_4

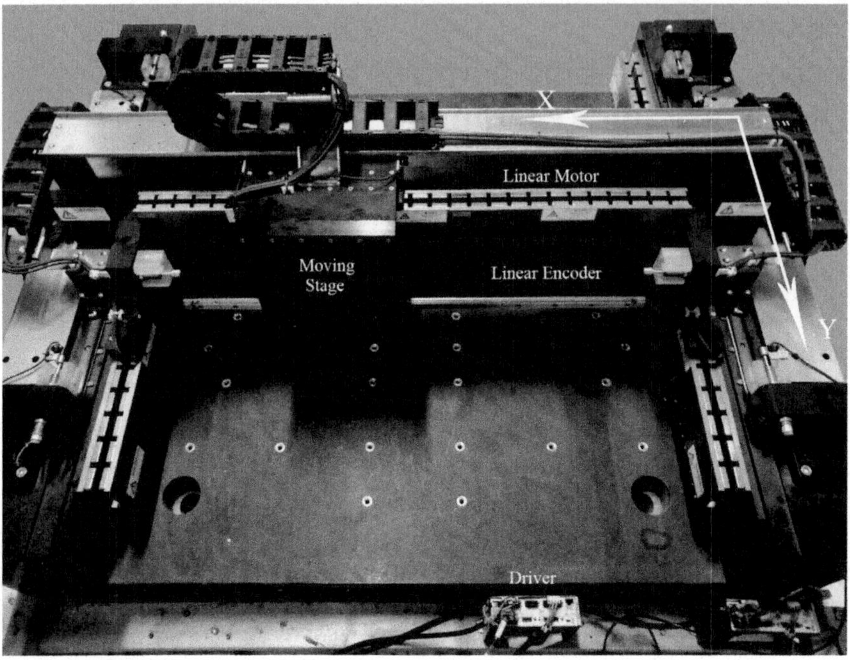

Fig. 4.1 Experimental facility for linear-motor-based motion control system

high-accuracy linear optical encoder (with a resolution of 50 nm). Based on the previous work in [15], the simplified continuous model can be presented as follows

$$\dot{p} = v, \tag{4.1}$$

$$m\dot{v} = (1 - k_m)(k_f u - f - d), \tag{4.2}$$

$$f = k_c \text{sgn}(v) + k_v v, \tag{4.3}$$

where m is the mass of the moving stage. p and v refer to the absolute position and the velocity of the moving stage, respectively. k_m stands for the gain of the variable load mass, k_f is the voltage-to-force amplifier gain and d denotes the lumped uncertainty that consists of unmodeled system dynamics and external disturbances. The simplified friction model f here consists of two items with Coulomb friction $k_c\text{sgn}(v)$ and viscous friction $k_v v$, where k_c is the Coulomb friction coefficient, k_v is the viscous friction coefficient and $\text{sgn}(\cdot)$ is the signum function.

By using the Euler's discretization method, the continuous model can be given as

$$\begin{aligned} \Delta p(k) &= v(k), \\ \Delta v(k) &= \frac{1 - k_m}{m}(k_f u(k) - f(k) - d(k)), \end{aligned}$$

where Δ denotes the forward difference operator, which can be obtained by

$$\Delta x(k) = \frac{x(k+1) - x(k)}{h}, \ \forall x(k) \in R,$$

and h is the sampling interval. Furthermore, the operator Δ^i satisfies $\Delta^i = \Delta(\Delta^{i-1})$, $i = 2, 3, 4, \ldots$. Thus, the natural discrete model of the linear motor platform can be rewritten as

$$p(k+1) = p(k) + hv(k), \tag{4.4}$$

$$v(k+1) = v(k) + h\frac{1-k_m}{m}(k_f u(k) - f(k) - d(k)), \tag{4.5}$$

$$f(k) = k_c \mathrm{sgn}(v(k)) + k_v v(k). \tag{4.6}$$

Considering the fact that the uncertainties always exist and are always bounded in practice, the following parametric uncertainties are reasonable to be assumed as

$$\begin{aligned} \frac{1}{\theta_m} \leq \frac{\bar{m}}{m} &\leq \theta_m, \\ \frac{1}{\theta_{k_m}} \leq \frac{1-\bar{k}_m}{1-k_m} &\leq \theta_{k_m}, \\ |\bar{k}_c - k_c| &\leq \theta_{k_c}, \\ |\bar{k}_v - k_v| &\leq \theta_{k_v}, \\ |d| &\leq \theta_d, \end{aligned} \tag{4.7}$$

where $\bar{\bullet}$ denotes the nominal model term, $\theta_\bullet$ stands for the boundary of the corresponding parametric uncertainty.

4.3 Preliminaries of Fractional-Order Calculus

To begin our discussion, some basic concepts and FO calculus will be presented, including the definition and practical stability results. In mathematics history, researchers have proposed many definitions in their own fields [9]. The most famous continuous definitions are the Riemann-Liouville definition, Caputo definition, and Grünwald-Letnikov (GL) definition [8]. Among them, the GL definition is usually adopted to describe the discrete-time case because of the clear expression.

Definition 4.1 ([9]) The FO difference of the function $x(k)$ in the discrete GL form is given by

$$\Delta^{\alpha}x(k) = \frac{1}{h^{\alpha}}\sum_{j=0}^{k}(-1)^{j}\binom{\alpha}{j}x(k-j), \tag{4.8}$$

where h is the sampling interval, $\alpha \in R$ is the order of fractional difference, and $k \in N_{+}$ is the number of samples for calculation. The binomial term $\binom{\alpha}{j}$ is given as

$$\binom{\alpha}{j} = \begin{cases} 1, & \text{for } j = 0, \\ \dfrac{\alpha(\alpha-1)\cdots(\alpha-j+1)}{j!}, & \text{for } j = 1, 2, \ldots. \end{cases} \tag{4.9}$$

From (4.8), it is clear that FO calculus can be regarded as a weighted sum of the function $x(k)$ where the number of items will become larger and larger as k increases. Moreover, the weighting is significant when its step is close to step k and small when it is far away from step k. In practical engineering applications, especially for digital control, the sum of the item involved is often reduced to a suitable length L [2]. Then, the $\Delta^{\alpha}x(k)$ can be represented as follows:

$$\Delta^{\alpha}x(k) = \frac{1}{h^{\alpha}}\sum_{j=0}^{L}(-1)^{j}\binom{\alpha}{j}x(k-j), \tag{4.10}$$

The weighted coefficients in (4.8) can be computed by the following simple algorithm [3].

$$C_k(\alpha) = (-1)^k\binom{\alpha}{k}, \quad k = 1, 2, 3, \ldots$$
$$C_{k+1}(\alpha) = C_k\frac{k+1-\alpha}{k+2}, \quad k = 1, 2, 3, \ldots$$

And the sum of all these coefficients can be written as

$$\sum_{j=0}^{L}(-1)^{j}\binom{\alpha}{j} = \frac{\Gamma(L+1-\alpha)}{\Gamma(1-\alpha)\Gamma(L+1)}, \tag{4.11}$$

where $\Gamma(\cdot)$ is the Gamma function. Moreover, from (4.9), there is a fact that

$$(-1)^{j}\binom{n}{j} > 0, \quad n < 0. \tag{4.12}$$

Based on (4.10)–(4.12), the following Lemmas are easy to be obtained.

Lemma 4.2 *For the fractional difference described by (4.10), the following inequality can be obtained*

$$K\frac{m}{h^\alpha} < |\Delta^\alpha x(k)| < K\frac{M}{h^\alpha},$$

where

$$K = \frac{\Gamma(L+1-\alpha)}{\Gamma(1-\alpha)\Gamma(L+1)},$$

and m, M are the minimum and maximum of $|x(i)|$, $i = k-L, k-L+1, \ldots, k$, respectively. Obviously, K is a positive constant for $\alpha < 0$.

Lemma 4.3 ([4]) *The discrete-time FO linear system described by*

$$\Delta^\alpha x(k+1) = x(k), \quad \alpha \in (0, 1), \tag{4.13}$$

is practically stable if and only if

$$\omega(z) \neq 0, \ |z| \geq 1,$$

where

$$\omega(z) = \det\left\{Iz - A_0 - \sum_{k=1}^{L} A_k z^{-k}\right\},$$

and L is the given length of practical implementation.

4.4 Fractional-Order Sliding Mode Controller Design

Considering the discrete model of the linear motor platform (4.4)–(4.6), define the tracking error $\check{p}(k)$ as

$$\check{p}(k) = p(k) - p_r(k), \tag{4.14}$$

where $p_r(k)$ is the reference signal, and $p(k)$ is the actual position.

Similar to their integer-order counterpart, the FO SMC approach comprises two phases.

4.4.1 Sliding Surface

The first phase is to design a stable sliding surface with some desired properties. Here, a new FO-type sliding surface is constructed as

$$s(k) = v(k) - \Delta p_r(k) + q\Delta^{\lambda-1}\check{p}(k), \tag{4.15}$$

where $q > 0$ and $0 < \lambda < 1$.

The following theorem is given to guarantee the proposed sliding surface's stability.

Theorem 4.4 *For the sliding surface (4.15), the tracking error $\check{p}(k)$ can always converge to the equilibrium point $\check{p}(k) = 0$ asymptotically if and only if the point $z_0 = 1 - \frac{qh}{h^{\lambda-1}}$ is located in the stability region $S(\lambda, L, q, h)$, which consists of the origin and is bounded by*

$$\rho(\omega) = e^{i\omega} - \sum_{j=1}^{L}(-\frac{qh}{h^{\lambda-1}})(-1)^j \binom{\lambda-1}{j} e^{ji\omega}.$$

Proof When $\check{p}(k)$ is on the sliding surface, it is easy to get that

$$v(k) - \Delta p_r(k) = -q\Delta^{\lambda-1}\check{p}(k).$$

Considering the definition of the operator Δ as well as part of the plant model (4.4), the following equation can be obtained

$$\begin{aligned}\check{p}(k+1) = &\sum_{j=1}^{L}(-\frac{qh}{h^{\lambda-1}})(-1)^j \binom{\lambda-1}{j} \check{p}(k-j) \\ &+(1 - \frac{qh}{h^{\lambda-1}})\check{p}(k).\end{aligned} \tag{4.16}$$

The above equation can be treated as a linear multi-delay system, which should be proved to be asymptotically stable. On the other hand, according to [4], the following system is asymptotically stable when the system (4.13) is practically stable.

$$x(k+1) = A_0x(k) + \sum_{j=1}^{L} A_j x(k-j).$$

Therefore, based on Lemma 4.3, the system (4.16) is asymptotically stable if and only if

$$\left| z - 1 + \frac{qh}{h^{\lambda-1}} + \sum_{k=1}^{L} \frac{qh}{h^{\lambda-1}}(-1)^j \binom{\lambda-1}{j} z^{-k} \right| \neq 0,$$

when $|z| \geq 1$. If $\Omega(z)$ is defined as

$$\Omega(z) = \left| z - 1 + \frac{qh}{h^{\lambda-1}} + \sum_{k=1}^{L} \frac{qh}{h^{\lambda-1}}(-1)^j \binom{\lambda-1}{j} z^{-k} \right|,$$

in other words, the roots of the equation $\Omega(z) = 0$ should be in the unit circle on the complex plane. However, the roots are difficult to obtain. Thus, we use the D-composition method [6] to simplify this problem. The constant term $1 - \frac{qh}{h^{\lambda-1}}$ is replaced by a symbol ρ.

According to the D-composition method, that the roots of $\Omega(z) = 0$ are in the unit circle is equal to that ρ is in the stable region, which is determined by

$$\rho(\omega) = e^{i\omega} - \sum_{j=1}^{L}(-\frac{qh}{h^{\lambda-1}})(-1)^j \binom{\lambda-1}{j} e^{ij\omega},$$

and consists of the origin. Thus, the proof of Theorem 4.4 is completed. ■

4.4.2 Discrete-Time Fractional-Order Sliding Mode Controller

For the second phase of sliding mode controller design, a control input should be constructed, which guarantees that $s(k)$ reaches the vicinity of the sliding surface in finite steps. A new sliding mode controller is presented as

$$u(k) = u_{eq}(k) + u_{sw}(k), \tag{4.17}$$

where the sliding mode control consists of two parts: the equivalent control $u_{eq}(k)$ and the switching control $u_{sw}(k)$. Based on the basic concepts of the equivalent control [11], the $u_{eq}(k)$ on the manifold $s(k) = 0$ can be obtained by solving $s(k+1) = 0$, i.e.,

$$\begin{aligned} s(k+1) - s(k) &= v(k+1) - v(k) - h\Delta^2 p_r(k) \\ &\quad + hq\Delta^{\lambda-1}(v(k) - \Delta p_r(k)) = 0, \end{aligned} \tag{4.18}$$

which implies

$$\begin{aligned} u_{eq}(k) = &\frac{\bar{m}}{k_f(1-\bar{k}_m)}\left[\Delta^2 p_r(k) - q\Delta^{\lambda-1}(v(k) - \Delta p_r(k))\right] \\ &+ \frac{1}{k_f}\left(\bar{k}_c \text{sgn}(v(k)) + \bar{k}_v v(k)\right). \end{aligned}$$

As for the switching control $u_{sw}(k)$, it is designed as

$$u_{sw}(k) = -\frac{\bar{m}}{k_f(1-\bar{k}_m)}(k_1 s^3(k) + k_2 \text{sig}^b(s(k))), \tag{4.19}$$

where $\mathrm{sig}^b(\cdot) = |\cdot|^b \mathrm{sgn}(\cdot)$, k_1 and k_2 are two tunable parameters whose value ranges will be discussed later.

Under the DFSM controller (4.17), and considering the discrete model (4.4)–(4.6), we can obtain the sliding mode dynamics as

$$s(k+1) = -\Phi_1 h s^3(k) - \Phi_2 h \mathrm{sig}^b(s(k)) + s(k) + \Phi_0(k)h, \tag{4.20}$$

where

$$\begin{aligned}
\Phi_0(k) &= \left[\frac{(1-k_m)\bar{m}}{(1-\bar{k}_m)m} - 1\right]\left[\Delta^2 p_r(k) - q\Delta^{\lambda-1}(v(k) - \Delta p_r(k))\right] - \frac{1-k_m}{m}\varrho(k),\\
\Phi_1 &= \frac{k_1(1-k_m)\bar{m}}{(1-\bar{k}_m)m},\quad \Phi_2 = \frac{k_2(1-k_m)\bar{m}}{(1-\bar{k}_m)m},\\
\varrho(k) &= (k_v - \bar{k}_v)v(k) + (k_c - \bar{k}_c)\mathrm{sgn}(v(k)) + d(k).
\end{aligned}$$

It is necessary to point out that k_1 and k_2 should be in a certain range satisfying $0 < \Phi_1 h < 1$ and $0 < \Phi_2 h < 1$. Furthermore, the following assumptions and propositions are also indispensable.

Assumption 4.5 For the tracking control system in (4.4)–(4.6), the sampling interval h is selected small enough to satisfy

$$|s(k)| < \left(\frac{1-\Phi_2 h}{\Phi_1 h}\right)^{\frac{1}{2}}. \tag{4.21}$$

Remark 4.6 In most practical discrete-time systems, if the sampling interval is too large, the discrete-time model may not describe the practical system precisely. Fortunately, the assumption can be easily achieved with the help of modern high-speed control processors.

Proposition 4.7 ([5]) $|\Phi_0^\star h| \le \Phi_2 h \Phi^b \le (1-\Phi_1 h)\Phi$, *where* Φ *is defined as*

$$\Phi = \max\left[\left(\frac{|\Phi_0^\star h|}{\Phi_2 h}\right)^{\frac{1}{b}}, \left(\frac{\Phi_2 h}{1-\Phi_1 h}\right)^{\frac{1}{1-b}}\right],$$

and $\Phi \ge 0$ *apparently.*

Proof If

$$\Phi = \left(\frac{\Phi_2 h}{1-\Phi_1 h}\right)^{\frac{1}{1-b}} \ge \left(\frac{|\Phi_0^\star h|}{\Phi_2 h}\right)^{\frac{1}{b}},$$

considering $\Phi_2 h > 0$, we have

$$|\Phi_0^\star h| \le \Phi_2 h \left(\frac{\Phi_2 h}{1-\Phi_1 h}\right)^{\frac{b}{1-b}} = \Phi_2 h \Phi^b = (1-\Phi_1 h)\Phi.$$

If

$$\Phi = \left(\frac{|\Phi_0^\star h|}{\Phi_2 h}\right)^{\frac{1}{b}} \geq \left(\frac{\Phi_2 h}{1-\Phi_1 h}\right)^{\frac{1}{1-b}},$$

then we can directly get that $|\Phi_0^\star h| = \Phi_2 h \Phi^b$. On the other hand, it can be obtained that

$$\left|\Phi_0^\star h\right|^{1-b} \geq \frac{\Phi_2 h}{(1-\Phi_1 h)^b}.$$

Taking $|\Phi_0^\star h| = \Phi_2 h \Phi^b$ into consideration, we can further get that $|\Phi_0^\star h)|^b \leq \Phi^b(1-\Phi_1 h)^b$, which means that

$$|\Phi_0^\star h| = \Phi_2 h \Phi^b \leq (1-\Phi_1 h)\left(\frac{|\Phi_0^\star h|}{\Phi_2 h}\right)^{1/b} = (1-\Phi_1 h)\Phi,$$

which completes the proof. ∎

Theorem 4.8 *For the discrete-time tracking control system (4.4)–(4.6) under Assumption 4.5, if the discrete-time FO sliding mode controller (4.17) is implemented, then the sliding mode state $s(k)$ from initial state converges into the region Ξ within at most K_m steps, where*

$$\begin{aligned}
\Xi &= \left\{s(k)\big||s(k)| \leq \xi\right\}, \\
\xi &= \Psi(b)\cdot \max\left[\left(\frac{|\Phi_0^\star h|}{\Phi_2 h}\right)^{\frac{1}{b}}, \left(\frac{\Phi_2 h}{1-\Phi_1 h}\right)^{\frac{1}{1-b}}\right], \\
K_m &= \left\lceil \frac{s^2(0)-\xi^2}{\mu^2}\right\rceil,
\end{aligned} \tag{4.22}$$

with

$$\Psi(b) = 1 + b^{\frac{b}{1-b}} - b^{\frac{1}{1-b}},\ \mu = \Phi_1 h \xi^3 + [\Psi^b(b)-1]|\Phi_0^\star| h,$$
$$|\Phi_0^\star| = \max\{|\Phi_0(k)|\},\quad k = 0, 1, 2, \ldots.$$

Proof Construct the following Lyapunov function

$$V(k) = s^2(k), \tag{4.23}$$

and then consider the cases under two circumstances separately.

Case 1. When $s(k) > \xi > 0$.
On the basis of (4.23), the difference between $V(k+1)$ and $V(k)$ is given as

$$V(k+1) - V(k) = [s(k+1) - s(k)][s(k+1) + s(k)].$$

On the one hand, taking Proposition 4.7 into consideration, we can easily get that

$$\begin{aligned} s(k) - s(k+1) &= \Phi_1 h s^3(k) + \Phi_2 h \operatorname{sig}^b(s(k)) - \Phi_0(k)h \\ &\geq \Phi_1 h s^3(k) + \Phi_2 h \operatorname{sig}^b(s(k)) - |\Phi_0^\star h| \\ &> \Phi_1 h \xi^3 + \left[\Psi^b(b) - 1\right] |\Phi_0^\star h| \\ &= \mu > 0. \end{aligned}$$

On the other hand, it is apparent that

$$\begin{aligned} s(k) + s(k+1) &= -\Phi_1 h s^3(k) - \Phi_2 h s^b(k) + 2s(k) + \Phi_0(k)h \\ &\geq -\Phi_1 h s^3(k) - \Phi_2 h s^b(k) + 2s(k) - |\Phi_0^\star h|. \end{aligned}$$

When $s(k) > 1$ and Assumption 4.5 is satisfied, by using Proposition 4.7, it follows that

$$\begin{aligned} s(k) + s(k+1) &\geq -\Phi_1 h s^3(k) - \Phi_2 h s^b(k) + 2[\Phi_1 h s^3(k) + \Phi_2 h s(k)] - |\Phi_0^\star h| \\ &> -\Phi_1 h s^3(k) - \Phi_2 h s^b(k) + 2[\Phi_1 h s^3(k) + \Phi_2 h s^b(k)] - |\Phi_0^\star h| \\ &> \Phi_1 h \xi^3 + \left[\Psi^b(b) - 1\right] |\Phi_0^\star h| \\ &= \mu > 0. \end{aligned}$$

When $s(k) \leq 1$, it is obvious that $s^3(k) \leq s(k)$. Consequently, by Proposition 4.7 again, we have

$$\begin{aligned} s(k) + s(k+1) &\geq -\Phi_1 h s^3(k) - \Phi_2 h s^b(k) + 2[\Phi_2 h s^b(k) + \Phi_1 h s(k)] - |\Phi_0^\star h| \\ &\geq -\Phi_1 h s^3(k) - \Phi_2 h s^b(k) + 2[\Phi_2 h s^b(k) + \Phi_1 h s^3(k)] - |\Phi_0^\star h| \\ &> \Phi_1 h \xi^3 + \left[\Psi^b(b) - 1\right] |\Phi_0^\star h| = \mu > 0. \end{aligned}$$

Thus we have

$$V(k+1) - V(k) = -[s(k) + s(k+1)][s(k) - s(k+1)] < -\mu^2 < 0.$$

Case 2. When $s(k) < -\xi < 0$.
The proof procedure of this case is similar to that of the former case. It can be proved that $V(k+1) - V(k) < -\mu^2 < 0$ is still true in this case.

According to the results above, we can conclude that $V(k+1) - V(k) = s^2(k+1) - s^2(k) < \mu^2$ is always valid for $k = 0, 1, 2, \ldots$. It can be concluded that $s^2(m) < s^2(0) - m\mu^2$.

By solving inequality $s^2(K_m) < s^2(0) - K_m\mu^2 \leq \xi^2$, we can find that $s(k)$ will converge into Ξ in K_m steps at most, where

$$K_m = \left\lceil \frac{s^2(0) - \xi^2}{\mu^2} \right\rceil.$$

So far, we have completed the proof. ■

The previous theorem shows that the sliding variable $s(k)$ converges into Ξ from any initial state. However, its dynamics in Ξ still need to be investigated. Here, we give the following Lemma and Theorem.

Lemma 4.9 ([7]) *If function* $\Psi(b) = 1 + b^{\frac{b}{1-b}} - b^{\frac{1}{1-b}}$ *and* $0 < b < 1$, *then* $1 < \Psi(b) < 2$ *and the following condition holds for* $\forall x \in [0, 1]$

$$x\Psi(b) - x^b\Psi^b(b) + \Psi(b) - 1 \geq 0.$$

Theorem 4.10 *For the discrete-time tracking control system (4.4)–(4.6) under Assumption 4.5 and the DFSM controller (4.17), once the sliding variable* $s(k)$ *enters the region* Ξ, *it cannot escape from it thereafter.*

Proof The basic idea proving the theorem is to show that if $s(k) \in \Xi$, then $s(k+1) \in \Xi$. As $-\xi \leq s(k) \leq \xi$, we can find a coefficient $\theta(-1 \leq \theta \leq 1)$ that satisfies $s(k) = \theta\xi = \theta\Psi(b)\Phi$.

Firstly, $s(k+1) \leq \xi$ will be proved, and two different cases are discussed as follows.

Case 1. When $\theta\Psi(b) \geq 0$, it is clear that $s(k) \geq 0$ and $s(k+1) = s(k) - \Phi_1 hs^3(k) - \Phi_2 hs^b(k) + \Phi_0(k)h$.

On this condition, by using Proposition 4.7, we can obtain

$$\begin{aligned} s(k+1) &\leq s(k) - \Phi_1 hs^3(k) - \Phi_2 h\theta^b\Psi^b(b)\Phi^b + |\Phi_0^\star h| \\ &\leq s(k) - \Phi_1 hs^3(k) + (1 - \Phi^b\Psi^b(b))|\Phi_0^\star h| \\ &\leq s(k) + (1 - \theta^b\Psi^b(b))|\Phi_0^\star h|. \end{aligned}$$

If $\theta\Psi(b) \geq 1$, it is obvious that $s(k+1) \leq s(k) \leq \xi$. If $\theta\Psi(b) < 1$, it is clear that $\theta\Psi(b) < \theta^b\Psi^b(b) < 1$. Therefore, considering Proposition 4.7, it can be deduced that

$$
\begin{aligned}
s(k+1) \leq & \theta\Psi(b)\Phi + (1-\theta^b\Psi^b(b))(1-\Phi_1 h)\Phi \\
= & (1 + (\theta\Psi(b) - \theta^b\Psi^b(b)) + (\theta^b\Psi^b(b) - 1)\Phi_1 h)\Phi \\
< & \Phi < \Psi(b)\Phi = \xi.
\end{aligned}
$$

Case 2. When $\theta\Psi(b) < 0$, it is clear that $s(k) < 0$ and $s(k+1) = -|s(k)| + \Phi_1 h|s(k)|^3 + \Phi_2 h|s(k)|^b + \Phi_0(k)h$.

If $-1 < s(k) < 0$ then $|s(k)|^b > |s(k)| > |s(k)|^3$, based on this and Proposition 4.7, we have

$$
\begin{aligned}
s(k+1) &\leq -|s(k)| + \Phi_1 h|s(k)|^3 + \Phi_2 h|s(k)|^b + |\Phi_0^\star h| \\
&\leq -(1-\Phi_1 h)\Phi[|\theta\Psi(b)| - |\theta\Psi(b)|^b - 1].
\end{aligned}
$$

According to Lemma 4.9, it is obvious that $|\theta\Psi(b)| - |\theta\Psi(b)|^b - 1 \geq -\Psi(b)$. Thus we can further get that

$$
s(k+1) \leq (1-\Phi_1 h)\Psi(b)\Phi < \Psi(b)\Phi = \xi.
$$

If $s(k) \leq -1$ then $|s(k)|^b \leq |s(k)| \leq |s(k)|^3$, thus based on this and Assumption 4.5, we can get that

$$
\begin{aligned}
s(k+1) &\leq -|s(k)| + \Phi_1 h|s(k)|^3 + \Phi_2 h|s(k)|^b + |\Phi_0^\star h| \\
&\leq -(1-\Phi_2 h)|\theta\Psi(b)\Phi| + \Phi_1 h|\theta\Psi(b)\Phi|^3 + |\Phi_0^\star h| \\
&\leq |\Phi_0^\star h| \leq (1-\Phi_1 h)\Phi \\
&< \Psi(b)\Phi = \xi.
\end{aligned}
$$

Secondly, $s(k+1) > -\xi$ will be proved. Similarly, two different cases are discussed as follows.

Case 1. When $\theta\Psi(b) \geq 0$, it is clear that $s(k) \geq 0$ and $s(k+1) = s(k) - \Phi_1 hs^3(k) - \Phi_2 hs^b(k) + \Phi_0(k)h$.

If $s(k) \geq 1$ then $-s(k)^3 \leq -s(k) \leq -s(k)^b$, based on this and Proposition 4.7, we have

$$
\begin{aligned}
s(k+1) &\geq s(k) - \Phi_1 hs^3(k) - \Phi_2 hs^b(k) - |\Phi_0^\star h| \\
&\geq (1-\Phi_2 h)s(k) - \Phi_1 hs^3(k) - |\Phi_0^\star h| \\
&\geq -|\Phi_0^\star h| \geq -(1-\Phi_1 h)\Phi > -\xi.
\end{aligned}
$$

If $0 \leq s(k) < 1$ then $-s^3(k) \geq -s(k) \geq -s^b(k)$, based on this and Proposition 4.7, we have

$$\begin{aligned} s(k+1) &\geq s(k) - \Phi_1 h s^3(k) - \Phi_2 h s^b(k) - |\Phi_0^\star h| \\ &\geq s(k) - \Phi_1 h s(k) - \Phi_2 h s^b(k) - |\Phi_0^\star h| \\ &\geq (1 - \Phi_1 h)\Phi \left[\theta\Psi(b) - \theta^b \Psi^b(b) - 1\right] \\ &\geq -\Psi(b)(1 - \Phi_1 h)\Phi \\ &> -\Psi(b)\Phi = -\xi. \end{aligned}$$

Case 2. When $\theta\Psi(b) < 0$, it is clear that $s(k) < 0$ and $s(k+1) = -|s(k)| + \Phi_1 h|s(k)|^3 + \Phi_2 h|s(k)|^b + \Phi_0(k)h$.

In this case, by Proposition 4.7, we have

$$\begin{aligned} s(k+1) &\geq -|s(k)| + \Phi_1 h|s(k)|^3 + \Phi_2 h|s(k)|^b - |\Phi_0^\star h| \\ &\geq -|\theta\Psi(b)\Phi| + (|\theta\Psi|^b - 1)|\Phi_0^\star h| + \Phi_1 h|\theta\Psi(b)\Phi|^3 \\ &\geq -\left[|\theta\Psi(b)\Phi| + (1 - |\theta\Psi(b)|^b)|\Phi_0^\star h|\right]. \end{aligned}$$

If $|\theta\Psi(b)| \geq 1$, it is obvious that $|\theta\Psi(b)|^b \geq -\xi$.

If $|\theta\Psi(b)| < 1$, it is clear that

$$\begin{aligned} s(k+1) &\geq -\Phi\left[|\theta\Psi(b)| + (1 - |\theta\Psi(b)|^b)(1 - \Phi_1 h)\right] \\ &= -\Phi\left[1 + (|\theta\Psi(b)|^b - 1)\Phi_1 h + |\theta\Psi(b)| - |\theta\Psi(b)|^b\right] \\ &\geq -\Phi > -\Psi(b)\Phi = -\xi. \end{aligned}$$

In summary, if $s(k) \in \Xi$ then $s(k+1) \in \Xi$ always holds. Here completes the proof. ■

4.4.3 Tracking Error Analysis

In Theorem 4.4, we have discussed that after the sliding surface (4.15) is reached, $\check{p}(k)$ will converge to zero if a certain condition is satisfied. However, as shown in Theorems 4.8 and 4.10, due to the existence of parametric uncertainties, $s(k)$ can only reach the region Ξ around the surface. Therefore, $\check{p}(k)$ can only converge to the vicinity of zero correspondingly. Therefore, the performance of the tracking error needs to be studied further.

Theorem 4.11 *For the discrete-time tracking control system (4.4)–(4.6) under Assumption 4.5 and the DFSM controller (4.17), the tracking error satisfies*

$$|\check{p}(k)|_{\max} \leq \frac{\xi\beta h^{\lambda-1}}{Kq}, \tag{4.24}$$

where $|\check{p}(k)|_{max}$ is a local maximum from $\check{p}(k-L)$ to $\check{p}(k)$ and β is a selected time variant parameter, which satisfies $\beta > 1$.

Proof As the sliding surface (4.15) shows that $s(k) = v(k) - \Delta p_r(k) + q\Delta^{\lambda-1}\check{p}(k)$, we have

$$
\begin{aligned}
|s(k)| &\geq |v(k) - \Delta p_r(k)| + |q\Delta^{\lambda-1}\check{p}(k)| \\
&\geq q|\Delta^{\lambda-1}\check{p}(k)|.
\end{aligned}
$$

Therefore,

$$|\Delta^{\lambda-1}\check{p}(k)| \leq \frac{|s(k)|}{q}, \tag{4.25}$$

gives the vicinity around zero that the tracking error $\check{p}(k)$ can reach.

On the other hand, according to Lemma 4.2, we can see that

$$|\Delta^{\lambda-1}\check{p}(k)| = \frac{KM}{h^{\lambda-1}\beta}. \tag{4.26}$$

Combining (4.25) and (4.26), it is easy to get

$$\frac{K|\check{p}(k)|_{\max}}{h^{\lambda-1}\beta} \leq \frac{|s(k)|}{q} \leq \frac{\xi}{q}.$$

Here, we can conclude that

$$|\check{p}(k)| \leq |\check{p}(k)|_{\max} \leq \frac{\xi\beta h^{\lambda-1}}{Kq}.$$

This completes the proof. ■

Remark 4.12 Through further research into (4.24), we can find approaches to tune DFSM parameters to get a smaller error in practice. Firstly, noticing (4.24) and the relationship between ξ and q, it is easy to find that a larger value of q leads to a smaller error as long as the stability of the sliding surface is guaranteed. Secondly, according to (4.22), we can find that a smaller k_1 and a larger k_2 within their limits result in a smaller ξ, which leads to a smaller error finally.

4.5 Simulations and Experiments

4.5.1 Simulations

In order to validate the stability analyzed in Section 4, two groups of simulations in the Matlab environment are carried out. The first group of simulations consists of three tests. They are implemented to study the dynamics of $\check{p}(k)$ on the sliding

Table 4.1 Parameters of the sliding mode surface

Parameter	Test 1	Test 2	Test 3
λ	0.5	0.5	0.5
q	100	1500	4000
h	0.01	0.01	0.01
L	20	20	20

surface. The parameters of different sliding surfaces (4.15) are listed in Table 4.1. The simulation results of Tests 1 and 2 are shown in (a1)–(b1) and (a2)–(b2) of Fig. 4.2, respectively. It proves that when z_0 is in the same region with origin (no matter z_0 is to the left or the right of origin), the state of $\check{p}(k)$ converges to zero asymptotically. Besides, we can also find that the closer z_0 is to the origin, the faster $\check{p}(k)$ converges to zero. The simulation result of Test 3 is illustrated in (a3) and (b3) of Fig. 4.2. In (a3) of the figure, z_0 is not in the same region as with the origin, and accordingly, the state of $\check{p}(k)$ diverges. In summary, all these simulation results accord well with Theorem 4.4.

The second group of simulations is implemented to study the trajectories of $s(k)$ under different controller parameters. We set the parameters of the practical model as $m = 11$ kg, $k_m = 0.03$, $k_c = 2.91$ N·s/mm, $k_v = 0.08$ N and $k_f = 10$ N/V. The parameters of controller (4.17) are listed in Table 4.2. Besides, $s(0) = 5$ and $h = 0.01$ s are selected to satisfy Assumption 4.5. Figure 4.3 demonstrates the simulation results of Tests 4–6. First of all, we design Test 4 to study the dynamics of $s(k)$ when the ideal model (without any uncertainties and disturbances) is considered. As shown in (a) of Fig. 4.3, the switching function converges to region Ξ in finite steps, and after the trajectory of $s(k)$ crosses the switching manifold for the first time, it crosses the switching manifold again at every successive sampling interval, which accords well with the conclusion in [10].

Secondly, based on Test 4, some parameters are changed and uncertainties are taken into consideration to get Test 5, the result of which can be seen in (b) of Fig. 4.3. Obviously, after several steps, $s(k)$ reaches region Ξ and keeps floating in Ξ, which accords well with Theorems 4.8 and 4.10.

In order to study further, Test 6 is conducted, in which more parameters are changed compared with Test 4. As shown in (c) of Fig. 4.3 although the curve of the switching function of Test 6 is not as smooth as that of Test 5, the curve still obeys the rules described by Theorems 4.8 and 4.10. So far, the stability of the system has been validated by simulations.

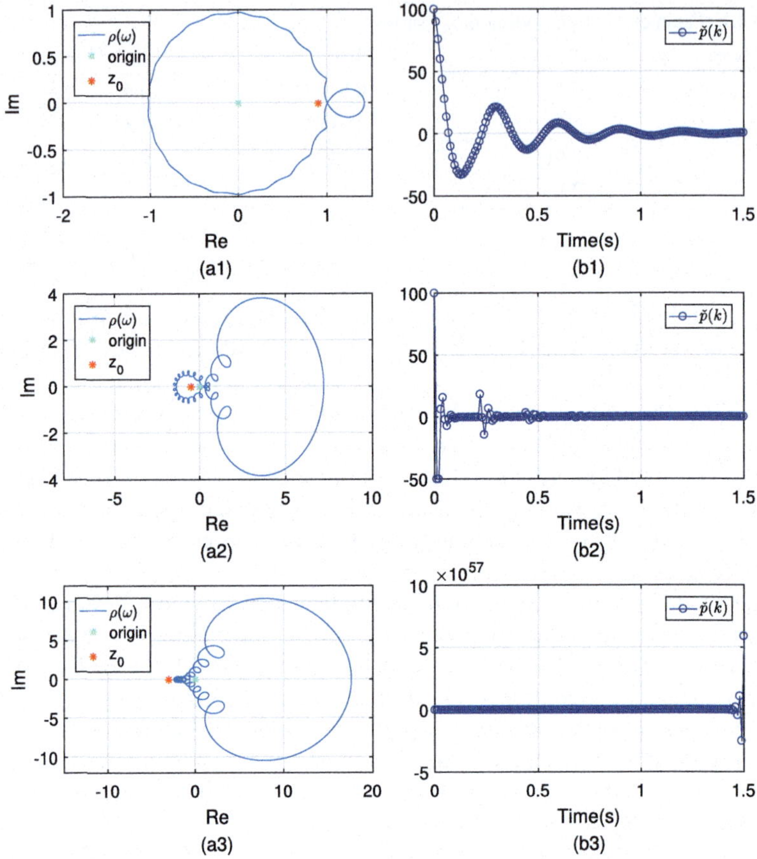

Fig. 4.2 The position of z_0 on the complex plane and the state of $\check{p}(k)$. (a·) The position of z_0; (b·) The state of $\check{p}(k)$; (·1) Test 1; (·2) Test 2; (·3) Test 3

4.5.2 Experiments

To verify the effectiveness of the proposed DFSM scheme, experiments of cosinoidal signal and triangular signal are implemented on the real linear motor platform. As few experiments on the linear motor system have been carried out in the field of discrete-time sliding mode control, we refer to the control approaches proposed in [1, 5], and apply them to the linear motor platform for comparison.

The basic physical parameters of the linear motor platform are $\bar{m} = 11\,\text{kg}$ and $\bar{k}_m = 0.03$, and the parameters about the friction can be obtained by uniform linear motion testing and utilizing particle swarm optimization: $\bar{k}_v = 0.032\,\text{N s/mm}$ and $\bar{k}_c = 9.738\,\text{N}$. As for the controllers, the sampling interval h is decided on 0.001 s to satisfy Assumption 4.5. In the DFSM controller, b and λ are selected as 0.5 without loss of generality, L is selected as 20 to balance the accuracy of the system and the

Table 4.2 Parameters of controller

Parameter	Test 4	Test 5	Test 6
$\bar{m}$	11	10.8	10.8
$\bar{k}_m$	0.03	0.032	0.032
$\bar{k}_c$	2.91	3	3
$\bar{k}_v$	0.08	0.085	0.085
q	100	10	110
k_1	1	0.1	0.1
k_2	1	3	3
λ	0.5	0.5	0.2
b	0.5	0.5	0.9
L	20	20	20

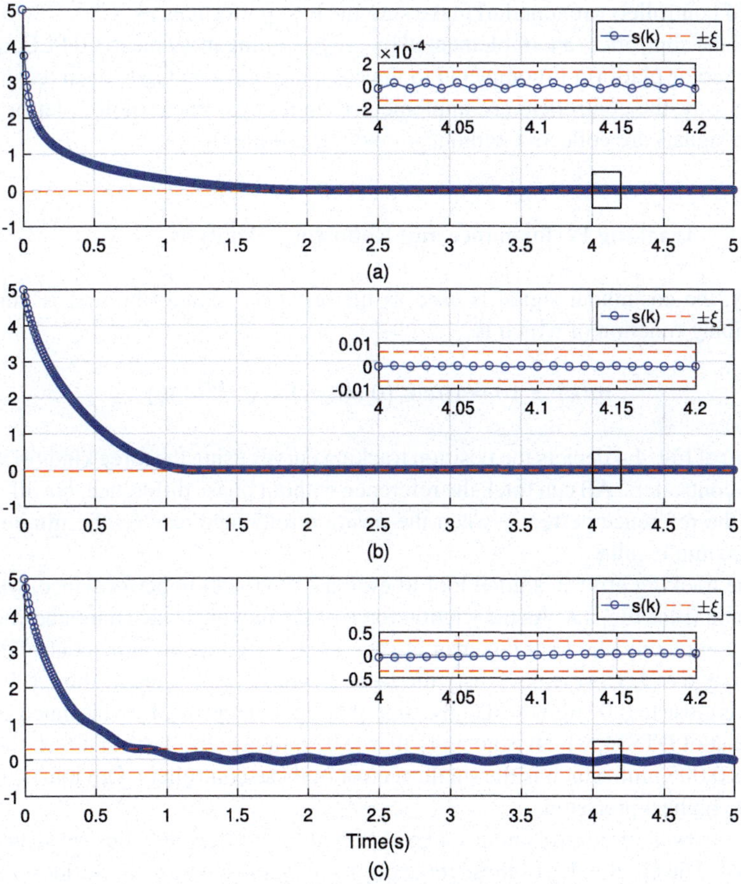

Fig. 4.3 Switching function and region ξ of **a** Test 4; **b** Test 5; **c** Test 6

Table 4.3 Parameters of experiment

Parameter	Cosinoidal tracking	Triangular tracking
q	3500	2000
k_1	2000	120
k_2	55000	70000
λ	0.5	0.5
b	0.5	0.5
L	20	20

amount of computation. Other parameters of the DFSM controller are tuned using the method in Remark 4.12 and the values are shown in Table 4.3. To guarantee the rationality and persuasion of our comparison, the parameters in the CDSM and the DTSM controllers are tuned to make sure the best performances.

All the controllers are implemented by programming in Twincat 3.0 PLC, and the linear motor platform is connected to the computer through CAN open technology. In this way, the controllers are implemented on the real linear motor platform, and all the signals are collected without smoothing or filtering.

4.5.2.1 Tracking Performance and Robustness Analysis

Firstly, the cosinoidal signal is used as the reference signal (the unit is mm), the describing function of which is

$$p_r(k) = 10 - 10\cos\left(0.001\pi(1 - e^{-0.001k})k\right).$$

Part (a) of Fig. 4.4 depicts the position tracking curves using the three kinds of sliding mode controllers. All can track the reference without phase difference, but all cannot track the reference perfectly when the position reference reaches the minimum or the maximum value.

The tracking error is a good tool to evaluate the tracking performance, which is shown in (b)of Fig. 4.4. As the friction force of the moving stage cannot be compensated completely, the tracking error always exists. Among the three controllers, the error range of DTSM is obviously the largest, whose peak value is about 0.07 mm, and it is close to 0.04 mm most of the time. The error range of CDSM is much smaller than that of DTSM, which is no more than 0.02 mm usually. As for DFSM, the error covers 0.005 mm, which is the smallest error of these controllers. Obviously, DFSM has the highest precision.

The curves of control input under different controllers are illustrated in **(c)** of Fig. 4.4. The amplitudes of these three curves are almost the same. Besides, through the enlarged area window, we find that the curve of DTSM keeps chattering with a large amplitude, which is caused by the control approach, while the curves of DFSM

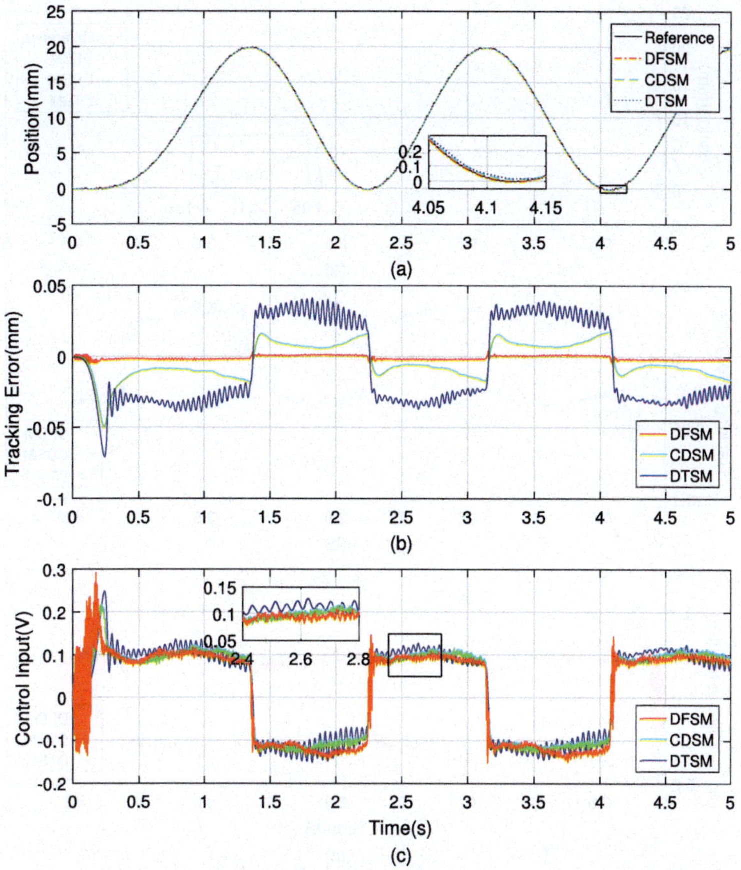

Fig. 4.4 Performances of tracking a cosinoidal reference signal when no extra payload is applied. **a** Position tracking responses; **b** Tracking errors; **c** Control inputs

and CDSM are only chattering with a rather small amplitude, which is mainly caused by measurement noises instead of control approach.

Since the payload change of the linear motor system is really common in practical industrial usage, we conducted experiments to investigate the robustness of our controller against the change of payload. Experiments are implemented with an extra 3.3 kg payload (accounts for 30% of the moving stage's mass) fixed on the stage and all the controller parameters remain the same as those before. As Fig. 4.5 shows all three controllers almost maintain the same performance. This phenomenon indicates that all of these three sliding mode controllers have good robustness.

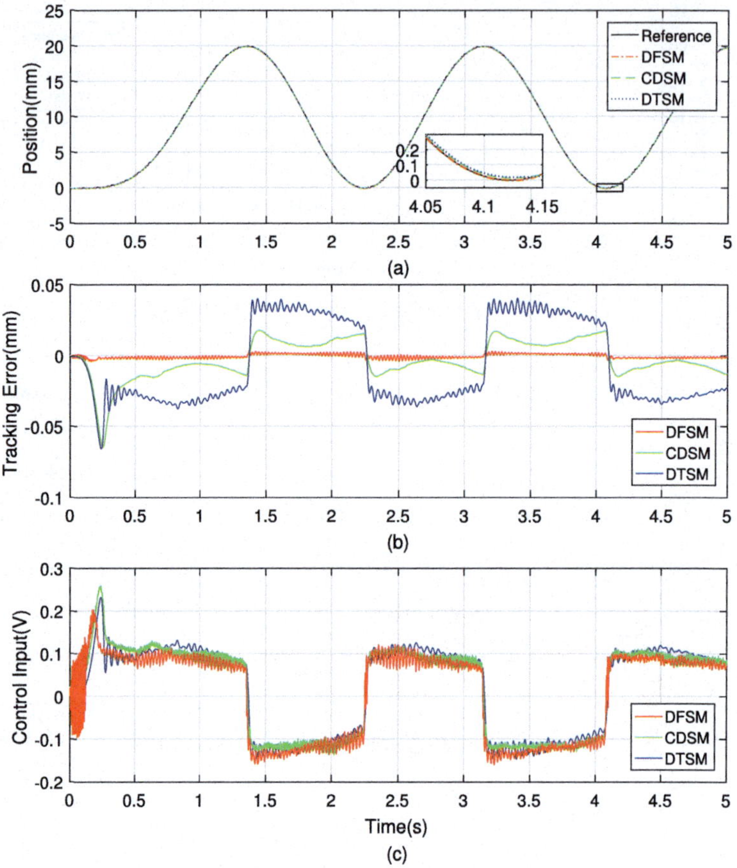

Fig. 4.5 Performances of tracking a cosinoidal reference signal when 3.3 kg extra payload is applied. **a** Position tracking responses; **b** Tracking errors; **c** Control inputs

4.5.2.2 Slope-Varying Triangular Tracking Performance and Robustness Analysis

To further verify the effectiveness, a slope-varying triangular reference signal (the unit is mm) is designed, of which the describing function is written as

$$p_r(k) = \begin{cases} 0.01k/1.6, & 0 \le k < 1600, \\ 10-10(0.001k-1.6)/1.4, & 1600 \le k < 3000, \\ 10(0.001k-3)/1.2, & 3000 \le k < 4200, \\ 10-10(0.001k-4.2), & 4200 \le k < 5000. \end{cases}$$

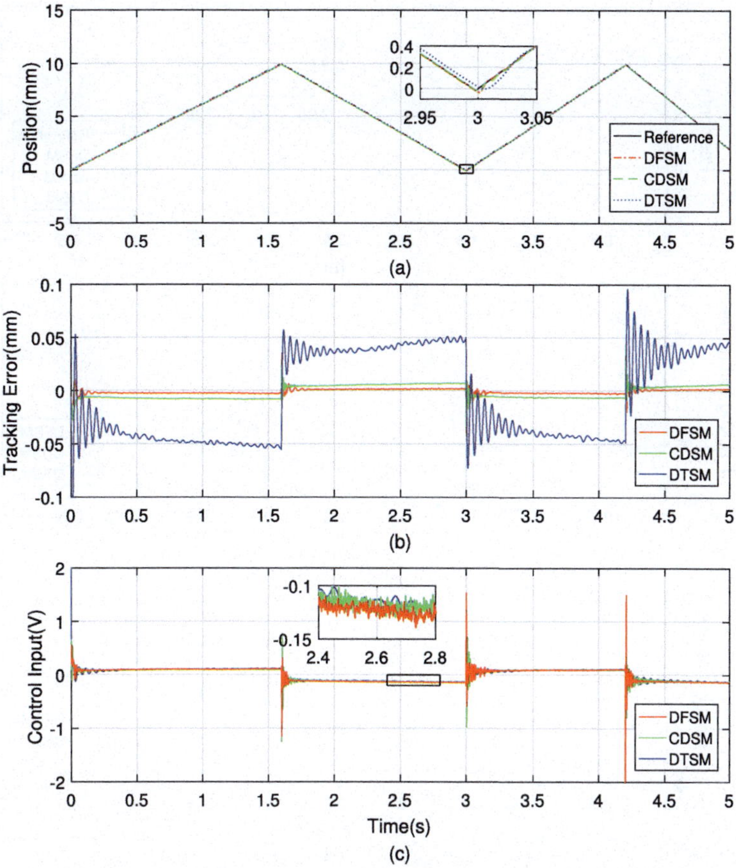

Fig. 4.6 Performances of tracking a slope-varying triangular reference signal when no extra payload is applied. **a** Position tracking responses; **b** Tracking errors; **c** Control inputs

When no extra payload is attached to the moving stage, Fig. 4.6 shows the position tracking curves in (a), the tracking errors in (b) and control inputs under different controllers in (c), respectively. We find that when the motion changes direction, the three approaches cannot track the reference signal very well, and the errors reach the peak value. When the stage moves in the same direction, the DFSM has the smallest error, whose value is about 0.002 mm. While CDSM and DTSM have larger errors, whose values are about 0.005 and 0.04 mm, respectively. As for the control input, the amplitudes of DFSM and CDSM are more stable than those of DTSM.

When the extra 3.3 kg load mass is attached to the moving stage, the position curve, the tracking error curve, and the control input curve are shown in Fig. 4.7. The results prove that the DFSM performs a noticeable robustness against the uncertainty once more.

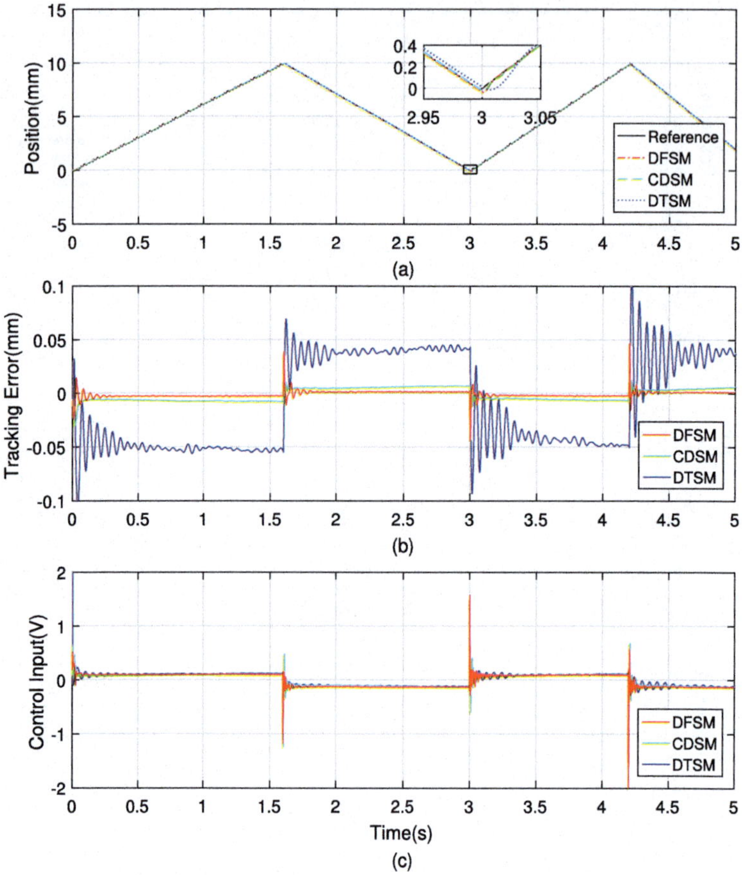

Fig. 4.7 Performances of tracking a slope-varying triangular reference signal when 3.3 kg extra payload is applied. **a** Position tracking responses; **b** Tracking errors; **c** Control inputs

4.5.2.3 Comparison Analysis and Summary

The qualitative analysis has shown that DFSM has a good performance when it is conducted on the linear motor platform. To make the conclusion more convincing, quantitative analysis and comparison will be conducted in this section.

First of all, the definition of root-mean-square (RMS) error is given as follows

$$RMS = \sqrt{\sum_{i=1}^{N} \frac{\check{p}^2(i)}{N}}, \tag{4.27}$$

where N denotes the sample number. Apparently, RMS errors describe the average precision of the tracking performance.

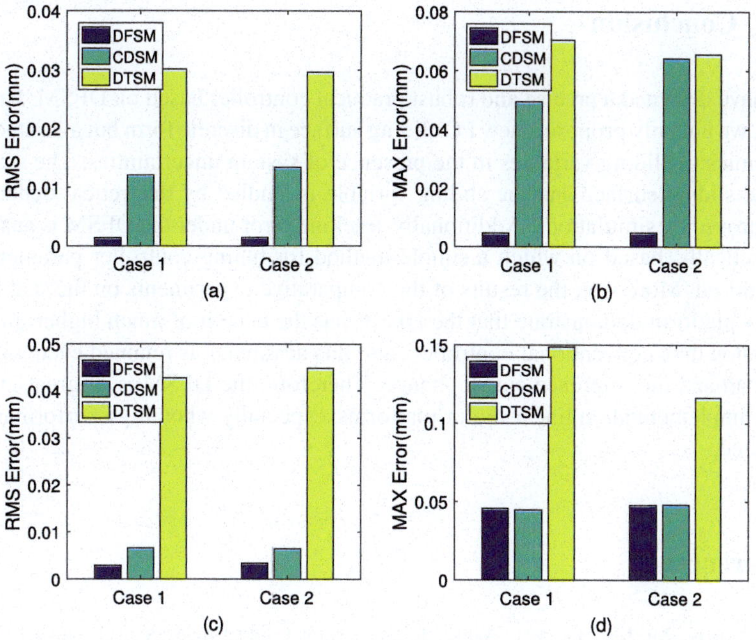

Fig. 4.8 Tracking error comparison between experiments (Case 1: without payload. Case 2: with 3.3 kg extra payload). **a** Root-mean-square errors when cosinoidal reference is tracked; **b** Maximum errors when cosinoidal reference is tracked; **c** Root-mean-square errors when slope-varying triangular reference is tracked; **d** Maximum errors when slope-varying triangular reference is tracked

To some extent, the maximum (MAX) error can also reflect the precision. Moreover, it plays a rather important role under certain circumstances in industry. Thus, we calculate all the statistics of RMS errors and MAX errors under the circumstances above. The results are displayed as bar charts in Fig. 4.8.

In Fig. 4.8a, b display that the RMS error and MAX error of DFSM is much smaller than those of CDSM and DTSM when the system is tracking a sinusoidal reference, no matter extra payload is attached or not. We can also find that when the mass of the stage changes, the tracking precision of DFSM changes little, which means that the robustness of DFSM is promising.

Analogously, (c) and (d) of Fig. 4.8 demonstrate that the RMS error of DFSM is the smallest when the system is tracking a triangular reference. The MAX error of DFSM is very close to that of CDSM, and they are much smaller than that of DTSM. In addition, the comparison between Cases 1 and 2 also proves the system under DFSM controller is robust.

In summary, no matter whether DFSM is implemented to track a cosinoidal signal or a slope-varying triangular signal on the real linear motor platform performs a high precision and a noticeable robustness.

4.6 Conclusion

We have designed a precise and robust practical controller based on DFSM. Particularly, we not only propose a new FO sliding surface in discrete form but also study the dynamics of sliding variables in the presence of system uncertainties. The stability of the sliding surface and the sliding variable is studied by theoretical deductions and proved by simulations. Additionally, tracking error under the DFSM is analyzed theoretically, based on which a simple method for tuning controller parameters is introduced. Moreover, the results of the comparative experiments on the real linear motor platform demonstrate that the DFSM has the benefit of much higher tracking precision than conventional controllers and this advantage is retained even with the payload and the reference signal change. Therefore, the DFSM is of great interest to be implemented on linear motor platforms, especially when high performance is required.

References

1. Abidi, K., Xu, J.X., She, J.H.: A discrete-time terminal sliding-mode control approach applied to a motion control problem. IEEE Trans. Ind. Electron. **56**(9), 3619–3627 (2008)
2. Bandyopadhyay, B., Kamal, S.: Stabilization and Control of Fractional Order Systems: A Sliding Mode Approach. Springer, Switzerland (2015)
3. Busłowicz, M., Kaczorek, T.: Simple conditions for practical stability of positive fractional discrete-time linear systems. Int. J. Appl. Math. Comput. Sci. **19**(2) (2009)
4. Busłowicz, M., Ruszewski, A.: Necessary and sufficient conditions for stability of fractional discrete-time linear state-space systems. Bull. Polish Acad. Sci. Tech. Sci. **61**(4), 779–786 (2013)
5. Du, H., Yu, X., Chen, M.Z., Li, S.: Chattering-free discrete-time sliding mode control. Automatica **68**, 87–91 (2016)
6. Gryazina, E.N., Polyak, B.T., Tremba, A.A.: D-decomposition technique state-of-the-art. Autom. Remote Control **69**(12), 1991–2026 (2008)
7. Li, S., Du, H., Yu, X.: Discrete-time terminal sliding mode control systems based on Euler's discretization. IEEE Trans. Autom. Control **59**(2), 546–552 (2013)
8. Li, Y., Chen, Y., Podlubny, I.: Mittag-leffler stability of fractional order nonlinear dynamic systems. Automatica **45**(8), 1965–1969 (2009)
9. Podlubny, I.: Fractional Differential Equations. Academic, New York, NY, USA (1999)
10. Qu, S., Xia, X., Zhang, J.: Dynamics of discrete-time sliding-mode-control uncertain systems with a disturbance compensator. IEEE Trans. Ind. Electron. **61**(7), 3502–3510 (2013)
11. Su, W.C., Drakunov, S.V., Ozguner, U.: An O (T^2) boundary layer in sliding mode for sampled-data systems. IEEE Trans. Autom. Control **45**(3), 482–485 (2000)
12. Wang, Y., Luo, G., Gu, L., Li, X.: Fractional-order nonsingular terminal sliding mode control of hydraulic manipulators using time delay estimation. J. Vib. Control **22**(19), 3998–4011 (2016)
13. Yin, C., Chen, Y., Zhong, S.M.: Fractional-order sliding mode based extremum seeking control of a class of nonlinear systems. Automatica **50**(12), 3173–3181 (2014)
14. Zhang, B., Pi, Y., Luo, Y.: Fractional order sliding-mode control based on parameters auto-tuning for velocity control of permanent magnet synchronous motor. ISA Trans. **51**(5), 649–656 (2012)
15. Zheng, J., Fu, M.: A reset state estimator using an accelerometer for enhanced motion control with sensor quantization. IEEE Trans. Control Syst. Technol. **18**(1), 79–90 (2009)

Chapter 5
Practical Tracking Control via Adaptive Fractional-Order Terminal SMC

5.1 Introduction

This chapter proposes a practical adaptive FO terminal SMC strategy for tracking control of the linear motor. Compared with the conventional fast nonsingular SMC, the proposed approach, with a FO integral sliding surface and the adaptive switching input, can obtain higher convergence precision, even though the motion control system suffers from system uncertainties. The adaptive term is designed to guarantee finite-time high-precision convergence of the sliding mode variable and to degenerate the effect of uncertainties by selecting the proper adaptive gain. Moreover, continuous input due to cancelling the signed term ensures that the motion control system is chattering-free. Finally, to further improve the precision, we introduce the super-twisting sliding mode disturbance observer for reducing unknown bounded disturbance, i.e., the quantization noise caused by velocity estimation. Experimental results indicate that the control system with the proposed controller is easily implemented and has higher tracking precision and considerable robustness to uncertainties compared with the existing controllers.

5.2 Model Review of X-Y Servo Platform

The experimental facility (by Akribis System Company) used for verifying the effectiveness of the designed linear motor based motion control system is shown in Fig. 4.1. An absolute position encoder is equipped on the moving stage, which is driven directly by the rotor of the linear motor and has a 50 cm travel range along the linear guide. The plant model of the linear motor can be given by [21]

$$m\ddot{p} = (1 - k_m)(k_f u - f - d), \tag{5.1}$$

$$f = k_c \mathrm{sgn}(\dot{p}) + k_v \dot{p}, \tag{5.2}$$

G. Sun et al., *Fractional-Order Sliding Mode Control: Methodologies and Applications*, Studies in Systems, Decision and Control 532,
https://doi.org/10.1007/978-3-031-60847-6_5

where m denotes the mass of the moving stage, p means the absolute position, k_m represents the gain of the variable load, k_f is the voltage-to-force amplifier gain, d stands for the bounded lumped uncertainty including disturbance and measurement noise. In (5.2), we formulate the simplified nonlinear friction model f, which regards the static force effect as an uncertainty. Hence, a certain part of the friction force contains Coulomb friction $k_c sgn(\dot{p})$ and viscous friction $k_v \dot{p}$.

In a practical motion control system, parametric uncertainty exists and satisfies the following conditions

$$\begin{aligned} &\frac{1}{\theta_m} \le \frac{\bar{m}}{m} \le \theta_m, \quad \frac{1}{\theta_{k_m}} \le \frac{1-\bar{k}_m}{1-k_m} \le \theta_{k_m}, \\ &|\bar{k}_c - k_c| \le \theta_{k_c}, \quad |\bar{k}_v - k_v| \le \theta_{k_v}, \quad |d| \le \theta_d, \end{aligned} \tag{5.3}$$

where $\bar{\bullet}$ stands for the nominal model term, and $\theta_\bullet$ is the positive scalar denoting the difference between the actual plant model and the nominal model.

5.3 Controller Design

This work aims to design a reliable motion control strategy that is insensitive to uncertainties and disturbances. To satisfy the requirements of tracking performance, an adaptive FO terminal SMC scheme is provided such that the moving stage can track the reference signal precisely and rapidly. In reality, The velocity signal is estimated by a numeric approach depending on the position measurement. Since the position measurement itself is corrupted by the quantization noise, the estimated velocity is also contaminated. To deal with this uncertainty and unmodeled disturbance, the SMDO is synthesized into the motion control system. The block diagram of the control strategy is shown in Fig. 5.1. This section will analyze the reachability of the adaptive switching control and the tracking precision involving the FO integral terminal sliding surface.

5.3.1 *Preliminaries*

To construct the adaptive FO terminal sliding mode controller, some preliminary knowledge of the FO calculus and terminal sliding mode theory is exhibited.

Definition 5.1 ([7]) The uniform FO integration operator is defined as

$${}_{t_0}D_t^{-\nu} f(t) = \frac{1}{\Gamma(\nu)} \int_{t_0}^{t} \frac{f(\varsigma)}{(t-\varsigma)^{1-\nu}} d\varsigma, \tag{5.4}$$

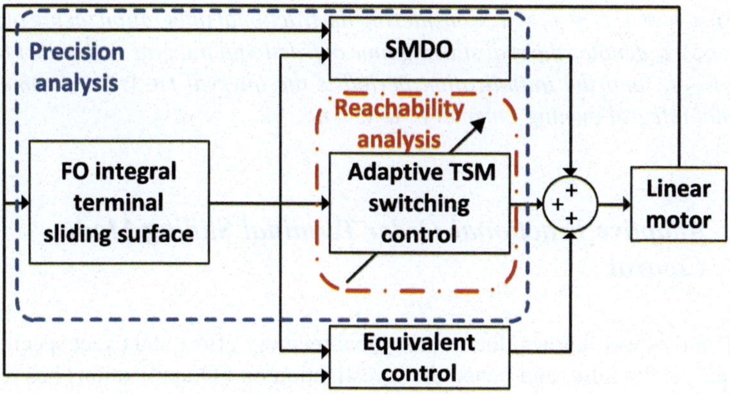

Fig. 5.1 Block diagram of control strategy

where $f(t)$ is locally integrable, $\Gamma(x) = \int_0^\infty \varsigma^{x-1} e^{-\varsigma} d\varsigma$ is the Gamma function. t_0 and t are the integral terminal values and $0 < \nu < 1$. Caputo fractional derivative can be calculated by the following expression:

$$ {}_{t_0}D_t^{\nu} f(t) = \frac{1}{\Gamma(1-\nu)} \int_{t_0}^{t} \frac{f'(\varsigma)}{(t-\varsigma)^{\nu}} d\varsigma. \tag{5.5}$$

Lemma 5.2 ([7]) *Denoting the real part of c as $\Re(c)$, if the fractional integration operators $D_{a+}^{-\nu}$ and $D_{b-}^{-\nu}$ with $\Re(c) > 0$ are bounded in $L_\iota(a, b)$, $1 \le \iota \le \infty$, $\iota \in N$, the following bounded conditions hold*

$$\|D_{a+}^{-\nu} f(x)\|_\iota \le K \|f(x)\|_\iota, \quad \|D_{b-}^{-\nu} f(x)\|_\iota \le K \|f(x)\|_\iota,$$
$$K = \frac{(b-a)^{\Re(c)}}{\Re(c)|\Gamma(c)|}.$$

Lemma 5.3 ([17]) *If the Lyapunov function $V(x)$ with the initial value V_0 satisfies*

$$\dot{V}(x) + \alpha V(x) + \beta V^{\gamma}(x) \le 0, \tag{5.6}$$

where α and β are positive scalars, $0 < \gamma < 1$, one can deduce the corresponding finite convergence time as

$$T \le \frac{\ln(\beta + \alpha V_0^{1-\gamma}) - \ln \beta}{\alpha(1-\gamma)}. \tag{5.7}$$

Property 5.4 ([20]) *The initialized fractional integral is defined as*

$$\begin{aligned} {}_{t_s}D_t^{-\nu} f(t) &= {}_{t_0}D_t^{-\nu} f(t) + \psi(f, -\nu, t_s, t_0, t), \\ \psi(f, -\nu, t_s, t_0, t) &= \frac{1}{\Gamma(\nu)} \int_{t_s}^{t_0} \frac{f(\varsigma)}{(t-\varsigma)^{1-\nu}} d\varsigma, \end{aligned} \tag{5.8}$$

where $0 < \nu < 1, t_s \leq t_0 < t$. *Considering the fractional integration taking place for* $t \in [t_0, \infty)$, t_s *denotes the real starting time of fractional integral such that* $f(t) = 0$ *for all* $t \leq t_s$, *and the initialization period is the interval* $t \in [t_s, t_0]$, *namely, the fractional integral owning "memory" at* $t = t_0$.

5.3.2 Adaptive Fractional-Order Terminal Sliding Mode Control

FO calculus is well known due to the global memory effect and exact specification of the physical nature, and hence the investigation on FO applications has recently become a hot issue [2, 8, 9, 11, 15, 18]. Besides the studies on FO systems, many FO controllers have been designed for FO and integer-order systems to obtain better performance than the conventional integer-order controllers [1, 3]. Regarding the FO sliding mode control, previous experimental results have verified the effectiveness of robotic manipulator maneuver [14] and velocity control of motor [19].

Motivated by the above factors, we regard the FO sliding mode technique as one essential component of the motion controller. However, the performance of the bare FO sliding mode control is just passable, and hence to achieve fast and precise tracking, a high-precision robust control law synthesized with the FO terminal sliding surface is also needed. Using the experience of the precision control scheme presented in [21], we further design the FO terminal sliding mode controller by introducing an iterative update term and FO terminal sliding surface in the conventional fast terminal sliding mode control law to improve the tracking performance directly. Moreover, the iterative update term contributes to the final tracking precision since it can be designed as an adaptive term to weaken the effect of uncertainties.

The adaptive FO terminal SMC strategy is designed based on the FO terminal sliding surface [13, 14], and the initialization function is taken into account:

$$s = \dot{\check{p}} + k \left[{}_{t_0}D_t^{\lambda-1} \lceil \check{p} \rfloor^a + \psi(\lceil \check{p} \rfloor^a, 1-\lambda, t_s, t_0, t) \right], \tag{5.9}$$

where $\check{p} = p - p_r$ is defined as the tracking error, p_r is the reference signal, p is the actual position, $\dot{\check{p}}$ is the velocity tracking error, $0 < \lambda < 1$, $k > 0$, and $\lceil \bullet \rfloor^a = |\bullet|^a \text{sgn}(\bullet)$ with $0 < a < 1$. The starting time of the fractional integral is t_s, namely, $\lceil \check{p} \rfloor^a = 0$ for all $t \leq t_s$, and the adaptive FO terminal SMC strategy begins working at instant $t = t_0$. $\psi(\lceil \check{p} \rfloor^a, 1-\lambda, t_s, t_0, t)$ represents the initialization function from t_s to t_0. These parameter selections make the motion on the sliding surface finite-time stable, which has been proved in [14], and the manipulator regulation based on this kind of sliding surface has also verified its engineering applicability [13].

Subjected to the nominal plant model, taking the derivative of s respective to time obtains

$$\begin{aligned} \dot{s} =& \frac{1-\bar{k}_m}{\bar{m}}(k_f u - \bar{k}_c \mathrm{sgn}(\dot{p}) - \bar{k}_v \dot{p} - d) \\ & - \ddot{p}_r + k\left({}_{t_s}D_t^{\lambda}\lceil \check{p} \rfloor^a + \frac{\lceil \check{p}(t_s) \rfloor^a}{\Gamma(1-\lambda)}(t-t_s)^{-\lambda}\right). \end{aligned} \tag{5.10}$$

Neglecting the lumped uncertainty, introducing $\dot{s} = 0$ into the above equation and replacing u by u_{eq}, gives rise to

$$\begin{aligned} u_{eq} =& \frac{1}{k_f}(\bar{k}_c \mathrm{sgn}(\dot{p}) + \bar{k}_v \dot{p}) + \frac{\bar{m}}{k_f(1-\bar{k}_m)}\left(\ddot{p}_r \right. \\ & \left. -k\left[{}_{t_s}D_t^{\lambda}\lceil \check{p} \rfloor^a + \frac{\lceil \check{p}(t_s) \rfloor^a}{\Gamma(1-\lambda)}(t-t_s)^{-\lambda}\right]\right). \end{aligned} \tag{5.11}$$

Furthermore, based on fast-terminal-sliding-mode-type reaching law [14, 21], we introduce the adaptive term to develop a novel adaptive terminal sliding mode switching control input u_{sw} expressed by

$$u_{sw} = -\frac{\bar{m}}{k_f(1-\bar{k}_m)}(k_1 s^3 + k_2 \lceil s \rfloor^{\alpha} + k_3 \xi,) \tag{5.12}$$

$$\dot{\xi} = -k_1 \xi^3 - k_2 \lceil \xi \rfloor^{\alpha} + k_3 s, \tag{5.13}$$

where $0 < \alpha < 1$, $k_1 > 0$ and $k_2 > 0$, k_3 can be designed as a constant or a time-varying parameter. The adaptive term ξ is introduced to suppress the adverse effect of uncertainties.

Theorem 5.5 *Consider the LM plant model (5.1) with the uncertainties satisfying conditions shown in (5.3), and under the regulation of the AFOTSM control strategy*

$$u = u_{eq} + u_{sw}, \tag{5.14}$$

where the form of u_{eq} and u_{sw} are given in (5.11) and (5.12), respectively, one can obtain the following tracking precision.

(1) The sliding mode dynamics are guaranteed to reach the following convergence region in finite time

$$\begin{aligned} |s| \le \Gamma = \min(\Gamma_1, \Gamma_2), \Gamma_1 = \left[\frac{\left|\theta_m \theta_{k_m} - 1\right| |\vartheta_d| + \frac{1-\bar{k}_m}{\bar{m}}\varrho}{k_1}\right]^{\frac{1}{3}}, \\ \Gamma_2 = \left[\frac{\left|\theta_m \theta_{k_m} - 1\right| |\vartheta_d| + \frac{1-\bar{k}_m}{\bar{m}}\varrho}{k_2}\right]^{\frac{1}{\alpha}}, \end{aligned} \tag{5.15}$$

where ϑ_d is a small positive scalar describing the desired boundary of the uncertainty gain. The boundary of the friction force f and the lumped uncertainty d is denoted as

$$\varrho = \theta_{k_v}|\dot{p}| + \theta_{k_c} + \theta_d \geq (k_v - \bar{k}_v)\dot{p} + (k_c - \bar{k}_c)sgn(\dot{p}) + d. \tag{5.16}$$

(2) The tracking error satisfies the following precisions

$$|\check{p}| \leq \left(\frac{\beta\Gamma}{kK}\right)^{\frac{1}{a}}, \ \left|\dot{\check{p}}\right| \leq 2\Gamma, \ \beta \geq 1. \tag{5.17}$$

Proof Consider the following Lyapunov candidate function

$$V = \frac{1}{2}s^2 + \frac{1}{2}\xi^2, \tag{5.18}$$

where the adaptive term (5.13), and the derivative of V are subjected to the plant model (5.1) with the input u in (5.14)

$$\dot{V} = \Phi_0 s - \Phi_1 s^4 - \Phi_2|s|^{\alpha+1} - k_1\xi^4 - k_2|\xi|^{\alpha+1}, \tag{5.19}$$

where

$$\begin{aligned}
\Phi_0 = &\left[\frac{(1-k_m)\bar{m}}{(1-\bar{k}_m)m} - 1\right]\left[\ddot{p}_r - k\left(\frac{\lceil\check{p}(t_s)\rfloor^a}{\Gamma(1-\lambda)}(t-t_s)^{-\lambda}\right.\right.\\
&\left.\left. + {}_{t_s}D_t^{\lambda}\lceil\check{p}\rfloor^a\right) - k_3\xi\right] - \frac{1-k_m}{m}\varrho\\
\Phi_1 = &\frac{k_1(1-k_m)\bar{m}}{(1-\bar{k}_m)m}, \quad \Phi_2 = \frac{k_2(1-k_m)\bar{m}}{(1-\bar{k}_m)m}.
\end{aligned}$$

Apparently, if the uncertainties vanish in the nominal plant model, certain constant parameters Φ_1, Φ_2, and $\Phi_0 = 0$ hold, which yields a classical adaptive sliding mode control. Considering the uncertainties existing in the nominal plant model, we will provide the analysis of the finite-time stability of the reaching phase under the control law in (5.14), and in what follows, the analysis is going to be divided into two similar cases.

Case 1 *The sliding mode dynamics are in the reaching phase with $s \neq 0$, and combining the biquadrate of s and $\Phi_0 s$ produces*

$$\dot{V} = -\left(\Phi_1 - \frac{\Phi_0}{s^3}\right)s^4 - \Phi_2|s|^{\alpha+1} - k_1\xi^4 - k_2|\xi|^{\alpha+1}. \tag{5.20}$$

Obviously, when $\Phi_1 - \frac{\Phi_0}{s^3} > 0$, we can choose

$$\varepsilon_1 + \varepsilon_2 = \min\left\{\left(\Phi_1 - \frac{\Phi_0}{s^3}\right), \Phi_2, k_1, k_2\right\}, \tag{5.21}$$

with $\varepsilon_1 > 0$ and $\varepsilon_2 > 0$ obtaining the finite-time stable form of Lyapunov candidate function (see Remark 5.6)

$$\begin{aligned}\dot{V} &\le -(\varepsilon_1 + \varepsilon_2)(s^4 + |s|^{\alpha+1} + \xi^4 + |\xi|^{\alpha+1}) \\ &\le -\varepsilon_1(s^2 + \xi^2) - \varepsilon_2(s^2 + \xi^2)^{\frac{\alpha+1}{2}} = -2\varepsilon_1 V - \varepsilon_2(2V)^{\frac{\alpha+1}{2}}.\end{aligned} \tag{5.22}$$

Remark 5.6 Typically, the exponential function $f(x) = a^x$ is monotonically increasing with $a > 1$, and conversely, it is monotonically decreasing with $0 < a < 1$. Hence, we have $x^4 + |x|^{\alpha+1} > x^2$ with $0 < \alpha < 1$, and we proceed to obtain

$$\varepsilon_1(s^4 + |s|^{\alpha+1} + \xi^4 + |\xi|^{\alpha+1}) > \varepsilon_1(s^2 + \xi^2). \tag{5.23}$$

According to the theory of basic inequality, we have

$$\begin{aligned}&\varepsilon_2\left(s^4 + |s|^{\alpha+1} + \xi^4 + |\xi|^{\alpha+1}\right) \\ \ge &\varepsilon_2\left(|s|^{\alpha+1} + |\xi|^{\alpha+1}\right) \ge \varepsilon_2\left(|s|^2 + |\xi|^2\right)^{\frac{\alpha+1}{2}},\end{aligned} \tag{5.24}$$

because $(b_1^2 + b_2^2 + \ldots + b_n^2)^{\alpha+1} \le (b_1^{\alpha+1} + b_2^{\alpha+1} + \ldots + b_n^{\alpha+1})^2$ holds when $0 < \alpha < 1$ and $b_i > 0$.

Hence one can calculate the finite convergence time as

$$t_r \le \frac{1}{(1-\alpha)\varepsilon_1} \ln\left[1 + \frac{\varepsilon_1}{\varepsilon_2}(2V_0)^{\frac{1-\alpha}{2}}\right], \tag{5.25}$$

where V_0 is the initial Lyapunov candidate function related to initializations of s and ξ.

After the discussion mentioned above, we shall seek the essential convergence region of s to guarantee finite-time stability. If and only if the following conditions hold, the analysis from (5.20) to (5.25) is valid.

$$\begin{aligned}|s| &> \left(\frac{|\Phi_0|}{\Phi_1}\right)^{\frac{1}{3}} = \left[\frac{\left|\left(1 - \frac{(1-\bar{k}_m)m}{(1-k_m)\bar{m}}\right)\vartheta - \frac{1-\bar{k}_m}{\bar{m}}\varrho\right|}{k_1}\right]^{\frac{1}{3}}, \\ \vartheta &= \ddot{p}_r - k\left[{}_{t_s}D_t^{\lambda}\lceil\check{p}\rfloor^a + \frac{\lceil\check{p}(t_s)\rfloor^a}{\Gamma(1-\lambda)}(t-t_s)^{-\lambda}\right] - k_3\xi.\end{aligned} \tag{5.26}$$

Considering the uncertainties existing in the nominal model, introducing (5.3) into (5.26) leads to

$$\left(\frac{|\Phi_0|}{\Phi_1}\right)^{\frac{1}{3}} \leq \left[\frac{\left|\theta_m \theta_{k_m} - 1\right| |\vartheta| + \frac{1-\bar{k}_m}{\bar{m}} \varrho}{k_1}\right]^{\frac{1}{3}} = \Gamma_1, \tag{5.27}$$

where ϑ depends on the selection of k_3. It is conservative to choose $|s| > \Gamma_1$ such that $\Phi_1 - \frac{\Phi_0}{s^3} > 0$ is guaranteed. Rephrasing that, the vicinity of sliding surface $|s| \leq \Gamma_1$ can be reached under the AFOTSM control in (5.14).

For reducing the vicinity Γ_1 to induce s to get closer to the origin, we design the adaptive gain k_3 as a time-varying parameter to degenerate the uncertainty gain ϑ in (5.27). Prescribe the desired uncertainty gain as ϑ_d, which is a small positive scalar, and the adaptive gain can be calculated by

$$k_3 = \begin{cases} \left(\ddot{p}_r - k\left[{}_{t_s}D_t^{\lambda}\lceil \check{p}\rfloor^a + \frac{\lceil \check{p}(t_s)\rfloor^a}{\Gamma(1-\lambda)}(t-t_s)^{-\lambda}\right] - \vartheta_d\right)/\xi, & \xi \neq 0, \\ \bar{k}_3, & \xi = 0, \end{cases} \tag{5.28}$$

where $\bar{k}_3 > 0$, and the convergence region Γ_1 will shrink down to

$$\Gamma_1 = \left[\frac{\left|\theta_m \theta_{k_m} - 1\right| |\vartheta_d| + \frac{1-\bar{k}_m}{\bar{m}} \varrho}{k_1}\right]^{\frac{1}{3}}. \tag{5.29}$$

Case 2 *Considering the Lyapunov candidate function in (5.19), combine $\Phi_0 s$ with fractional power of s to obtain*

$$\dot{V} = -\left(\Phi_2 - \frac{\Phi_0}{sgn(s)|s|^{\alpha}}\right)|s|^{\alpha+1} - \Phi_1 s^4 - k_1 \xi^4 - k_2 |\xi|^{\alpha+1}.$$

Correspondingly, the finite-time convergence of V is guaranteed when $\Phi_2 - \frac{\Phi_0}{sgn(s)|s|^{\alpha}} > 0$, and utilizing the similar analyses in Case 1, the convergence region can be expressed as follows:

$$|s| \leq \Gamma_2 = \left[\frac{\left|\theta_m \theta_{k_m} - 1\right| |\vartheta_d| + \frac{1-\bar{k}_m}{\bar{m}} \varrho}{k_2}\right]^{\frac{1}{\alpha}}. \tag{5.30}$$

At this point, we can conclude that under the control law in (5.14), the sliding mode dynamics can be guaranteed to converge to the following region in finite time

$$|s| \leq \Gamma = \min\left(\Gamma_1, \Gamma_2\right). \tag{5.31}$$

Substituting (5.31) into (5.9) yields

$$s = \dot{\check{p}} + k \left({}_{t_s}D_t^{\lambda-1} \lceil \check{p} \rfloor^a \right), |s| \leq \Gamma. \tag{5.32}$$

When $D^{\lambda-1}\lceil \check{p} \rfloor^a \neq 0$, merging it with s yields

$$\dot{\check{p}} + \left(k - \frac{s}{{}_{t_s}D_t^{\lambda-1} \lceil \check{p} \rfloor^a} \right) {}_{t_s}D_t^{\lambda-1} \lceil \check{p} \rfloor^a = 0. \tag{5.33}$$

For satisfying the condition of the sliding surface in (5.9) with $|s| \leq \Gamma$, the gain of the FO integral term needs to be a positive scalar, namely,

$$k - \frac{s}{{}_{t_s}D_t^{\lambda-1} \lceil \check{p} \rfloor^a} > 0. \tag{5.34}$$

Furthermore, the finite-time stability of the equilibrium point along the desired sliding surface in (5.9) can be guaranteed by

$$\left| {}_{t_s}D_t^{\lambda-1} \lceil \check{p} \rfloor^a \right| > \frac{\Gamma}{k}, \tag{5.35}$$

in other words,

$$\left| {}_{t_s}D_t^{\lambda-1} \lceil \check{p} \rfloor^a \right| \leq \frac{\Gamma}{k} \tag{5.36}$$

implies a maximum vicinity where the tracking error $\check{p}$ can reach. Based on Lemma 5.2, we can choose a time-varying parameter $\beta \geq 1$ to get [13]

$$\left| {}_{t_s}D_t^{\lambda-1} \lceil \check{p} \rfloor^a \right| = \frac{K |\check{p}|_{max}^a}{\beta}, \tag{5.37}$$

where K can be calculated from Lemma 5.2. Combining (5.37) and (5.36), we have

$$\frac{K |\check{p}|_{max}^a}{\beta} \leq \frac{\Gamma}{k}. \tag{5.38}$$

Thus, we obtain the convergence precision of the tracking error $\check{p}$ as

$$|\check{p}| \leq |\check{p}|_{max} \leq \left(\frac{\beta \Gamma}{k K} \right)^{\frac{1}{a}}, \tag{5.39}$$

and the precision of the velocity error $\dot{\check{p}}$ follows

$$\left| \dot{\check{p}} \right| = \left| s - {}_{t_s}D_t^{\lambda-1} \lceil \check{p} \rfloor^a \right| \leq 2\Gamma \tag{5.40}$$

When ${}_{t_s}D_t^{\lambda-1}\lceil \check{p} \rfloor^a = 0$, no matter whether the tracking error leaves or stays at the origin, it will still satisfy the precision shown in (5.39) using the analyses from (5.32) to (5.40), which completes the proof. ∎

Remark 5.7 Theoretically, the time-varying k_3 can be regulated to achieve a small enough ϑ_d according to (5.28). However, ξ converging to the vicinity of origin will result in an enormous adaptive gain k_3. This enormous adaptive gain is frequently invalid for digital computer control, and it further induces the actual motion control system to be unstable.

For stable system performance in realistic engineering applications, we adopt the saturation function to avoid the undesired enormous gain. Hence in our experiment (5.28) is applied in the following form

$$k_3 = \bar{k}_3 \text{sat}\left(\frac{\ddot{p}_r - k\left[{}_{t_s}D_t^{\lambda}\lceil \check{p} \rfloor^a + \frac{\lceil \check{p}(t_s) \rfloor^a}{\Gamma(1-\lambda)}(t-t_s)^{-\lambda}\right] - \vartheta_d}{\bar{k}_3 \xi}\right),$$

where sat(•) is the normalized saturation function.

Once the saturation of the expression mentioned above occurs, it will not keep quality like (5.28). However, it creates a smaller uncertainty gain boundary layer ϑ_s depending on $\bar{k}_3$ to reduce Γ_1.

$$\begin{aligned}\vartheta_s &= \ddot{p}_r - k\left[{}_{t_s}D_t^{\lambda}\lceil \check{p} \rfloor^a + \frac{\lceil \check{p}(t_s) \rfloor^a}{\Gamma(1-\lambda)}(t-t_s)^{-\lambda}\right] \\ &\quad - |\bar{k}_3|\text{sgn}\left(-k\left[{}_{t_s}D_t^{\lambda}\lceil \check{p} \rfloor^a + \frac{\lceil \check{p}(t_s) \rfloor^a}{\Gamma(1-\lambda)}(t-t_s)^{-\lambda}\right] + \ddot{p}_r - \vartheta_d\right)|\xi|.\end{aligned} \tag{5.41}$$

Neglecting the effect of the small prescribed ϑ_d leads to

$$\begin{aligned}|\vartheta_s| &= \Bigg|\ddot{p}_r - k\left[{}_{t_s}D_t^{\lambda}\lceil \check{p} \rfloor^a + \frac{\lceil \check{p}(t_s) \rfloor^a}{\Gamma(1-\lambda)}(t-t_s)^{-\lambda}\right] \\ &\quad -|\bar{k}_3\xi|\text{sgn}\left(\ddot{p}_r - k\left[{}_{t_s}D_t^{\lambda}\lceil \check{p} \rfloor^a + \frac{\lceil \check{p}(t_s) \rfloor^a}{\Gamma(1-\lambda)}(t-t_s)^{-\lambda}\right]\right)\Bigg| \\ &\le |\vartheta|.\end{aligned}$$

This compromise therefore makes a weaker effect in the uncertainty gain than (5.28) when the calculated gain is greater than the prescribed one, nevertheless, the convergence region is more narrow than the strategy with the constant gain in (5.27).

Remark 5.8 The proposed control strategy has outstanding advantages on the position and velocity tracking precision with the same finite convergence time since there exists the adaptive term ξ reducing the gain of the uncertainty.

A practical integer-order FNTSM control strategy has a distinctly different precision form, which is hard to compare with the proposed precisions in (5.39) and (5.40). However, in adaptive FO terminal SMC, unlike the FNTSM, the uncertainty gains ϑ

is adaptive, and the adaptive effect happens on the convergence region of s directly, which leads to a higher precision of the sliding mode variable.

These mentioned methods are all continuous and nonsingular, and they can therefore be applied in the control system of the linear motor. In theory, the real decisions depend on the parameter selection of the respective strategy.

Remark 5.9 Using Property 5.4, (5.9) can be rewritten as [20]

$$
\begin{aligned}
s &= \dot{\check{p}} + k \left({}_{t_s}D_t^{\lambda-1} \lceil \check{p} \rfloor^a \right) \\
&= \dot{\check{p}} + k \left({}_{t_s}D_{t_0}^{\lambda-1} \lceil \check{p} \rfloor^a + {}_{t_0}D_t^{\lambda-1} \lceil \check{p} \rfloor^a \right)
\end{aligned}
\tag{5.42}
$$

which is a generalization of the FO integral sliding surface involving the initialization function.

It is worth noting that $t_s = t_0 = 0$, namely, zero initial condition, is widely considered to be an essential condition in engineering applications because the efforts of controllers are always accompanied by system states with time varying [3, 13, 14, 16, 18]. Without loss of generality, $t_s = t_0 = 0$ is satisfied in this work.

5.3.3 Sliding Mode Disturbance Observer Design

The previous sections have designed the adaptive FO terminal SMC strategy based on the nominal plant model for the real linear motor-based motion control system. The proposed method abandons the high-gain sign function to guarantee the continuous control law, and conversely, the robustness to the disturbance is sacrificed. Owing to the unknown difference between the nominal model and the actual one, the control performance can just be evaluated by the precision of the convergence region rather than converging to the origin.

Although the adaptive term ξ reduces the uncertainty gain, there also exist some unforeseen disturbances in the lumped uncertainty, i.e., for the real LM, the velocity signal is estimated by the backward differentiator of the position signals, because our experimental facility only equips with the absolute position encoder. The backward differentiator estimation is widely used in engineering applications, but the drawback is that the measurement noise will be introduced into the obtained velocity signal. Hence in [21], the authors employ the robust exact differentiator to suppress the noise. Alternatively, in our work, we adopt the sliding mode disturbance observer (SMDO) to suppress the noise due to the noise being regarded as one part of the uncertainties. Compared with the robust exact differentiator method, the SMDO method not only takes into account the measured noise caused by using a differentiator but also suppresses the uncertainties involving parametric uncertainty and some disturbances.

The basic SDMO can be constructed as [4]

$$
\dot{s} = \Psi + \Delta\Psi + u, \ \Delta\Psi = d + e_m, \tag{5.43}
$$

where Ψ denotes the known terms, namely, the switching control, $\Delta\Psi$ represents the unknown terms, including the lumped uncertainty d and the model error e_m. Rephrasing that, $\Delta\Psi$ can be regarded as a bounded nonlinear function,

$$|\Delta\Psi| = |f(d, e_m, t)| < L, \tag{5.44}$$

where $L > 0$ and the actual form of $f(d, e_m, t)$ is unknown.

For estimating $\Delta\Psi$ by utilizing SMDO, the auxiliary parameters and the robustness compensation follow:

$$\begin{aligned} \sigma &= s + z, \\ \dot{z} &= -\Psi - u - v, \\ v &= (L + \rho)\text{sgn}(\sigma), \end{aligned} \tag{5.45}$$

where $\rho > 0$. Considering the Lyapunov candidate function $V_1 = \frac{1}{2}\sigma^2$, we have

$$\dot{V}_1 = \sigma \left[\Delta\Psi - (L + \rho)\text{sgn}(\sigma)\right] \leq -\rho|\sigma| \leq -\sqrt{2}\rho(V_1)^{\frac{1}{2}} \leq 0. \tag{5.46}$$

It is clear from the above deduction that once the sliding surface σ is stable, namely, $\sigma = 0$, we have $s + z = 0$, which implies $\dot{s} = \Psi + u$, meaning the disturbance and uncertainty have been eliminated theoretically. The apparent drawback is that the sign function is used in the observer, which may cause the chattering.

To avoid the chattering, we implement the super-twisting compensation [4] to guarantee that all the alternative auxiliary sliding mode variables and high-order derivatives converge to the origin. What's more, the chattering is eliminated by hiding it in the high-order derivative. To design the practical continuous disturbance observer, the compensation in (5.45) is replaced by $v = \dot{\sigma} + \varpi_1|\sigma|^{\frac{1}{2}}\text{sgn}(\sigma) + \varpi_2 \int \text{sgn}(\sigma)d\tau$. Here the parameters in the above super-twisting disturbance observer satisfy the below sufficient conditions [4]

$$\varpi_1 > C,\ \varpi_2^2 \geq \frac{4C(\varpi_1 + C)}{\varpi_1 - C},\ C > \left|\Delta\dot{\Psi}\right|. \tag{5.47}$$

The super-twisting sliding mode technique is mature, and the proof of the stability of this observer can be seen in articles [4], which is omitted here.

5.4 Experiments

To validate the effectiveness of the proposed control strategy, we will apply the proposed controller with SMDO on the real linear motor platform.

To carry out the practical experiments, some nominal parameters and controller design parameters shall be given out ahead. For the nominal plant model, the mass m is

11 kg, and $\bar{k}_m = 0.03$ means a constant load. The user parameter k_f is determined by configuring the gain of the drive. Without loss of generality, we assume the following uncertainty boundedness: $\theta_m = 1.3$, $\theta_{k_v} = 3$, $\theta_{k_c} = 5$, $\theta_{k_m} = 1.2$ and $\theta_d = 40$. The parameters $\bar{k}_c$ and $\bar{k}_v$ are related to sliding friction, and we carry out a set of constant velocity experiments to determine their selections as $\bar{k}_c = 7.847$ and $\bar{k}_v = 0.019$ based on PSO-based evaluation with 300 particles by iterating 300 generations [5].

In terms of the adaptive FO terminal SMC, the parameter λ determines the fractional order of the sliding surface. If $\lambda = 0$ holds, the finite-time stability of the motion on the sliding surface vanishes theoretically. Selecting $\lambda = 1$ gives the FTSM controller involving the singular issue, and the nonsigularity variant has been introduced in detail in [21]. To emphasize the characteristics of the FO controller, we select $\lambda = 0.5$ in this work. The parameter a, the gain k_1, k_2 and k together determine the tracking precision directly. Since the singularity issue is avoided due to the FO operator, the parameter a can be selected freely in (0, 1). A large gain k gives a large bandwidth, namely a fast response, but too large k will cause undesired overshoots or even oscillation, and we select $k = 260$ based on the experimental performance in our work. It is clear from (5.15) that larger k_1 and k_2 result in a higher precision. Moreover, the positive gain k_1 and k_2 can be designed as the adaptive terms as follows:

$$k_1 = k_2 = 10\left[\left|\theta_m\theta_{k_m} - 1\right| |\vartheta_d| + \frac{1 - \bar{k}_m}{\bar{m}}\varrho\right],$$

such that the tracking precision of 40 μm can be obtained. The time-varying gain k_3 works to generate the small uncertainty gain ϑ. Its value relies on the selection of gain $\bar{k}_3$, which can be freely chosen theoretically. Another parameter of the switching input is carefully set to $\alpha = 0.5$, since a large α will result in the chattering, and conversely, too small α yields a slow convergence of the sliding mode variable theoretically. Through trial and error, the parameters $\bar{k}_3 = 1$ and $a = 0.6$ are chosen such that a high-precision performance is achieved. From (5.47), we have a couple of valid parameters, $\varpi_1 = 400$ and $\varpi_2 = 14$, which can compensate the unknown disturbance satisfying $\left|\Delta\dot{\Psi}\right| < C = 40$. We approximately calculate the FO numerical expression by Charef's method [10], in which the high and low transitional frequencies are $\omega_h = 100$ rad/s and $\omega_b = 0.01$ rad/s, respectively, and furthermore, the Runge-Kutta method with the fixed step of 0.25 ms is employed for solving the differential equation.

Neglecting the adaptive term ξ in the AFOTSM controller produces a degenerated FO terminal sliding mode controller [13]. For comparison, FOTSM, FNTSM, PID, and an observer-based sliding mode control (OSMC) [12] are tested on the linear motor. According to the investigation of [21], the controller parameters can be designed as the time-varying parameters to achieve 40 μm of tracking precision theoretically. Other comparisons are also well tuned, i.e., PID controller is tuned by Ziegler-Nichols method [22] to obtain coarse parameters, then fine-tuning is applied to seek a high-precision performance by trial and error.

All the implementation of the proposed strategy is based on Twincat 3.0 PLC, and the sample period of our experimental facility is 0.25 ms. The communication between the upper computer and the drive depends on CANopen technology.

A cosinusoidal signal and a triangular signal are presented as reference signals to examine the tracking performance. The cosinusoidal reference signal has the following form:

$$p_{cr}(t) = 10^{-2}\left[1 - \cos(2\pi t)\right](1 - e^{-2t}), \tag{5.48}$$

and the form of the triangular reference is

$$p_{tr}(t) = 10^{-3}(-1)^{c_0}(c_1 t - c_2), \tag{5.49}$$

where $c_1 = 1.6$, and c_0 is the integer quotient of t and c_1, which drops the remainder. $c_2 = 2n$ is an even number satisfying $2n - 1 \leq c_0 \leq 2n, n \in N$.

We also take into account the load $\bar{m}$ in two cases, since the variant load affects the stability of the control system [6]. Case 1 explores the situation that there is not any load added on the moving stage, and Case 2 discusses the situation that the stage is moving with extra load. For each case, the symbol a and b stand for cosinusoidal reference and triangular reference, respectively, i.e., Case 1a means that the stage is tracking the cosinusoidal reference without extra load.

The performances related to different λ are shown in the end of this section, where λ are selected as 0.2, 0.5 and 0.9 to construct different FO integral sliding surfaces.

5.4.1 Experimental Results Without Load

Figures 5.2 and 5.4 show the position tracking trajectories. It is clear from the enlarged figures that PID and OSMC perform worse than the others, especially at the moments that the motion changes direction, i.e., at 1 s of Case 1a and 1.25 s of Case 1b. The trajectories of the FO controller are closer to the reference signal than that of PID, OSMC, and FNTSM controllers. The tracking errors shown in Figs. 5.3 and 5.5 can evaluate the performances intuitively. The errors of PID and OSMC are always larger than others, and the max error is over 0.03 mm, but the other three methods hold a max error of less than 0.02 mm, which can be observed from the enlarged figures. Noting the errors shown in Fig. 5.3, the AFOTSM controller makes the errors lower than 0.005 mm and more smooth and steady than other methods. The max errors occur when the motion direction changes because this change results in a transient approximating zero velocity, magnifying the influence of inadequate estimation for the real friction model and the actual velocity. The max errors of FO controllers are smaller than the integer-order ones', since the FO terminal sliding surface depends on the velocity signal slighter than the FNTSM controller. Although the sliding mode disturbance observer is utilized in these experiments, there exist some minor oscillations of errors caused by uncertainties and disturbances.

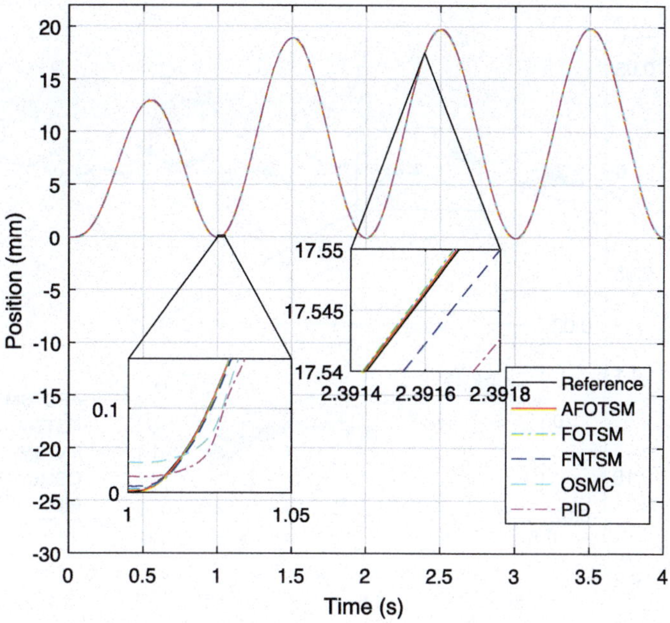

Fig. 5.2 Case 1a: tracking trajectories of cosinusoidal reference without extra load

The excellent performance of the proposed method for tracking the triangle signal can be also seen from the experimental results shown in Fig. 5.5. The AFOTSM controller has the smallest max error at the moment of direction change among these methods and also maintains the smallest absolute average tracking error intuitively. The AFOTSM and FOTSM controllers have tracking precisions of 0.005 mm similar to Case 1a, which is better than FNTSM, PID, and OSMC.

5.4.2 Experimental Results with Extra 30% Load

In Case 2, the extra 30% load is attached to the moving stage for validating the robustness against uncertainties. Figures 5.6 and 5.7 show the tracking errors. These experimental results demonstrate that the AFOTSM and FOTSM controllers can still guarantee better tracking performance than the integer-order controllers, and the tracking errors are no more than 0.01 mm, which is similar to the cases without extra load. One can compare Fig. 5.6 with Fig. 5.3 and find that the oscillations and average errors of the FNTSM and PID controller in Case 2a are worse than in Case 1a. However the FO controllers perform high-quality control, and it is worth noting that the AFOTSM controller does not seem to be influenced by the extra payload. Similar results can also be obtained from the experimental results in Figs. 5.5 and 5.7.

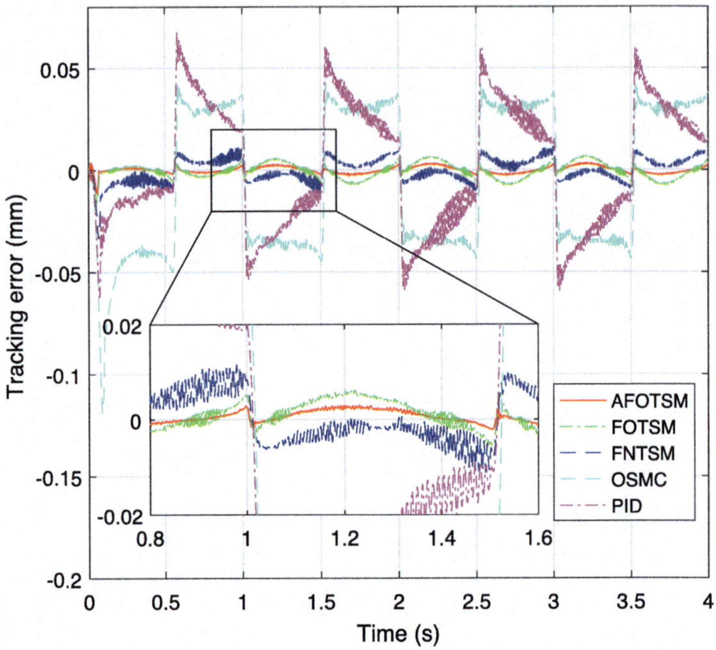

Fig. 5.3 Case 1a: tracking errors of AFOTSM, FOTSM, FNTSM, and PID

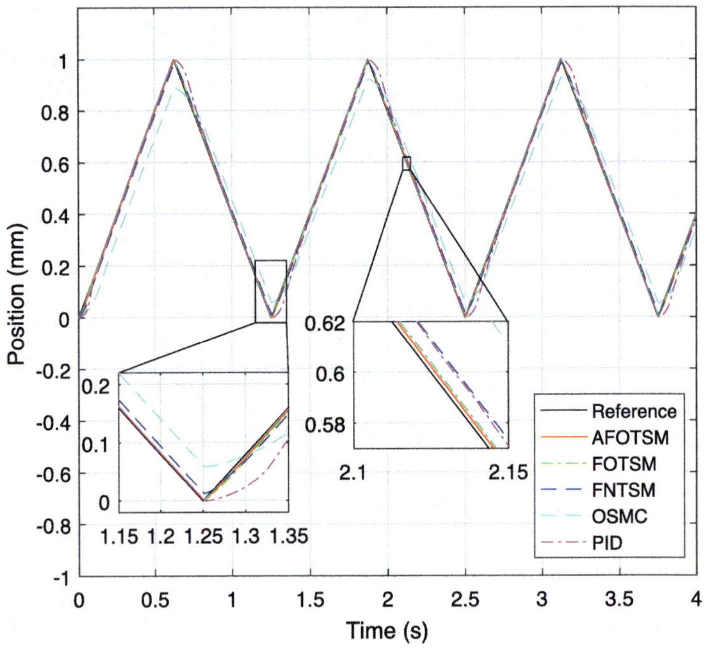

Fig. 5.4 Case 1b: tracking trajectories of triangular reference without extra load

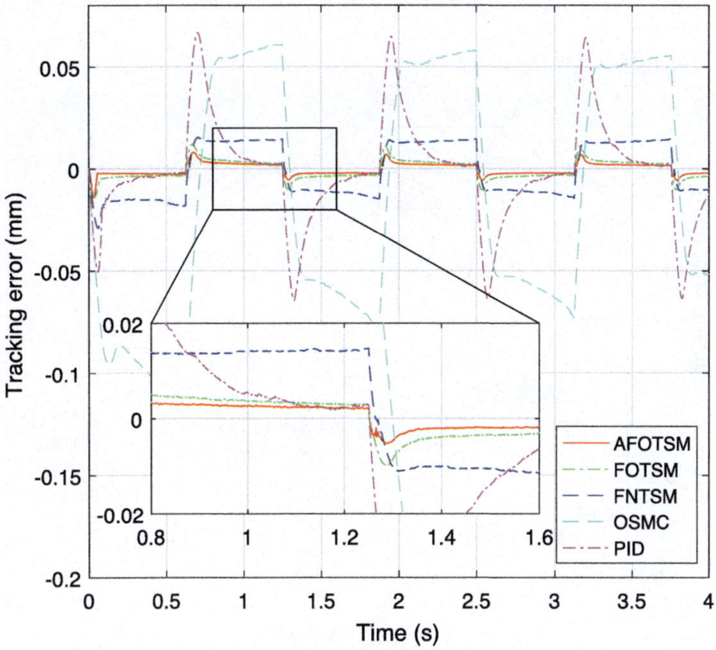

Fig. 5.5 Case 1b: tracking errors of AFOTSM, FOTSM, FNTSM, and PID

The control inputs are drawn in Fig. 5.8. Compared with FOTSM, FNTSM, PID, and OSMC, the inputs of the proposed controller for tracking the reference signal have the smallest range of variation, and minor chattering is caused by disturbance and noise. Moreover, the PID control inputs oscillate most drastically, and these oscillations get worse in Case 2b, which indicates that the PID controller relies on parameters deeply. Conversely, the proposed controller can always maintain the proper control performance and is robust to this extra payload due to the adaptive algorithm.

Hence, based on the results shown in Figs. 5.2, 5.3, 5.4, 5.5, 5.6 and 5.7, we conclude that the three sliding mode controllers can maintain the tracking performance with the precision of 0.02 mm under different reference, and among these controllers, the AFOTSM controller has the highest tracking precision. Moreover, these experimental results are coincident with the theoretical analyses. Similar to FNTSM and FOTSM, the finite-time convergence guarantees a faster response compared with the asymptotic stability of PID and OSMC, which can be easily observed from the time evolution of tracking error in Fig. 5.5. Compared with the integer-order controllers, the proposed controller utilizes FO sliding surface to reduce the steady-state error and meanwhile avoid a magnitude of overshoot, i.e., in Fig. 5.5, PID achieves a satisfactory precision at about 1.1 s, but the overshoot is undesired. However, the AFOTSM

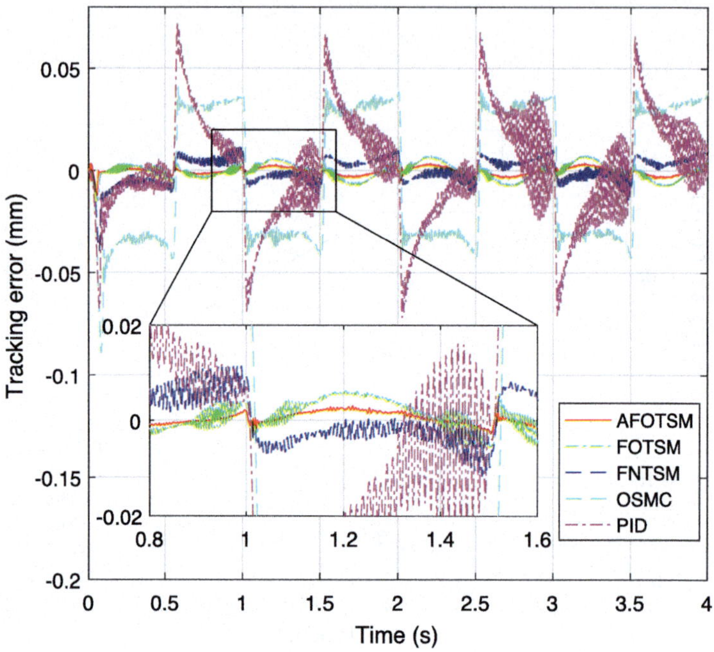

Fig. 5.6 Case 2a: tracking errors of AFOTSM, FOTSM, FNTSM, and PID

controller realizes a smooth transition with smaller overshoot and higher steady-state error, and the adaptive algorithm allows the strategy to get higher tracking precision.

To further improve the performance, λ can be tuned to construct different sliding surfaces obtaining higher precision. Figure 5.9 shows the tracking errors with different λ. When $\lambda = 0.9$, the errors become flatter and smaller than the other two's, meanwhile, a smaller absolute average error can be obtained, which indicates that increasing λ leads to better performance. Conversely, when $\lambda = 0.2$, the performance is unsatisfactory, and large errors can be observed. The essential reason for this phenomenon is that the motion on the sliding surface is governed by $\dot{\check{p}} = -k\left({}_{t_s}D_t^{\lambda-1}\lceil \check{p} \rfloor^{a}\right)$. When $\lambda \to 1$, $\check{p}$ performs a fast response, and when $\lambda \to 0$, $\check{p}$ performs a damping oscillation with a large overshoot. Particularly, if $\lambda = 1$, the singular issue will appear in u_{sw}, resulting in an enormous control effort to destroy the control system; if $\lambda = 0$, the stability of the sliding surface can not be guaranteed.

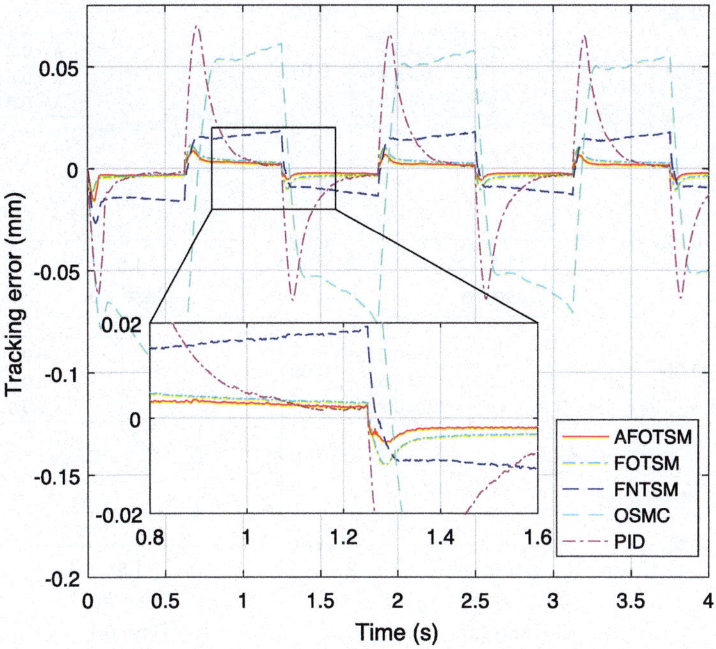

Fig. 5.7 Case 2b: tracking errors of AFOTSM, FOTSM, FNTSM, and PID

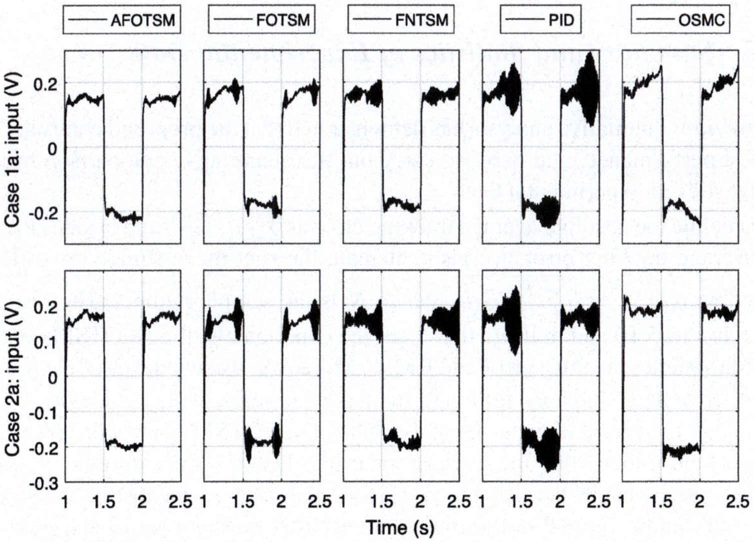

Fig. 5.8 Control inputs in Case 1a and Case 2a

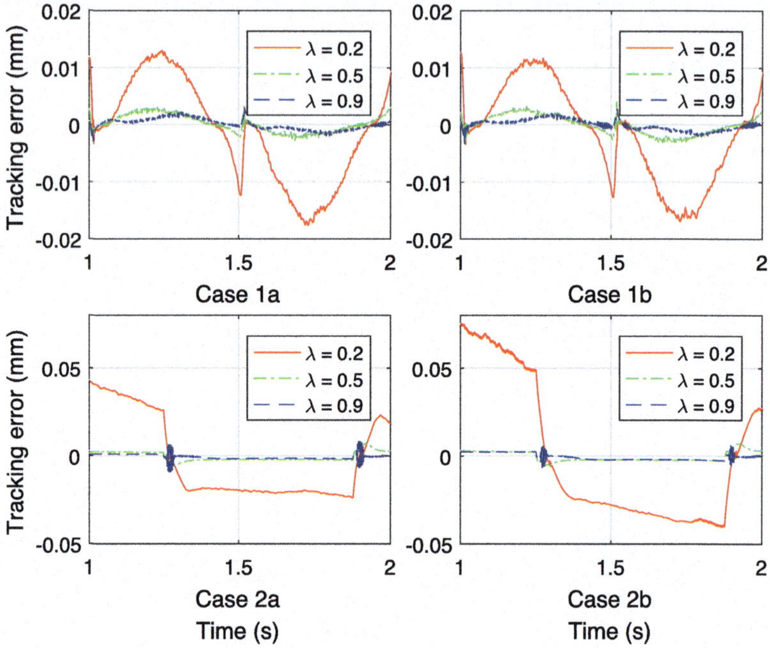

Fig. 5.9 Tracking error with $\lambda = 0.2, 0.5$ and 0.9

5.4.3 *Summary and Statistics of Experimental Data*

The previous qualitative analysis has demonstrated that the proposed controller owns the best performance, and here we carry out the quantitative comparison based on the statistics of experimental data.

We define the absolute average tracking error as $\sum_{i=1}^{N} \frac{|\check{p}(i)|}{N}$, and besides the absolute average tracking error, we also calculate the root-mean-square error (RMSe) defined as $RMSe = \sqrt{\sum_{i=1}^{N} \frac{\check{p}^2(i)}{N}}$, where N is the sample number. The results are shown in Fig. 5.10 and indicate that in all the experiments, the AFOTSM controller owns the smallest average error and RMSe. Moreover, these indexes of the proposed controller remain stable even though the payload varies, which coincides with the theoretical analyses and experimental results. The FOTSM controller takes second place, which implies that the tracking precision benefits from introducing the FO integral into the controller design. Therefore, the results indicate that the proposed controller can be applied to the linear motor-based motion control and guarantees fast response and high-precision tracking performance with robustness against uncertainties.

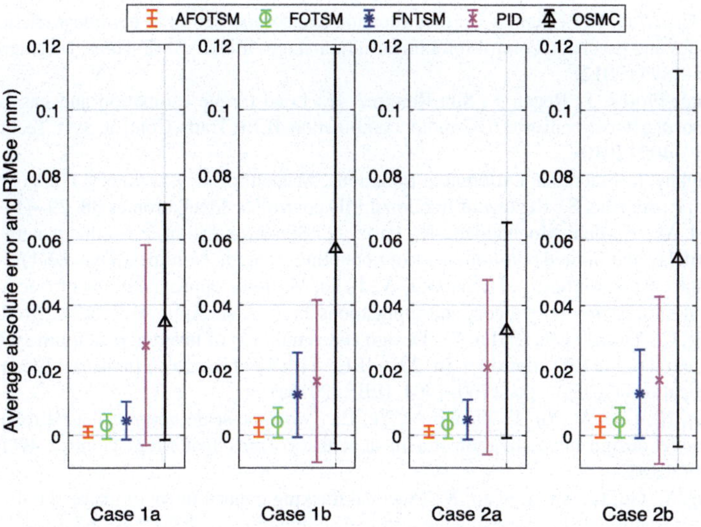

Fig. 5.10 Average absolute error and RMSe

5.5 Conclusion

In this chapter, we have designed a practical AFOTSM controller adopted on the linear motor-based motion control. The FO calculus and the adaptive algorithms guarantee high-precision and fast-response performance with robustness against uncertainties. The sign does not exist in the switching input; hence, the proposed approach is continuous and chattering-free. The SMDO is also introduced into the control strategy to improve the tracking precision further, even though the linear motor-based motion suffers from unknown uncertainties. The experimental results demonstrate that the AFOTSM controller has the benefits of higher tracking precision and faster response than the conventional FNTSM and FOTSM controllers. The proposed approach is also effective when the liner motor has an extra load.

References

1. Chen, D., Liu, Y., Ma, X., Zhang, R.: Control of a class of fractional-order chaotic systems via sliding mode. Nonlinear Dyn. **67**(1), 893–901 (2012)
2. Efe, M.Ö.: Fractional order systems in industrial automation-a survey. IEEE Trans. Ind. Inform. **7**(4), 582–591 (2011)
3. Erenturk, K.: Fractional-order $PI^{\lambda}D^{\mu}$ and active disturbance rejection control of nonlinear two-mass drive system. IEEE Trans. Ind. Inform. **60**(9), 3806–3813 (2012)
4. Levant, A.: Universal single-input-single-output (SISO) sliding-mode controllers with finite-time convergence. IEEE Trans. Autom. Control. **46**(9), 1447–1451 (2001)

5. Lin, C.J., Yau, H.T., Tian, Y.C.: Identification and compensation of nonlinear friction characteristics and precision control for a linear motor stage. IEEE/ASME Trans. Mechatron. **18**(4), 1385–1396 (2012)
6. Linares-Flores, J., Reger, J., Sira-Ramírez, H.: Load torque estimation and passivity-based control of a boost-converter/DC-motor combination. IEEE Trans. Control. Syst. Technol. **18**(6), 1398–1405 (2010)
7. Podlubny, I.: Fractional Differential Equations. Academic, New York, NY, USA (1999)
8. Shah, P., Agashe, S.: Review of fractional PID controller. Mechatronics **38**, 29–41 (2016)
9. Tousi, M., Kardehi Moghaddam, R., Pariz, N.: Synchronization in oscillator networks with time delay and limited non-homogeneous coupling strength. Nonlinear Dyn. **82**(1), 1–8 (2015)
10. Vinagre, B., Podlubny, I., Hernandez, A., Feliu, V.: Some approximations of fractional order operators used in control theory and applications. Fract. Calc. Appl. Anal. **3**(3), 231–248 (2000)
11. Wang, C., Yin, G., Liu, C., Fu, W.: Design and simulation of inverted pendulum system based on the fractional PID controller. In: 2016 IEEE 11th Conference on Industrial Electronics and Applications (ICIEA), pp. 1760–1764. IEEE (2016a)
12. Wang, N., Lin, W., Yu, J., Zhang, A., Ye, C.: Two time-scale observer-based robust motion controller design and realization of a linear actuator. Trans. Inst. Meas. Control. **40**(11), 3241–3251 (2018)
13. Wang, Y., Gu, L., Xu, Y., Cao, X.: Practical tracking control of robot manipulators with continuous fractional-order nonsingular terminal sliding mode. IEEE Trans. Ind. Electron. **63**(10), 6194–6204 (2016)
14. Wang, Y., Luo, G., Gu, L., Li, X.: Fractional-order nonsingular terminal sliding mode control of hydraulic manipulators using time delay estimation. J. Vib. Control **22**(19), 3998–4011 (2016)
15. Yin, C., Sm, Zhong, Wf, Chen: Design of sliding mode controller for a class of fractional-order chaotic systems. Commun. Nonlinear Sci. Numer. Simul. **17**(1), 356–366 (2012)
16. Yin, C., Stark, B., Chen, Y., Sm, Zhong, Lau, E.: Fractional-order adaptive minimum energy cognitive lighting control strategy for the hybrid lighting system. Energy Build. **87**, 176–184 (2015)
17. Yu, S., Yu, X., Shirinzadeh, B., Man, Z.: Continuous finite-time control for robotic manipulators with terminal sliding mode. Automatica **41**(11), 1957–1964 (2005)
18. Zeng, G.Q., Chen, J., Dai, Y.X., Li, L.M., Zheng, C.W., Chen, M.R.: Design of fractional order PID controller for automatic regulator voltage system based on multi-objective extremal optimization. Neurocomputing **160**, 173–184 (2015)
19. Zhang, B., Pi, Y., Luo, Y.: Fractional order sliding-mode control based on parameters auto-tuning for velocity control of permanent magnet synchronous motor. ISA Trans. **51**(5), 649–656 (2012)
20. Zhang, F., Li, C.: Remarks on the initialization of caputo derivative. In: Proceedings of 2012 IEEE/ASME 8th IEEE/ASME International Conference on Mechatronic and Embedded Systems and Applications, pp. 325–329. IEEE (2012)
21. Zheng, J., Wang, H., Man, Z., Jin, J., Fu, M.: Robust motion control of a linear motor positioner using fast nonsingular terminal sliding mode. IEEE/ASME Trans. Mechatron. **20**(4), 1743–1752 (2014)
22. Ziegler, J.G., Nichols, N.B.: Optimum settings for automatic controllers. Trans. Am. Soc. Mech. Eng. **64**(8), 759–765 (1942)

Chapter 6
Discrete-Time Fractional-Order Terminal Sliding Mode Tracking Control

6.1 Introduction

This chapter presents a practical discrete-time FO terminal sliding mode control strategy for high-precision tracking tasks based on linear motors. In particular, the practical parametric uncertainties involving sliding friction, uncertain payload, and disturbance in tracking tasks are considered in this work. Combining Grünwald-Letnikov FO definition and terminal sliding mode technique, the proposed method synthesizes a novel discrete-time FO terminal SMC law to drive the sliding mode dynamics into the stable region in finite steps theoretically even though the system is suffering from uncertainties and disturbances, and the motion on the surface can guarantee higher tracking precision than the conventional discrete-time terminal sliding surface by selecting suitable controller parameters. The theoretical analyses give out the guideline of parameter selection, and experiments are carried out on the linear-motor-based test platform to demonstrate that the proposed controller is easily implemented and can achieve high-precision tracking, fast response, and considerable robustness to uncertainties.

6.2 Model Review of X-Y Servo Platform

The experimental setup used in this work is the linear motor-based motion control system (by Akribis System Company) shown in Fig. 4.1. The moving stage consists of the iron-core translator and the inertia load and travels along the linear guidance with the travel range of 700 mm. The continuous-time dynamic equations are given by [7]

G. Sun et al., *Fractional-Order Sliding Mode Control: Methodologies and Applications*, Studies in Systems, Decision and Control 532,
https://doi.org/10.1007/978-3-031-60847-6_6

$$\begin{aligned} \dot{p} &= v, \\ m\dot{v} &= u - f - d, \\ f &= k_c \text{sgn}(\dot{p}) + k_v \dot{p}, \end{aligned} \tag{6.1}$$

where p, v, and $\dot{v}$ represent the absolute position, velocity, and acceleration of the moving stage, respectively. The system input u is the force generated by the linear motor. The mass of the iron-core translator and inertia load is denoted as m, and the lumped uncertainty d contains bounded unknown disturbance and measurement noise. The nonlinear friction f is related to the velocity of the moving stage, and the viscous friction coefficient and Coulomb friction coefficient are denoted by k_v and k_c, respectively. To design the digital motion control system, it is essential to discretize the plant model to yield a nominal discrete-time plant model [1]

$$\Delta p(k) = v(k), \tag{6.2}$$

$$m\Delta v(k) = u(k) - (f(k) + \delta f(k)) - d(k) - \delta d(k), \tag{6.3}$$

$$f(k) = k_c \text{sgn}(v(k)) + k_v v(k), \tag{6.4}$$

where $\delta f(k)$ means the friction uncertainty, and $\delta d(k)$ describes the possible parameter variation, discretization errors, and measure errors.

Assumption 6.1 ([1]) The uncertain terms $d(k)$, $\delta f(k)$, and $\delta d(k)$ are assumed to be bounded and own the following boundary.

$$|d(k)| \leq d_m, \quad |\delta f(k)| \leq \theta_f, \quad |\delta d(k)| \leq \theta_d. \tag{6.5}$$

Using Euler's discretization in [2], the forward difference of position signal at sampling point k denoted by $\Delta p(k)$ can be expressed by

$$\Delta p(k) = \frac{p(k+1) - p(k)}{h}, \tag{6.6}$$

where h represents the sampling interval and $k \in \mathbb{N}$.

6.3 Discrete-Time Fractional-Order Terminal Sliding Mode Controller Design

6.3.1 Preliminaries

To design the discrete-time FO terminal sliding mode (DFOTSM) controller for linear motor-based motion, some basic notions and definitions for FO operators are recalled, and some essential lemmas are also given out.

Definition 6.2 The Grünwald-Letnikov FO difference with arbitrary order $\alpha \in \mathbb{R}$ is described by

$$\Delta^{\alpha} y(k) = \frac{1}{h^{\alpha}} \sum_{j=0}^{k} (-1)^j \binom{\alpha}{j} y(k-j), \tag{6.7}$$

where h represents the sampling interval and the number of samples is denoted as k. $\binom{\alpha}{j}$ is the notation of the binomial coefficients, which can be calculated according to

$$\binom{\alpha}{j} = \begin{cases} 1, & j = 0, \\ \frac{\alpha(\alpha-1)\cdots(\alpha-j+1)}{j!} & , j = 1, 2, 3, \ldots \end{cases} \tag{6.8}$$

It is inefficient for the real-time motion controller to record and calculate all the sampling data to obtain Grünwald-Letnikov FO difference in engineering applications. To conserve computing resources, the finite recording number L is introduced into the definition of FO difference given by

$$\Delta^{\alpha} y(k) = \frac{1}{h^{\alpha}} \sum_{j=0}^{L} (-1)^j \binom{\alpha}{j} y(k-j). \tag{6.9}$$

Lemma 6.3 *The relation between* $\binom{\alpha}{j}$ *and* $\Gamma(x)$ *can be established by*

$$\sum_{j=0}^{L} (-1)^j \binom{\alpha}{j} = \frac{\Gamma(L+1-\alpha)}{\Gamma(1-\alpha)\Gamma(L+1)}, \tag{6.10}$$

where $\Gamma(x) = \int_0^{\infty} \tau^{x-1} e^{-\tau} d\tau$.

Considering the bounded sequence $y(k)$, $k = 0, 1, \ldots$ with $\max\{|y(k-1)|\} \le \rho_y$, the corresponding FO difference is bounded by

$$\frac{1}{h^{\alpha-1}} \sum_{j=1}^{L} (-1)^j \binom{\alpha-1}{j} y(k-j) \le \frac{(K-1)\rho_y}{h^{\alpha-1}}, \tag{6.11}$$

where $K = \frac{\Gamma(L+2-\alpha)}{\Gamma(2-\alpha)\Gamma(L+1)} \ge 1$.

6.3.2 Control System Design

To realize digital engineering application, the discrete-time FO terminal SMC is constructed in this section. Different from the continuous-time controller, some properties of the DFOTSM controller cater to digital systems, and will be analyzed in this section.

The continuous-time FOTSM scheme for tracking control of robotic manipulator has been investigated in [5], and the discrete-time FOTSM surface can be given by

$$z(k) = \Delta e(k) + \xi_1 \Delta^{\alpha-1} \lceil e(k) \rfloor^{\lambda_1}, \tag{6.12}$$

where $e(k) = p(k) - p_r(k)$ is the error between the position signal $p(k)$ and the reference signal $p_r(k)$. The notation $\lceil e(k) \rfloor^{\lambda_1} \triangleq |e(k)|^{\lambda_1} \mathrm{sgn}[e(k)]$ with $0 < \lambda_1 < 1$ denoting a ratio of odd integers, and $\xi_1 \in \mathbb{R}^+$. Taking the difference of $z(k)$ yields

$$\begin{aligned} z(k+1) =& z(k) + h\Delta \left[v(k) - \Delta p_r(k) + \xi_1 \Delta^{\alpha-1} \lceil e(k) \rfloor^{\lambda_1} \right] \\ =& z(k) + h\Delta v(k) - h\Delta \left[\Delta p_r(k) - \xi_1 \Delta^{\alpha-1} \lceil e(k) \rfloor^{\lambda_1} \right] \\ =& z(k) + \frac{h}{m} \big[u(k) - k_c \mathrm{sgn}(v(k)) - k_v v(k) - d(k) \\ & - \delta f(k) - \delta d(k) \big] - h \left[\Delta^2 p_r(k) - \xi_1 \Delta \left(\Delta^{\alpha-1} \lceil e(k) \rfloor^{\lambda_1} \right) \right] \end{aligned} \tag{6.13}$$

and based on the nominal model, the equivalent control input u_{eq} can be given by

$$\begin{aligned} u_{eq}(k) =& m \left[\Delta^2 p_r(k) - \xi_1 \Delta \left(\Delta^{\alpha-1} \lceil e(k) \rfloor^{\lambda_1} \right) \right] \\ & + k_c \mathrm{sgn}(v(k)) + k_v v(k). \end{aligned} \tag{6.14}$$

Furthermore, we introduce a novel terminal-sliding-mode-type reaching law given by

$$u_{sw}(k) = -m \left[\xi_2 z^3(k) + \xi_3 \lceil z(k) \rfloor^{\lambda_2} \right], \tag{6.15}$$

where $0 < \lambda_2 < 1$, $\xi_2 \in \mathbb{R}^+$ and $\xi_3 \in \mathbb{R}^+$. Combining the equivalent control and the reaching law produces the actual control law

$$u(k) = u_{eq}(k) + u_{sw}(k), \tag{6.16}$$

and substituting it into (6.13) gives rise to

$$z(k+1) = z(k) + h\rho_0(k) - h\xi_2 z^3(k) - h\xi_3 \lceil e(k) \rfloor^{\lambda_2}, \tag{6.17}$$

where

$$\rho_0(k) = \frac{1}{m}\left[\delta d(k) + \delta f(k) + d(k)\right].$$

To complete the stability analyses of our discrete-time control strategy, the following assumptions about the sliding surface are required.

Assumption 6.4 For a practice discrete-time control system, the sampling interval and sliding surface should satisfy

$$z^2(k) < \frac{1 - h\xi_3}{h\xi_2},$$

where $1 - h\xi_3 > 0$.

At this point, the DFOTSM controller can be constructed by the following theorem.

Theorem 6.5 *Considering the nominal discrete-time plant model subjected to uncertainties shown in (6.3) and sliding surface in (6.12) with Assumption 6.4, we can employ the novel control law in (6.16) to guarantee the following sliding mode dynamics.*

(1) The discrete-time sliding mode dynamics can be driven into a region Ω within η steps, and the region Ω can be described by

$$\Omega = \{z(k) \| z(k)| < \gamma\},$$

where

$$\gamma = \tilde{\rho}\Lambda(\lambda_2),\ \eta = \left\lfloor \frac{z^2(0) - \gamma^2}{\mu^2} \right\rfloor + 1,\ \tilde{\rho} = \max\left\{ \left(\frac{\hat{\rho}_0}{\xi_3}\right)^{\frac{1}{\lambda_2}}, \left(\frac{h\xi_3}{1 - h\xi_2}\right)^{\frac{1}{1-\lambda_2}} \right\}$$

$$\Lambda(y) = 1 + y^{\frac{y}{1-y}} - y^{\frac{1}{1-y}},\ \mu = h\xi_2\gamma^3 + \left[\Lambda^{\lambda_2}(\lambda_2) - 1\right]|h\hat{\rho}_0|,$$

and $\hat{\rho}_0 = d_m + \theta_f + \theta_d$.

(2) Once the sliding mode dynamics enter the region Ω, it will stay there and not escape in the subsequent time, which means $|z(k+1)| \leq \gamma$ if $|z(k)| \leq \gamma$.

Proof Before giving out the proof, we have to review a critical property of $\Lambda(y)$, that is, for any $0 \leq x \leq 1$ and $0 < y < 1$, we have the following relations [3]

$$1 < \Lambda(y) < 2, x\Lambda(y) - x^y\Lambda^y(y) + \Lambda(y) \geq 1. \tag{6.18}$$

The following properties will also be frequently used in this proof

$$|h\hat{\rho}_0| \le h\xi_3\tilde{\rho}^{\lambda_2} \le (1 - h\xi_2)\tilde{\rho}, \tag{6.19}$$

the proof of which is as follows.

[Proof of $|h\hat{\rho}_0| \le h\xi_3\tilde{\rho}^{\lambda_2} \le (1 - h\xi_2)\tilde{\rho}$ in (6.19)] According to the definition of γ and Assumption 6.4, considering the situation that $\left(\frac{\hat{\rho}_0}{\xi_3}\right)^{\frac{1}{\lambda_2}} \ge \left(\frac{h\xi_3}{1-h\xi_2}\right)^{\frac{1}{1-\lambda_2}}$, and since $\hat{\rho}_0$, ξ_3, h and $1 - h\xi_2$ are all positive, we have

$$(h\hat{\rho}_0)^{1-\lambda_2} \ge \frac{h\xi_3}{(1 - h\xi_2)^{\lambda_2}}. \tag{A.1}$$

Using the above expression, we can obtain

$$\begin{aligned} \tilde{\rho} = \left(\frac{\hat{\rho}_0}{\xi_3}\right)^{\frac{1}{\lambda_2}} &\Rightarrow |h\hat{\rho}_0| = h\xi_3\tilde{\rho}^{\lambda_2} \\ &\Rightarrow (h\hat{\rho}_0)^{\lambda_2} \le \tilde{\rho}^{\lambda_2}(1 - h\xi_2)^{\lambda_2} \\ &\Rightarrow |h\hat{\rho}_0| \le \tilde{\rho}(1 - h\xi_2). \end{aligned} \tag{A.2}$$

On the contrary, if $\left(\frac{\hat{\rho}_0}{\xi_3}\right)^{\frac{1}{\lambda_2}} < \left(\frac{h\xi_3}{1-h\xi_2}\right)^{\frac{1}{1-\lambda_2}}$ established, we have $\tilde{\rho} = \left(\frac{h\xi_3}{1-h\xi_2}\right)^{\frac{1}{1-\lambda_2}}$, and

$$h\xi_3\tilde{\rho}^{\lambda_2} = (1 - h\xi_2)\tilde{\rho}, \quad |h\hat{\rho}_0| < h\xi_3\left(\frac{h\xi_3}{1 - h\xi_2}\right)^{\frac{\lambda_2}{1-\lambda_2}},$$

which implies $|h\hat{\rho}_0| < h\xi_3\tilde{\rho}^{\lambda_2}$.

Hence, we have obtained $|h\hat{\rho}_0| \le h\xi_3\tilde{\rho}^{\lambda_2} \le (1 - h\xi_2)\tilde{\rho}$.

Remark 6.6 It is obvious from the result of this lemma that one can also get $|h\hat{\rho}_0| \le h\xi_3\xi^{\lambda_2} \le (1 - h\xi_2)\xi$ with $\xi > \tilde{\rho}$.

(1) Select the discrete-time Lyapunov function $V(k) = z^2(k)$, and we have

$$V(k + 1) - V(k) = -\left[z(k) - z(k + 1)\right]\left[z(k) + z(k + 1)\right]. \tag{6.20}$$

Then, we make classified discussions to prove that the sliding mode dynamics can reach the region Ω with the condition mentioned in Assumption 6.4 if $z(k) \notin \Omega$.

Case 1. Consider $z(k) > \gamma > 0$, and it is clear that

$$\begin{aligned} z(k) - z(k + 1) =& h\xi_2 z^3(k) + h\xi_3\lceil z(k)\rfloor^{\lambda_2} - h\rho_0(k) \\ \ge& h\xi_2\gamma^3 + h\xi_3\tilde{\rho}^{\lambda_2}\Lambda^{\lambda_2}(\lambda_2) - |h\hat{\rho}_0|. \end{aligned} \tag{6.21}$$

Using (6.19), we can obtain

$$z(k) - z(k+1) \geq h\xi_2\gamma^3 + \left[\Lambda^{\lambda_2}(\lambda_2) - 1\right]\left|h\hat{\rho}_0\right| \geq \mu. \tag{6.22}$$

Next, we proceed to examine the property of $z(k) + z(k+1)$

$$\begin{aligned} z(k) + z(k+1) =& h\rho_0(k) + 2z(k) - h\xi_2 z^3(k) - h\xi_3\lceil z(k)\rfloor^{\lambda_2} \\ \geq& 2z(k) - h\xi_2 z^3(k) - h\xi_3\lceil z(k)\rfloor^{\lambda_2} - |h\hat{\rho}_0|. \end{aligned} \tag{6.23}$$

Firstly, we consider the situation that $z(k) \geq 1$, hence we have $z(k) \geq \lceil z(k)\rfloor^{\lambda_2}$, and based on Assumption 6.4, we have

$$z(k) > h\xi_2 z^3(k) + h\xi_3\lceil z(k)\rfloor^{\lambda_2}, \tag{6.24}$$

and along the line from (6.21, 6.22), (6.23) becomes

$$z(k) + z(k+1) \geq h\xi_2 z^3(k) + h\xi_3\lceil z(k)\rfloor^{\lambda_2} - |h\hat{\rho}_0| \geq \mu.$$

Then, we consider the situation that $0 < z(k) < 1$, which implies $z^3(k) < z(k)$, and based on (6.19) and Remark 6.6, (6.23) becomes

$$\begin{aligned} z(k) + z(k+1) \geq& 2\left[h\xi_3\lceil z(k)\rfloor^{\lambda_2} + h\xi_2 z(k)\right] - h\xi_2 z^3(k) - h\xi_3\lceil z(k)\rfloor^{\lambda_2} - |h\hat{\rho}_0| \\ \geq& h\xi_3\lceil z(k)\rfloor^{\lambda_2} + h\xi_2 z^3(k) - |h\hat{\rho}_0| \geq \mu. \end{aligned} \tag{6.25}$$

Hence, we have $V(k+1) - V(k) \leq -\mu^2$ when $z(k) > \gamma$.

Case 2. Considering the situation that $z(k) < -\gamma < 0$ with Assumption 6.4, along the same line from (6.20) to (6.25), we can also prove that $V(k+1) - V(k) \leq -\mu^2$.

According to the above analyses, the discrete-time sliding mode dynamics can converge into the region Ω, and it is reasonable to assume that the sliding mode dynamics reach the region at the η step, which implies $z^2(\eta) \leq z^2(0) - \eta\mu^2 \leq \gamma^2$. Moreover, $\eta \in \mathbb{N}^+$ results in $\eta = \left\lfloor \frac{z^2(0)-\gamma^2}{\mu^2} \right\rfloor + 1$.

At this point, we have proved that the discrete-time sliding mode dynamics can converge into the region Ω within η steps.

(2) Once the sliding mode dynamics enter the region Ω, we have $|z(k)| < \gamma$ and define $0 \leq \beta \leq 1$ to obtain $|z(k)| = \gamma\beta$.

To prove $|z(k+1)| \leq \gamma$, it is essential to divide the analyses due to the location of $z(k)$.

Case 3. Proof of $-\gamma \leq z(k+1) \leq \gamma$ with $z(k) \geq 0$.

Firstly, the situation $0 \leq z(k) \leq \gamma$ is considered and it implies that $\beta\Lambda(\lambda_2) > 0$. Equation (6.21) becomes

$$\begin{aligned} z(k+1) =& z(k) - h\xi_2 z^3(k) - h\xi_3\lceil z(k)\rfloor^{\lambda_2} + h\rho_0(k) \\ \leq& z(k) - h\xi_2 z^3(k) - h\xi_3(\beta\tilde{\rho}\Lambda(\lambda_2))^{\lambda_2} + |h\hat{\rho}_0|. \end{aligned}$$

By $h\xi_2 z(k) > 0$ and (6.19), the above expression becomes

$$
\begin{aligned}
z(k+1) \leq & z(k) - h\xi_3 \tilde{\rho}^{\lambda_2} \beta^{\lambda_2} \Lambda^{\lambda_2}(\Lambda) + |h\hat{\rho}_0| \\
\leq & z(k) - |h\hat{\rho}_0| \beta^{\lambda_2} \Lambda^{\lambda_2}(\Lambda) + |h\hat{\rho}_0| \\
\leq & z(k) + \left[1 - \beta^{\lambda_2} \Lambda^{\lambda_2}(\Lambda)\right] |h\hat{\rho}_0|.
\end{aligned} \tag{6.26}
$$

If $\beta\Lambda(\lambda_2) \geq 1$, it is obvious that $z(k+1) \leq z(k) \leq \gamma$, and $\beta\Lambda(\lambda_2) < 1$ leads to $\beta^{\lambda_2}\Lambda^{\lambda_2}(\lambda_2) < 1$. Utilizing (6.19) and $z(k) = \beta\tilde{\rho}\Lambda(\lambda_2)$, (6.26) becomes

$$
\begin{aligned}
z(k+1) \leq & \beta\tilde{\rho}\Lambda(\lambda_2) + \left[1 - \beta^{\lambda_2}\Lambda^{\lambda_2}(\Lambda)\right](1 - h\xi_2)\tilde{\rho} \\
\leq & \tilde{\rho}\left[\beta\Lambda(\lambda_2) - \beta^{\lambda_2}\Lambda^{\lambda_2}(\lambda_2) + h\xi_2\left(\beta^{\lambda_2}\Lambda^{\lambda_2}(\lambda_2) - 1\right) + 1\right] \\
\leq & \tilde{\rho} \leq \tilde{\rho}\Lambda(\lambda_2) \leq \gamma.
\end{aligned}
$$

Now, we have proved that $z(k+1) \leq \gamma$ with $z(k) \geq 0$, and we proceed to analyze $z(k+1) \geq -\gamma$ with $z(k) \geq 0$.

Considering $z(k) \geq 1$, we have $\lceil z(k) \rfloor^{\lambda_2} \leq z(k) \leq z^3(k)$ and utilize Assumption 6.4 and (6.19) to get

$$
\begin{aligned}
z(k+1) \geq & z(k) - h\xi_3 z(k) - h\xi_2 z^3(k) - |h\hat{\rho}_0| \\
\geq & (1 - h\xi_3) z(k) - h\xi_2 z^3(k) - |h\hat{\rho}_0| \\
\geq & -|h\hat{\rho}_0| \geq -(1 - h\xi_2)\tilde{\rho} > -\Lambda(\lambda_2)\tilde{\rho} \geq -\gamma.
\end{aligned} \tag{6.27}
$$

Contrarily, considering $0 < z(k) < 1$, we have $z^3(k) \leq z(k) \leq \lceil z(k) \rfloor^{\lambda_2}$ and utilize Assumption 6.4 and (6.19) to get

$$
\begin{aligned}
z(k+1) \geq & (1 - h\xi_2)\beta\Lambda(\lambda_2)\tilde{\rho} - h\xi_3\left(\beta\Lambda(\lambda_2)\tilde{\rho}\right)^{\lambda_2} - |h\hat{\rho}_0| \\
\geq & \tilde{\rho}(1 - h\xi_2)\left(\beta\Lambda(\lambda_2) - \beta^{\lambda_2}\Lambda^{\lambda_2}(\lambda_2) - 1\right) \\
\geq & -\tilde{\rho}(1 - h\xi_2)\Lambda(\lambda_2) > -\tilde{\rho}\Lambda(\lambda_2) \geq -\gamma.
\end{aligned} \tag{6.28}
$$

Case 4. Here we will give out the proof about $-\gamma \leq z(k+1) \leq \gamma$ with $z(k) < 0$. Firstly, we will prove $z(k+1) \leq \gamma$ with $z(k) < 0$.

When $-1 < z(k) < 0$, utilizing $|z(k)|^{\lambda_2} > |z(k)| > |z(k)|^3$ and (6.19), (6.21) becomes

$$
\begin{aligned}
z(k+1) = & z(k) + h\xi_2 |z|^3(k) + h\xi_3 |z(k)|^{\lambda_2} + h\rho_0(k) \\
\leq & -|\beta\Lambda(\lambda_2)\tilde{\rho}| + h\xi_2|\beta\Lambda(\lambda_2)\tilde{\rho}|^3 + h\xi_3|\beta\Lambda(\lambda_2)\tilde{\rho}|^{\lambda_2} + |h\hat{\rho}_0| \\
\leq & \tilde{\rho}\left[(h\xi_2 - 1)|\beta\Lambda(\lambda_2)| + (1 - h\xi_2)|\beta\Lambda(\lambda_2)|^{\lambda_2} + (1 - h\xi_2)\right] \\
\leq & \tilde{\rho}(1 - h\xi_2)\left[1 - |\beta\Lambda(\lambda_2)| + |\beta\Lambda(\lambda_2)|^{\lambda_2}\right].
\end{aligned}
$$

By using (6.18) and Assumption 6.4, the following inequality holds.

$$
z(k+1) \leq (1 - h\xi_2)\tilde{\rho}_0\Lambda(\lambda_2) \leq \gamma. \tag{6.29}
$$

When $z(k) \leq -1$, the relation $|z(k)|^{\lambda_2} \leq |z(k)| \leq |z(k)|^3$ can be satisfied, and (6.21) becomes

$$\begin{aligned} z(k+1) \leq &- |\beta \Lambda(\lambda_2)\tilde{\rho}| + h\xi_2|\beta \Lambda(\lambda_2)\tilde{\rho}|^3 + h\xi_3|\beta \Lambda(\lambda_2)\tilde{\rho}|^{\lambda_2} + |h\hat{\rho}_0| \\ \leq &- (1 - h\xi_3)|\beta \Lambda(\lambda_2)\tilde{\rho}| + h\xi_2|\beta \Lambda(\lambda_2)\tilde{\rho}|^3 + |h\hat{\rho}_0|. \end{aligned}$$

By Assumption 6.4 and (6.19), we can obtain

$$z(k+1) \leq |h\hat{\rho}_0| \leq (1 - h\xi_2)\tilde{\rho} < \Lambda(\lambda_2)\tilde{\rho} \leq \gamma. \tag{6.30}$$

After $z(k+1) \leq \gamma$ with $z(k) < 0$ proved, we will discuss $z(k+1) > -\gamma$ with $z(k) < 0$.

Considering $z(k) < 0$, (6.21) can be rewritten as

$$\begin{aligned} z(k+1) \geq &- |z(k)| + h\xi_2|z(k)|^3 + h\xi_3|z(k)|^{\lambda_2} - |h\hat{\rho}_0| \\ \geq &- |\beta\tilde{\rho}\Lambda(\lambda_2)| + \left(|\beta\Lambda(\lambda_2)|^{\lambda_2} - 1\right)|h\hat{\rho}_0| + h\xi_2|\beta\tilde{\rho}_0\Lambda(\lambda_2)|^3 \\ \geq &- \left[|\beta\tilde{\rho}\Lambda(\lambda_2)| - \left(|\beta\Lambda(\lambda_2)|^{\lambda_2} - 1\right)|h\hat{\rho}_0|\right]. \end{aligned}$$

If $|\beta\Lambda(\lambda_2)| \geq 1$, we can directly get

$$z(k+1) \geq -|\beta\tilde{\rho}\Lambda(\lambda_2)| \geq -\gamma.$$

If $|\beta\Lambda(\lambda_2)| < 1$, using (6.19), we have

$$\begin{aligned} z(k+1) \geq &- \left[|\beta\tilde{\rho}\Lambda(\lambda_2)| - \left(|\beta\Lambda(\lambda_2)|^{\lambda_2} - 1\right)(1 - h\xi_2)\tilde{\rho}\right] \\ \geq &- \tilde{\rho}\left[1 - h\xi_2\left(1 - |\beta\Lambda(\lambda_2)|^{\lambda_2}\right) + \left(|\beta\Lambda(\lambda_2)| - |\beta\Lambda(\lambda_2)|^{\lambda_2}\right)\right] \\ \geq &- \tilde{\rho} > -\tilde{\rho}\Lambda(\lambda_2) \geq -\gamma. \end{aligned}$$

To this end, we have finished the analyses about $|z(k+1)| \leq \gamma$, that is the discrete-time sliding mode dynamics will enter the region Ω and not escape in the subsequent time by using the proposed control law. ■

Remark 6.7 This control law can accelerate the convergence to Ω when the system is with a large uncertainty leading to $\gamma > 1$.

Remark 6.8 In practice, Assumption 6.4 is easy to satisfy. According to the requirement of control performance, a sampling interval h is set firstly, i.e., for this work $h = 1$ ms, and we proceed to select suitable ξ_2 and ξ_3 to make $z(0)$ initially satisfy Assumption 1. The principle of selection of ξ_2 and ξ_3 is to make $\gamma^2 < \frac{1-h\xi_3}{h\xi_2}$, this is because this assumption is proposed to guarantee reaching condition to hold. If $\gamma^2 < z^2(0) < \frac{1-h\xi_3}{h\xi_2}$, we can activate the reaching phase, which leads to a decreasing $z^2(k)$, and the subsequent $z(k)$ satisfy the assumption directly. If $z^2(0) < \gamma^2$, the sliding phase works, and $|z^2(k)| < \gamma^2 < \frac{1-h\xi_3}{h\xi_2}$ can also be guaranteed.

Next, we will discuss the tracking precision when the sliding mode dynamics stay in the region Ω. Before moving on, a necessary lemma about discrete-time terminal sliding surfaces is exhibited.

Lemma 6.9 ([3]) *Consider the dynamics on the discrete-time terminal sliding surface $\Delta e(k) + \xi_1 \lceil e(k) \rfloor^{\lambda_1} = g(k)$ with $|g(k)| \leq \gamma$, and there is a finite step to guarantee*

$$|e(k)| \leq \Lambda(\lambda_1) \cdot \max\left\{ \left(\frac{\gamma}{\xi_1}\right)^{1/\lambda_1}, (h\xi_1)^{\frac{1}{1-\lambda_1}} \right\}. \tag{6.31}$$

According to the analyses in Theorem 6.5, the proposed control law can drive the system states into a certain region, and combined with the definition of the DFOTSM surface, we have $\Delta e(k) + \xi_1 \Delta^{\alpha-1} \lceil e(k) \rfloor^{\lambda_1} = g(k)$. Using Grünwald-Letnikov definition, the above expression becomes

$$\Delta e(k) + \frac{\xi_1}{h^{\alpha-1}} \lceil e(k) \rfloor^{\lambda_1} + \frac{\xi_1}{h^{\alpha-1}} \sum_{j=1}^{L} (-1)^j \binom{\alpha-1}{j} \lceil e(k-j) \rfloor^{\lambda_1} = g(k).$$

By the note of Lemma 6.3, it follows from the above expression

$$\Delta e(k) + \frac{\xi_1}{h^{\alpha-1}} \lceil e(k) \rfloor^{\lambda_1} \leq \gamma + \frac{\rho_e(K-1)}{h^{\alpha-1}}, \tag{6.32}$$

where $\rho_e \geq \left\{ \lceil e(k-1) \rfloor^{\lambda_1} \right\}$.

Based on the result in Lemma 6.9, we can obtain

$$|e(k)| \leq \Lambda(\lambda_1) \cdot \max\left\{ \left(\frac{h^{\alpha-1}\gamma_0}{\xi_1}\right)^{1/\lambda_1}, (h^{2-\alpha}\xi_1)^{\frac{1}{1-\lambda_1}} \right\}, \tag{6.33}$$

where $\gamma_0 = \gamma + \frac{\rho_e(K-1)}{h^{\alpha-1}}$. Compared with the precision in Lemma 6.9, because h is the sampling interval with $0 < h \ll 1s$ for digital computer system, we have

$$(h^{2-\alpha}\xi_1)^{\frac{1}{1-\lambda_1}} < (h\xi_1)^{\frac{1}{1-\lambda_1}},$$

which means that the proposed tracking precision is higher than the discrete-time terminal sliding mode control if $|e(k)|$ in (6.33) is determined by $(h^{2-\alpha}\xi_1)^{\frac{1}{1-\lambda_1}}$, i.e.,

$$(h^{2-\alpha}\xi_1)^{\frac{1}{1-\lambda_1}} \geq \left(\frac{h^{\alpha-1}\gamma_0}{\xi_1}\right)^{1/\lambda_1} \quad \Rightarrow \quad \xi_1 \geq \frac{\gamma_0^{1-\lambda_1}}{h^{1+\lambda_1-\alpha}} \tag{6.34}$$

6.4 Experiments

6.4.1 Experimental Setup

The proposed control scheme is examined on an iron-core linear motor shown in Fig. 4.1 and implemented in Twincat 3.0 with real-time code running in an industrial PLC. The communication protocol of the entire control system is CANopen, and the sampling interval for the performance test is decided to be 1 millisecond, which covers the sampling time of the position sensor and the communication period. The setup is equipped with an absolute position sensor with a resolution of 50 nm, and the velocity signal is estimated by using the backward difference technique. The mass of the nominal model is provided by the manufacturer $m = 11$ kg, and the values of parameters involving the sliding friction $k_c = 7.847$ N and $k_v = 0.019$ Ns/m are identified by using a PSO-based evaluation with 300 particles iterating 300 generations [4]. Conservatively, we set the bounds of parameters that indicate the parametric uncertainties as follows: $\theta_d = 2$, $\theta_f = 5$, and $d_m = 10$.

6.4.2 Controller Parameter Selections

Next, we will discuss the selection of the controller parameters. For satisfying Assumption 6.4, ξ_3 is conservatively set as $\xi_3 = 450$, and we set $\xi_2 = 10$ to obtain a broad selection of the initial sliding mode dynamics $z(0)$. It is worth noting that there exists a contradiction between the selections of ξ_1 and α if λ_1 is fixed. A small α results in a higher tracking precision and meanwhile requires a large ξ_1 to guarantee a better performance as Eqs. (6.33, 6.34) shown. However, too large ξ_1 may not meet the requirement of Assumption 6.4. Through trial and error, we set $\xi_1 = 500$, $\lambda_2 = 0.6$, $\lambda_1 = 0.6$ and $\alpha = 0.5$, and among these selections $\lambda_1 = 0.6$ and $\alpha = 0.5$ are frequently used in the FO or integer-order terminal sliding mode controllers [5, 6, 8]. These parameters can yield a theoretical tracking precision of 31.44 μm.

6.4.3 Experimental Results

In the previous sections, we have analyzed the tracking precision and corresponding parameter selections, and here will employ the real linear motor to verify the effectiveness. As a comparison, the discrete-time terminal sliding mode controller presented in [3] is also tested on the experimental setup, and for a fair comparison, the controller parameters of the comparison are the same as the proposed controller.

In this work, we assign two experiments as follows:

(1) Exp1: Controllers are applied on the experimental setup without extra payload to test the tracking performances.

(2) Exp2: An extra payload of 3 kg is added on the moving stage to test the robustness against the parametric uncertainties.

In both experiments, the controllers are required to track two kinds of reference signals to inspect the corresponding tracking performances. The first reference signal is a conoidal function with the final frequency of 0.5 Hz and the amplitude of 1 cm: $p_r(k) = 10^{-2}(1 - e^{-hk})\,[0.5 - \cos(\pi hk)]$, and the other one is a triangle signal with varying slopes. For clarity, we denote Exp1 with the cosinoidal reference as Case1A, and denote Exp1 with the triangle reference as Case1B. Accordingly, Case2A and Case2B are denoted to represent Exp2 with different references. The discrete-time FO terminal and discrete-time sliding mode controllers are marked as C1 and C2, respectively.

The experimental results of Exp1 are shown in Figs. 6.1, 6.2 and 6.3. It is obvious from Fig. 6.1 that the tracking trajectory of C1 is closer to the reference than C2s.

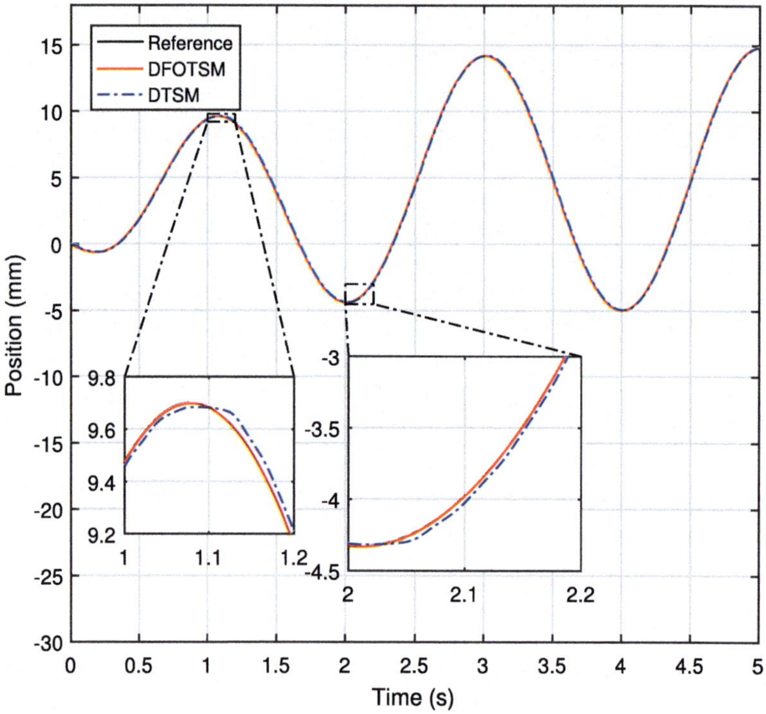

Fig. 6.1 Exp1 with the cosinoidal reference: tracking trajectories

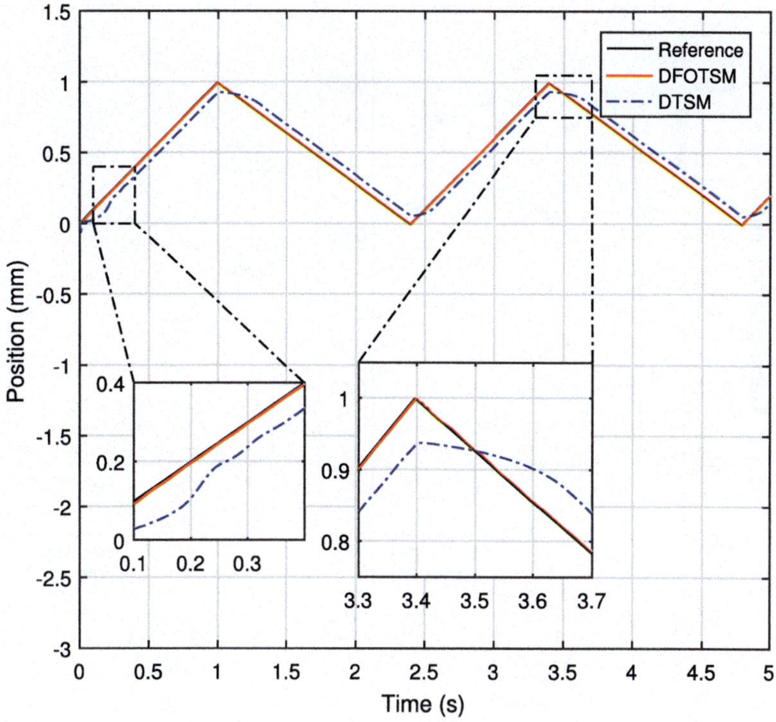

Fig. 6.2 Exp1 with the triangle reference: tracking trajectories

From the enlarged figures, one can see that there exist some overshoots when C2s motion direction changes. Meanwhile, C1s trajectory is smooth and almost coincides with the reference everywhere.

Case1B is assigned to test the low-speed tracking performance, and the experimental results are shown in Fig. 6.2. Because the amplitude of the reference is 1mm, which is smaller than that of Case1A, slight steady-state errors can be found in C1s trajectories, which are far smaller than C2s.

The tracking errors of Exp1 are given out in Fig. 6.3. Intuitively, in Case1A, C2s errors are about 40 μm, much larger than C1s errors that are less than 5 μm. These actual tracking precisions are much higher than the theoretical precisions because the bounds of the physical parameters are chosen conservatively in the analyses. In addition, C2s errors fluctuate more widely than C1s, and the enlarged figures indicate that C1s maximum errors appear at the moment of motion direction change and fall back in no more than 0.1 s. Moreover, the maximum errors are not big enough to cause the overshoots, which coincides with the aforementioned analyses.

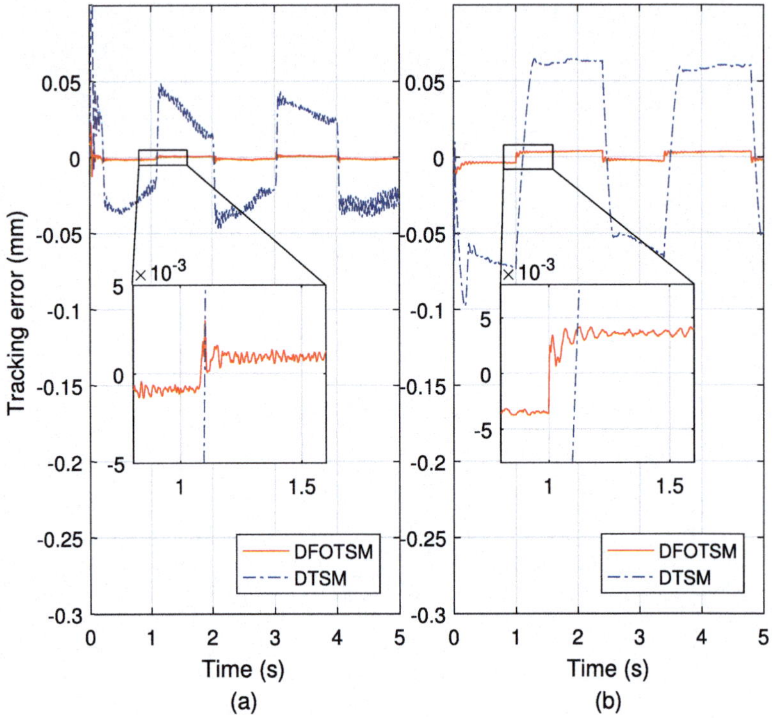

Fig. 6.3 Exp1: **a** tracking errors of Case1A **b** tracking errors of Case1B

Exp2 is carried out to examine the tracking performance of the proposed control scheme when the moving stage is subjected to severe parametric uncertainty. The tracking errors are shown in Fig. 6.4. Although the fluctuation of C1s errors becomes more frequent, and the maximum error also increases, C1 can still guarantee the tracking precision of approximating 5 μm.

The typical control inputs are illustrated by Fig. 6.5. It is clear to see that the adjustment range of C1s inputs is around 10 N, which is similar to C2s. Some fine adjustments can also be found in C1s inputs, and C2s inputs are more smooth but superficial. Considering the corresponding tracking errors, this contrast means that C1 is sensitive to the reference signal change, that is the control effort can be regulated in time to guarantee high-precision tracking when the reference signal is time-varying. This phenomenon is more conspicuous in Fig. 6.5c, the motion direction changes at 1 s and 2.4 s, and C1s control effort responds quickly to this change and turns steady at about 1.1 s; whereas C2 regulates the input more slowly and becomes steady after 1.3 s.

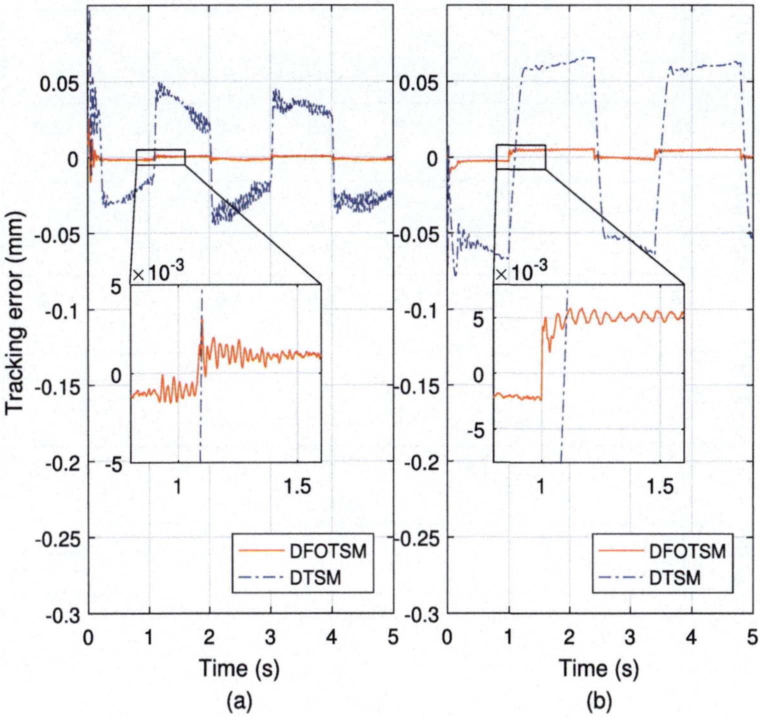

Fig. 6.4 Exp2: **a** tracking errors of Case2A **b** tracking errors of Case2B

6.4.4 Summary and Statistics of Experimental Results

The previous qualitative analysis has demonstrated that the DFOTSM controller has better performance than the DFTSM controller, and here we carry out the quantitative comparison based on the statistics of experimental results.

We define the average tracking error as $\sum_{k=1}^{N} \frac{|e(k)|}{N}$, and besides the average tracking error, we also calculate the root-mean-square error (RMSe) defined as $RMSe = \sqrt{\sum_{k=1}^{N} \frac{e^2(k)}{N}}$, where N is the sampling number. The results are shown in Fig. 6.6 and indicate that the DFOTSM controller owns the better average error and RMSe. Moreover, these indexes of the DFOTSM controller maintain stability even though the payload varies, which coincides with the theoretical analyses and experimental results. Compared to the statistical results of Exp1 and Exp2, it is apparent that the varying payload doesn't affect the stability of the control system and the proposed controller can retain a satisfactory robust performance.

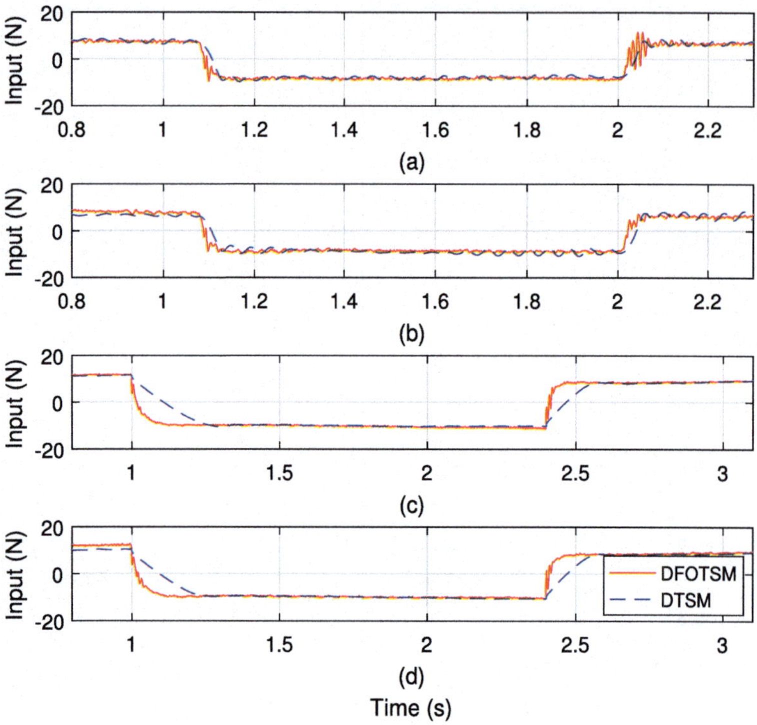

Fig. 6.5 Control inputs: **a** Case1A **b** Case2A **c** Case1B **d** Case2B

6.5 Conclusion

A discrete-time FO terminal SMC strategy has been developed for the practical linear motor-based high-precision motion control system. Grünwald-Letnikov FO difference is introduced into integer-order discrete-time sliding mode controller to synthesize a novel sliding surface. The control scheme takes into bounded parametric uncertainties including sliding friction, varying payload, and unknown disturbance, and the corresponding reaching law guarantees the sliding mode dynamics to converge into a certain region in finite steps. The theoretical analyses have proved that the proposed sliding surface can increase the tracking precision drastically compared with the discrete-time sliding mode controller. Experiments are carried out on an industrial linear motor-based setup with different payloads, and the proposed controller performs high-precision, fast response, and robustness to uncertainties, which coincides with the theoretical analyses.

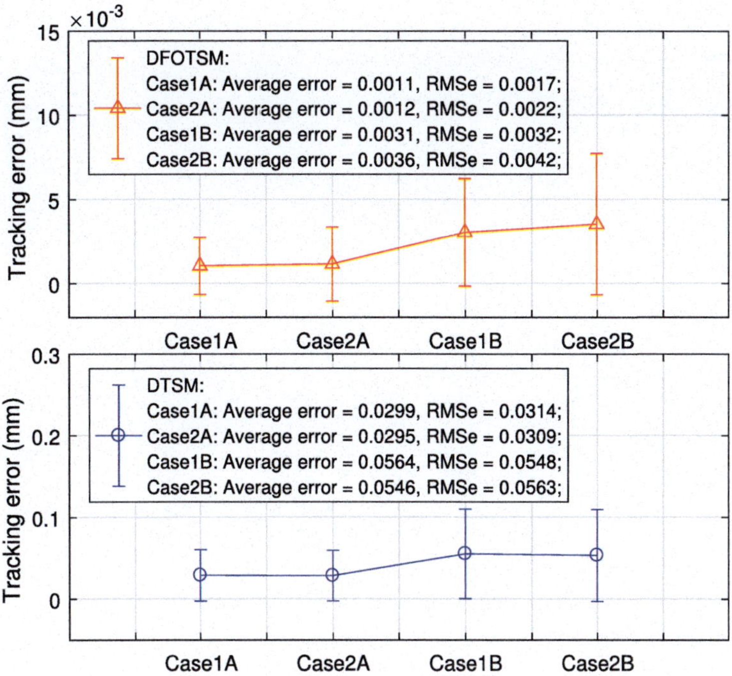

Fig. 6.6 Average errors and RMSes

References

1. Corradini, M.L., Fossi, V., Giantomassi, A., Ippoliti, G., Longhi, S., Orlando, G.: Discrete time sliding mode control of robotic manipulators: development and experimental validation. Control Eng. Pract. **20**(8), 816–822 (2012)
2. Galias, Z., Yu, X.: Euler's discretization of single input sliding-mode control systems. IEEE Trans. Autom. Control **52**(9), 1726–1730 (2007)
3. Li, S., Du, H., Yu, X.: Discrete-time terminal sliding mode control systems based on Euler's discretization. IEEE Trans. Autom. Control **59**(2), 546–552 (2013)
4. Lin, C.J., Yau, H.T., Tian, Y.C.: Identification and compensation of nonlinear friction characteristics and precision control for a linear motor stage. IEEE/ASME Trans. Mechatron. **18**(4), 1385–1396 (2012)
5. Wang, Y., Gu, L., Xu, Y., Cao, X.: Practical tracking control of robot manipulators with continuous fractional-order nonsingular terminal sliding mode. IEEE Trans. Ind. Electron. **63**(10), 6194–6204 (2016)
6. Wang, Y., Luo, G., Gu, L., Li, X.: Fractional-order nonsingular terminal sliding mode control of hydraulic manipulators using time delay estimation. J. Vib. Control **22**(19), 3998–4011 (2016)
7. Zheng, J., Fu, M.: A reset state estimator using an accelerometer for enhanced motion control with sensor quantization. IEEE Trans. Control Syst. Technol. **18**(1), 79–90 (2009)
8. Zheng, J., Wang, H., Man, Z., Jin, J., Fu, M.: Robust motion control of a linear motor positioner using fast nonsingular terminal sliding mode. IEEE/ASME Trans. Mechatron. **20**(4), 1743–1752 (2014)

Chapter 7
Fractional-Order Sliding Mode Contouring Error Control

7.1 Introduction

In order to achieve better performance of contouring control, not only an efficient contouring error estimation (CEE) but also a valid contouring error control (CEC) is necessary. In sight of existing practical contouring control approaches, most of them are designed as a continuous controller but utilized as a discrete-time form, which lacks theoretical support [7]. In application, the approximations caused by computer numerical techniques were not well taken into account, leading to a degradation of theoretical contouring performance. Moreover, nonlinear disturbances and uncertainties will aggravate the degradation [12], and hence it is imperative to design an effective computer numerical control strategy for contouring control of high-precision servo systems.

Inspired by the shortcomings of the existing CEC above, the discrete-time discrete-time sliding mode sliding mode (DFSMC) CEC for multidimensional systems is proposed in this chapter. In the process of designing the controller, a novel FO sliding surface is addressed. Generally, FO calculus can describe the system more accurately in comparison with integer-order calculus, as it has a weighted memory effect. Moreover, the sliding mode controller has a good robustness performance which has been proved by extensive theoretical and experimental results. Thus, the uncertainties and disturbances have few influences on the system under DFSMC. To validate the effectiveness of the proposed control strategy, groups of comparative experiments are implemented on a 2-DOF linear motor table, whose results validate the efficiency of the proposed strategy.

G. Sun et al., *Fractional-Order Sliding Mode Control: Methodologies and Applications*, Studies in Systems, Decision and Control 532,
https://doi.org/10.1007/978-3-031-60847-6_7

7.2 Simplified Newton-Based Contouring Error Estimation

For the n-dimensional contouring control problem, the reference contour map can be parameterized by time as

$$\boldsymbol{p}_r(t) = [p_{r1}(t), p_{r2}(t), \ldots, p_{rn}(t)]^T, \tag{7.1}$$

where t denotes the time, which indicates the reference position, and in the industrial application, $p_{ri}(t), i = 1, 2, \ldots, n$ is always a continuous function.

The actual contour map can be expressed as

$$\boldsymbol{p}(t) = [p_1(t), p_2(t), \ldots, p_n(t)]^T. \tag{7.2}$$

Due to the development of computer numerical control (CNC), $\boldsymbol{p}(t)$ is sampled as $\boldsymbol{p}(k) = [p_1(k), p_2(k), \ldots, p_n(k)]^T$ in the controller, where $t = kh$, h is the sampling interval of CNC.

Define a error vector $\boldsymbol{\varepsilon}(\hat{t})$ as follows

$$\boldsymbol{\varepsilon}(\hat{t}) = \boldsymbol{p}_r(\hat{t}) - \boldsymbol{p}(k), \tag{7.3}$$

which stands for the vector from the actual position $\boldsymbol{p}(k)$ to the reference contour map at time $\hat{t}$. The contouring error is defined as the shortest distance between the current position and the reference contour [4], thus the contouring error vector $\boldsymbol{\varepsilon}^*$ can be defined as

$$\boldsymbol{\varepsilon}^* = \boldsymbol{p}_r(t^*) - \boldsymbol{p}(k), \tag{7.4}$$

where $\boldsymbol{\varepsilon}^*$ happens at time t^*.

To make the definitions above clearer, the example of a 3-dimensional case is given out and the diagram is displayed in Fig. 7.1.

To complete the CEE, for an n-dimensional system, we propose a cost function as

$$J(\hat{t}) = |\boldsymbol{\varepsilon}(\hat{t}) \cdot (\boldsymbol{p}_r(\hat{t} + t_\delta) - \boldsymbol{p}_r(\hat{t}))|, \tag{7.5}$$

where t_δ is a tiny time interval. Proper $t_{\delta i}$ should be selected to satisfy that

$$\boldsymbol{p}_r(\hat{t} + t_\delta) - \boldsymbol{p}_r(\hat{t}) \neq \boldsymbol{0}. \tag{7.6}$$

The solution of the equation $J(\hat{t}) = 0$ is t^*. However, $J(\hat{t})$ is usually a nonlinear equation, which makes it difficult to give out the analytical solution. Thanks to the development of digital processors, the arithmetic solution can be obtained. Therefore, we apply the simplified Newton-based algorithm to estimate t^* in [8]. It is presented as

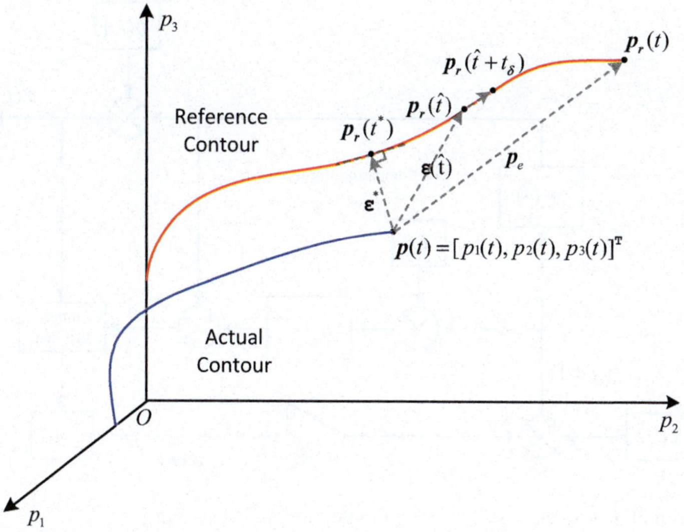

Fig. 7.1 Contouring error of 3-dimensional systems

$$\hat{t}(j+1) = \hat{t}(j) - \frac{J(\hat{t}(j))}{C}, \tag{7.7}$$

where j denotes the times of iteration, C is a positive constant. Obviously, C is the only adjustable parameter, and a smaller value of C results in a faster rate of convergence in this method.

It is necessary to point out that the results of SNE are not global but local, which means the selection of the initial value for the iteration must be proper to prevent diverging or converging to wrong results. A general approach to solve this problem is to use the estimated value $\hat{t}(j)$ as the initial value for the iteration of the next step. The process of the iteration is demonstrated in Fig. 7.2. The acquisition of $\boldsymbol{\varepsilon}(\hat{t})$ in one step is summarized as follows:

1. Initialize $\hat{t}$.
2. Use (7.4) to get $\boldsymbol{\varepsilon}$.
3. Set t_δ satisfying (7.6) and use (7.5) to obtain J.
4. Substitute $\hat{t}$ and J into (7.7) and get a new $\hat{t}$ which is noted as $\hat{t}(j+1)$.
5. Repeat the process 2-4 until the times of iterations are enough. Then output $\varepsilon(\hat{t})$ as the result of CEE.

Remark 7.1 If the t_δ is selected small enough, the result of the estimation is only affected by the constant C and the times of iterations. Therefore, compared with static CEE algorithms addressed in [2, 15], the simplified Newton-based CEE has a higher precision, especially for high-feed rate or highly curved contouring applications.

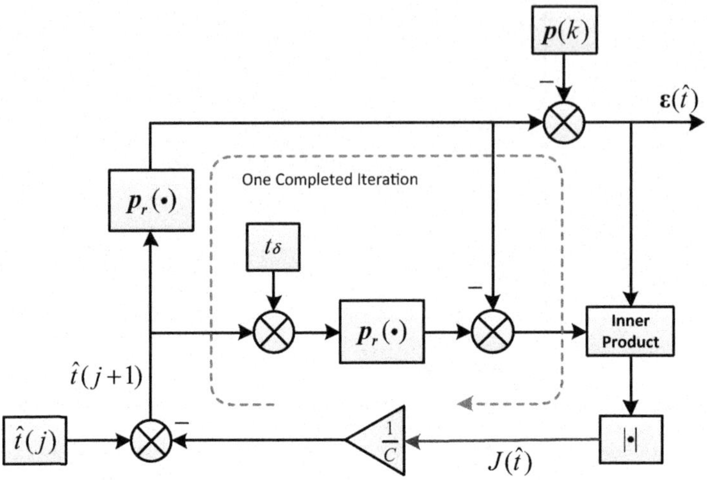

Fig. 7.2 Simplified Newton-based CEE for multidimensional systems

Remark 7.2 Ghaffari and Ulsoy presented a Newton-based CEE (NE) in [4]. To get the estimation of t^*, the Newton-based extremum-seeking algorithm is implemented, thus the gradient and Hessian of the cost function need to be given out. The calculation of the gradient and the Hessian requires the derivative of $p_{ri}(t)$. However, in industrial applications, the $p_{ri}(t)$ is not always derivable everywhere. The calculation of SNE in this chapter demands no derivatives, thus it has a wider range of applications. Moreover, the calculation in the iteration of SNE is much simpler, especially when $n \geq 3$, which is helpful to decrease the delay during the processing and promotes real-time performance.

7.3 Discrete-Time Fractional-Order Sliding Mode Contouring Error Control Design

7.3.1 Plant Modeling

The contouring control is usually implemented on electromechanical systems, such as biaxial linear-motor-driving tables and ball screw motor-driving table. These systems can be described as an n-dimensional dynamic model uniformly, which can be represented as [5]

$$\boldsymbol{M}\ddot{\boldsymbol{p}} + \boldsymbol{B}\dot{\boldsymbol{p}} + \boldsymbol{A}\boldsymbol{S}_f(\dot{\boldsymbol{p}}) = \boldsymbol{u} + \boldsymbol{d}, \tag{7.8}$$

where $\boldsymbol{p}$ is explained in (7.2) stands for the position vector, $\dot{\boldsymbol{p}} = [\dot{p}_1(t), \dot{p}_2(t), \ldots, \dot{p}_n(t)]^T = [v_1(t), v_2(t), \ldots, v_n(t)]^T = \boldsymbol{v}$ is the velocity vector, $\ddot{\boldsymbol{p}} = [\ddot{p}_1(t),$

$\ddot{p}_2(t), \ldots, \ddot{p}_n(t)]^T$ represents the acceleration vector, $\boldsymbol{M} = diag[M_1, M_2, \ldots, M_n]$ is the diagonal inertia matrix, $\boldsymbol{B} = diag[B_1, B_2, \ldots, B_n]$ denotes the damping matrix, $\boldsymbol{A} = diag[A_1, A_2, \ldots, A_n]$ stands for the friction coefficient matrix, $\boldsymbol{S_f}(\dot{\boldsymbol{p}}) = [\text{sgn}\,(v_1(t))\,, \text{sgn}\,(v_2(t))\,, \ldots, \text{sgn}\,(v_n(t))]$, $\boldsymbol{d}$ is the external and unmodelled disturbances vector.

By using Euler's discretization methods, the continuous model can be discretized as

$$\boldsymbol{p}(k+1) = \boldsymbol{p}(k) + h\boldsymbol{v}(k), \tag{7.9}$$

$$\begin{aligned} \boldsymbol{v}(k+1) = \boldsymbol{v}(k) + h\boldsymbol{M}^{-1}\boldsymbol{u}(k) + h\boldsymbol{M}^{-1}\boldsymbol{d}(k) \\ - h\boldsymbol{M}^{-1}\boldsymbol{B}\boldsymbol{v}(k) - h\boldsymbol{M}^{-1}\boldsymbol{A}\boldsymbol{S_f}(\boldsymbol{v}(k)). \end{aligned} \tag{7.10}$$

It is necessary to point out that in a practical engineering facility, because of the parametric uncertainties, the exact values of $\boldsymbol{A}$, $\boldsymbol{B}$, and $\boldsymbol{M}$ are unknown to us. Thus, we use $\bar{\bullet}$ to stand for the nominal matrices and $\tilde{\bullet} = \bullet - \bar{\bullet}$ to stand for the parametric uncertainties. During the process of designing a controller, the nominal model should be used instead of the actual model.

The position tracking error vector $\mathbf{p_e}(k)$ and the velocity tracking error vector $\mathbf{v_e}(k)$ are defined as $\mathbf{p_e}(k) = \mathbf{p_r}(k) - \mathbf{p}(k)$ and $\mathbf{v_e}(k) = \mathbf{v_r}(k) - \mathbf{v}(k)$ respectively, where $\mathbf{p_r}(k)$ is the reference position vector, $\mathbf{v_r}(k) = \frac{1}{h}(\mathbf{p_r}(k+1) - \mathbf{p_r}(k))$ is the reference velocity vector. As for the contouring error vector, the analysis in Section II shows that it is a posteriori vector, which cannot be obtained directly but can be estimated by the SNE. We use $\boldsymbol{\varepsilon}(k)$ to express the result of the estimation in discrete-time cases.

Taking positioning error and contouring error into consideration, define the position synthetical error vector and the velocity synthetical error vector respectively as

$$\check{\boldsymbol{p}}(k) = \boldsymbol{p}_e(k) + \eta \cdot \boldsymbol{\varepsilon}(k), \tag{7.11}$$

$$\check{\boldsymbol{v}}(k) = \boldsymbol{v}_e(k) + \eta \cdot \Delta\boldsymbol{\varepsilon}(k), \tag{7.12}$$

where $\Delta\boldsymbol{\varepsilon}(k) = [\Delta\varepsilon_1, \Delta\varepsilon_2, \ldots, \Delta\varepsilon_n]^T$ and η is a tunable parameter.

7.3.2 Control System Design

Before the designed controller is given out, the definition of discrete-time FO calculus in Grünwald-Letnikov (GL) form should be firstly presented as

Definition 7.3 ([9]) The FO difference of the function $x(k)$ in the discrete GL form is given by

$$\Delta^{\alpha}x(k) = \frac{1}{h^{\alpha}}\sum_{j=0}^{k}(-1)^{j}\binom{\alpha}{j}x(k-j), \tag{7.13}$$

where $\alpha \in \mathbb{R}$ stands for the order of fractional difference, h is the sampling interval, $k \in \mathbb{N}_{+}$ denotes the number of steps and the $\binom{\alpha}{j}$ is a binomial term, which is defined as

$$\binom{\alpha}{j} = \begin{cases} 1 & \text{for } j = 0, \\ \frac{\alpha(\alpha-1)\ldots(\alpha-j+1)}{j!} & \text{for } j = 1, 2, 3, \ldots \end{cases} \tag{7.14}$$

However, it is realized that Definition 7.3 is not convenient to be utilized to calculate FO calculus directly, because the sum goes to infinity as time goes on. In practice, the approximation $\Delta^{\alpha}x(k) = \frac{1}{h^{\alpha}}\sum_{j=0}^{l}(-1)^{j}\binom{\alpha}{j}x(k-j)$ is more useful, where l is a selected suitable length [1]. Through further calculation of the binomial term, it is easy to find that the weighting decreases with j increasing. In other words, the value of the fractional difference is mainly decided by several $x(k-j)$ where $j = 0, 1, \ldots, l$. Therefore, this approximation is rational intuitively.

The discrete-time FO sliding surface is designed as follows:

$$\boldsymbol{s}(k) = \boldsymbol{c}_1\check{\boldsymbol{p}}(k) + \boldsymbol{c}_2\check{\boldsymbol{v}}(k) + \boldsymbol{c}_3\Delta^{\lambda-1}\check{\boldsymbol{p}}(k), \tag{7.15}$$

where $\boldsymbol{s}(k) = [s_1(k), s_2(k), \ldots, s_n(k)]^{T}$ and $\Delta^{\lambda-1}\check{\boldsymbol{p}}(k) = \left[\Delta^{\lambda-1}\check{p}_1(k),\ \Delta^{\lambda-1}\check{p}_2(k), \ldots, \Delta^{\lambda-1}\check{p}_n(k)\right]^{T}$, λ is a constant satisfying $0 < \lambda < 1$, $\boldsymbol{c}_1 = diag[c_{11}, c_{12}, \ldots, c_{1n}]$, $\boldsymbol{c}_2 = diag[c_{21}, c_{22}, \ldots, c_{2n}]$ and $\boldsymbol{c}_3 = diag\,[c_{31}, c_{32}, \ldots,\ c_{3n}]$ are three constant diagonal matrices satisfying $c_{ij} > 0, \forall i, j$, and we can select $\boldsymbol{c}_2 = I_{n\times n}$ for the sake of simplicity.

It is necessary to investigate the stability of the surface, which is shown in the following theorem. Before giving such a theorem, two lemmas are provided.

Lemma 7.4 *If Λ_i is defined as (7.26), k_i and β are selected to satisfy $0 < k_i h < 1$ and $0 < \beta < 1$ respectively, then it can be obtained that*

- $D_i h \leq k_i h \Lambda_i^{\frac{1+\beta}{2}} \leq \Lambda_i^{\frac{3-\beta}{2}}$.
- $D_i h < \Lambda_i$ *and* $k_i h \leq \Lambda_i^{1-\beta}$.

Lemma 7.5 ([6]) *If $\Psi(\beta)$ is defined as $\Psi(\beta) = 1 + \beta^{\frac{\beta}{1-\beta}} - \beta^{\frac{1}{1-\beta}}, 0 < \beta < 1$, then $1 < \Psi(\beta) < 2$ and $x\Psi(\beta) - x^{\beta}\Psi^{\beta}(\beta) + \Psi(\beta) - 1 \geq 0, \forall x \in [0, 1]$ hold.*

Theorem 7.6 *The discrete-time FO sliding surface (7.15) is stable if there are a series of positive symmetric matrices $\boldsymbol{P}_0, \boldsymbol{P}_1, \ldots, \boldsymbol{P}_n \in \mathbb{R}^{n\times n}$ satisfying the LMI condition (7.16), in which*

$$\begin{bmatrix} \mathbf{A}_0^{\mathbf{T}}\mathbf{P}_0\mathbf{A} - \mathbf{P}_0 + \sum_{j=1}^{l}\mathbf{P}_j & \mathbf{A}_0^{\mathbf{T}}\mathbf{P}_0\mathbf{A}_1 & \mathbf{A}_0^{\mathbf{T}}\mathbf{P}_0\mathbf{A}_2 & \cdots & \mathbf{A}_0^{\mathbf{T}}\mathbf{P}_0\mathbf{A}_l \\ \mathbf{A}_1^{\mathbf{T}}\mathbf{P}_0\mathbf{A}_0 & \mathbf{A}_1^{\mathbf{T}}\mathbf{P}_0\mathbf{A}_1 & \mathbf{A}_1^{\mathbf{T}}\mathbf{P}_0\mathbf{A}_2 & \cdots & \mathbf{A}_1^{\mathbf{T}}\mathbf{P}_0\mathbf{A}_l \\ \mathbf{A}_2^{\mathbf{T}}\mathbf{P}_0\mathbf{A}_0 & \mathbf{A}_2^{\mathbf{T}}\mathbf{P}_0\mathbf{A}_1 & \mathbf{A}_2^{\mathbf{T}}\mathbf{P}_0\mathbf{A}_2 & \cdots & \mathbf{A}_2^{\mathbf{T}}\mathbf{P}_0\mathbf{A}_l \\ \vdots & \vdots & \vdots & \ddots & \vdots \\ \mathbf{A}_r^{\mathbf{T}}\mathbf{P}_0\mathbf{A}_0 & \mathbf{A}_r^{\mathbf{T}}\mathbf{P}_0\mathbf{A}_1 & \mathbf{A}_r^{\mathbf{T}}\mathbf{P}_0\mathbf{A}_2 & \cdots & \mathbf{A}_l^{\mathbf{T}}\mathbf{P}_0\mathbf{A}_l - \sum_{j=1}^{l}\mathbf{P}_j \end{bmatrix} \leq 0, \quad (7.16)$$

$$\boldsymbol{A}_0 = \boldsymbol{c}_2 - h\boldsymbol{c}_1 - h^{2-\lambda}\boldsymbol{c}_3, \tag{7.17}$$

$$\boldsymbol{A}_j = -h^{2-\lambda}\boldsymbol{c}_3(-1)^j\binom{\alpha}{j}, \; j = 1, 2, ..., l. \tag{7.18}$$

Proof Assuming that the sliding surface has been reached, we can get $\boldsymbol{s}(k) = \mathbf{0}$, i.e.,

$$\check{\boldsymbol{p}}(k+1) = (\boldsymbol{c}_2 - h\boldsymbol{c}_1 - h^{2-\lambda}\boldsymbol{c}_3)\check{\boldsymbol{p}}(k) - h^{2-\lambda}\boldsymbol{c}_3\sum_{j=1}^{l}(-1)^j\binom{\alpha}{j}\check{\boldsymbol{p}}(k-j).$$

If $\boldsymbol{A}_0$ and $\boldsymbol{A}_j$ are defined as (7.17) and (7.18) respectively, then

$$\check{\boldsymbol{p}}(k+1) = \boldsymbol{A}_0\check{\boldsymbol{p}}(k) + \sum_{j=1}^{l}\boldsymbol{A}_j\check{\boldsymbol{p}}(k-j). \tag{7.19}$$

After that, a Lyapunov function is constructed as

$$V[\check{\mathbf{p}}(k)] = \check{\mathbf{p}}^{\mathbf{T}}(k)\mathbf{P}_0\check{\mathbf{p}}(k) + \sum_{j=1}^{l}\sum_{i=1}^{l}\check{\mathbf{p}}^{\mathbf{T}}(k-i)\mathbf{P}_j\check{\mathbf{p}}(k-i).$$

Further deduction of $\hat{\Delta}V\left[\check{\mathbf{p}}(k)\right] = V\left[\check{\mathbf{p}}(k+1)\right] - V\left[\check{\mathbf{p}}(k)\right]$ is displayed in (7.20). It is apparent that if (7.16) is satisfied, then $\hat{\Delta}V\left[\check{\mathbf{p}}(k)\right] \leq 0$.

$$\begin{aligned} \Delta V\left[\check{\mathbf{p}}(k)\right] &= \check{\mathbf{p}}^{\mathbf{T}}(k+1)\mathbf{P}_0\check{\mathbf{p}}(k+1) + \sum_{j=1}^{l}\sum_{i=1}^{l}\check{\mathbf{p}}^{\mathbf{T}}(k+1-i)\mathbf{P}_j\check{\mathbf{p}}(k+1-i) \\ &\quad - \check{\mathbf{p}}^{\mathbf{T}}(k)\mathbf{P}_0\check{\mathbf{p}}(k) - \sum_{j=1}^{l}\sum_{i=1}^{l}\check{\mathbf{p}}^{\mathbf{T}}(k-i)\mathbf{P}_j\check{\mathbf{p}}(k-i) \\ &= \check{\mathbf{p}}^{\mathbf{T}}(k+1)\mathbf{P}_0\check{\mathbf{p}}(k+1) - \check{\mathbf{p}}^{\mathbf{T}}(k)\mathbf{P}_0\check{\mathbf{p}}(k) \end{aligned}$$

$$
\begin{aligned}
&+\sum_{j=1}^{l}\check{\mathbf{p}}^{\mathbf{T}}(k)\mathbf{P}_j\check{\mathbf{p}}(k)-\check{\mathbf{p}}^{\mathbf{T}}(k-l)\sum_{j=1}^{l}\mathbf{P}_j\check{\mathbf{p}}(k-l)\\
&=\left[\mathbf{A}_0\check{\mathbf{p}}(k)-\sum_{j=1}^{l}\mathbf{A}_j\check{\mathbf{p}}(k-j)\right]^{\mathbf{T}}\mathbf{P}_0\left[\mathbf{A}_0\check{\mathbf{p}}(k)-\sum_{j=1}^{l}\mathbf{A}_j\check{\mathbf{p}}(k-j)\right]\\
&-\check{\mathbf{p}}^{\mathbf{T}}(k)\mathbf{P}_0\check{\mathbf{p}}(k)+\check{\mathbf{p}}^{\mathbf{T}}(k)\sum_{j=1}^{l}\mathbf{P}_j\check{\mathbf{p}}(k)-\check{\mathbf{p}}^{\mathbf{T}}(k-l)\sum_{j=1}^{l}\mathbf{P}_j\check{\mathbf{p}}(k-l)\\
&=\begin{bmatrix}\check{\mathbf{p}}(k)\\ \check{\mathbf{p}}(k-1)\\ \check{\mathbf{p}}(k-2)\\ \vdots\\ \check{\mathbf{p}}(k-l)\end{bmatrix}^{\mathbf{T}}\\
&\times\begin{bmatrix}\mathbf{A}_0^{\mathbf{T}}\mathbf{P}_0\mathbf{A}-\mathbf{P}_0+\sum_{j=1}^{l}\mathbf{P}_j & \mathbf{A}_0^{\mathbf{T}}\mathbf{P}_0\mathbf{A}_1 & \cdots & \mathbf{A}_0^{\mathbf{T}}\mathbf{P}_0\mathbf{A}_l\\
\mathbf{A}_1^{\mathbf{T}}\mathbf{P}_0\mathbf{A}_0 & \mathbf{A}_1^{\mathbf{T}}\mathbf{P}_0\mathbf{A}_1 & \cdots & \mathbf{A}_1^{\mathbf{T}}\mathbf{P}_0\mathbf{A}_l\\
\mathbf{A}_2^{\mathbf{T}}\mathbf{P}_0\mathbf{A}_0 & \mathbf{A}_2^{\mathbf{T}}\mathbf{P}_0\mathbf{A}_1 & \cdots & \mathbf{A}_2^{\mathbf{T}}\mathbf{P}_0\mathbf{A}_l\\
\vdots & \vdots & \ddots & \vdots\\
\mathbf{A}_l^{\mathbf{T}}\mathbf{P}_0\mathbf{A}_0 & \mathbf{A}_l^{\mathbf{T}}\mathbf{P}_0\mathbf{A}_1 & \cdots & \mathbf{A}_l^{\mathbf{T}}\mathbf{P}_0\mathbf{A}_l-\sum_{j=1}^{l}\mathbf{P}_j\end{bmatrix}\\
&\times\begin{bmatrix}\check{\mathbf{p}}(k)\\ \check{\mathbf{p}}(k-1)\\ \check{\mathbf{p}}(k-2)\\ \vdots\\ \check{\mathbf{p}}(k-l)\end{bmatrix}.
\end{aligned}
\tag{7.20}
$$

■

To design the discrete-time sliding mode controller, we refer to the equivalent-control-based DSMC [13], of which procedures contain two steps.

The first step is to obtain the equivalent control

$$
\begin{aligned}
\boldsymbol{u}_{\boldsymbol{eq}}(k)=&\bar{\boldsymbol{M}}(\Delta\boldsymbol{v}_{\boldsymbol{r}}(k)+\boldsymbol{c}_3\Delta^{\lambda-1}(\Delta\check{\boldsymbol{p}}(k))+\boldsymbol{c}_1\check{\boldsymbol{v}}(k))\\
&+\bar{\boldsymbol{B}}\boldsymbol{v}_{\boldsymbol{r}}(k)-\bar{\boldsymbol{B}}\check{\boldsymbol{v}}(k)+\bar{\boldsymbol{A}}\boldsymbol{S}_f(\boldsymbol{v}(k)).
\end{aligned}
\tag{7.21}
$$

The second step is to design the switching control, which is presented as

$$
\boldsymbol{u}_{\boldsymbol{sw}}(k)=\boldsymbol{k}\bar{\boldsymbol{M}}\cdot\mathrm{sig}^{\gamma(k)}(\boldsymbol{s}(k)),
\tag{7.22}
$$

where $\boldsymbol{k} = diag\,[k_1, k_2, \ldots, k_n]$, $\mathrm{sig}^{\gamma(k)}(\boldsymbol{s}(k)) = \left[\mathrm{sig}^{\gamma_1(k)}(s_1(k)), \mathrm{sig}^{\gamma_2(k)}(s_2(k)), \ldots, \mathrm{sig}^{\gamma_n(k)}(s_n(k))\right]^T$, k_i is a positive constant satisfying $0 < k_i h < 1$, $\mathrm{sig}^{\bullet}(\cdot) = |\cdot|^{\bullet}\mathrm{sgn}(\cdot)$, $\mathrm{sgn}(\cdot)$ is the signal function and $\gamma_i(k) = \frac{|s_i(k)|+\beta}{|s_i(k)|+1}$, $0 < \beta < 1$. Obviously, $\gamma_i(k)$ satisfies $\beta < \gamma_i(k) < 1$.

The integrated controller consists of the equivalent control and the switching control, i.e.,

$$\boldsymbol{u}(k) = \boldsymbol{u}_{eq}(k) + \boldsymbol{u}_{sw}(k). \tag{7.23}$$

To study the dynamic of the system, considering (7.11)–(7.15), we substitute the control law (7.23) into the model (7.9) and (7.10), it can be obtained that

$$s_i(k+1) = -k_i h \mathrm{sig}^{\gamma_i(k)}(s_i(k)) + s_i(k) + h\tilde{d}_i(k), \tag{7.24}$$

where $\tilde{d}_i(k) = \frac{\tilde{M}_i}{M_i}\Delta v_{ri}(k) + \frac{\tilde{M}_i}{M_i}c_{3i}\Delta^{\lambda-1}(\Delta\check{p}_i(k)) + \frac{\tilde{M}_i}{M_i}c_{1i}\check{v}_i(k) + \frac{\tilde{B}_i}{M_i}v_i(k) + \frac{\tilde{A}_i}{M_i}S_{fi}(v_i(k)) + \frac{\tilde{B}_i}{M_i}\alpha\Delta\varepsilon_i(k) - k_i\frac{\tilde{M}}{M}\mathrm{sig}^{\gamma(k)}(s(k)) + \alpha\Delta^2\varepsilon(k) - \frac{1}{M_i}d_i(k)$. According to [3], it is reasonable to assume that the $\tilde{d}_i$ is bounded by a known constant, i.e., $|\tilde{d}_i(k)| < D_i$. Then the dynamic behavior of $\boldsymbol{s}(k)$ can be concluded by the following two theorems.

Theorem 7.7 *For the n-dimensional discrete-time system (7.9) (7.10) under the DFSMC (7.23), the sliding variable $s_i(k)$ converges to a boundary layer Z_i with the thickness of ζ_i after K_i steps, where*

$$Z_i = \{s_i(k)\big||s_i(k)| \le \zeta_i\},\ \zeta_i = \Psi(\beta)\cdot\Lambda_i, \tag{7.25}$$

$$\Lambda_i = \max\left\{\left(\frac{D_i}{k_i}\right)^{\frac{2}{1+\beta}}, (k_i h)^{\frac{1}{1-\beta}}\right\}, \tag{7.26}$$

$$K_i = \left\lceil \frac{s_i^2(0) - {\zeta_i}^2}{{\omega_i}^2} \right\rceil, \tag{7.27}$$

where $\Psi(\beta) = 1 + \beta^{\frac{\beta}{1-\beta}} - \beta^{\frac{1}{1-\beta}}$, $\omega_i = k_i h[\Psi(\beta)\Lambda_i]^{\frac{1+\beta}{2}} - D_i h$ and $\lceil\cdot\rceil$ standing for rounding up.

Proof A Lyapunov function in the discrete-time form is conducted as $V_i(k) = [s_i(k)]^2$ and $\hat{\Delta}V_i(k)$ is represented as

$$\begin{aligned}\hat{\Delta}V_i(k) =& V_i(k+1) - V_i(k)\\ =& -[s_i(k) - s_i(k+1)][s_i(k) + s_i(k+1)].\end{aligned} \tag{7.28}$$

As the case $s_i(k) > \zeta_i > 0$ is symmetrical to the case $s_i(k) < -\zeta_i < 0$, we only state the detailed proof procedure of the former case for the sake of simplicity.

Under the case $s_i(k) > \zeta_i > 0$, based on Lemma 7.4, we have

$$\begin{aligned} s_i(k) - s_i(k+1) &= k_i h[s_i(k)]^{\gamma_i(k)} - \tilde{d}_i(k)h \\ &\geq k_i h[s_i(k)]^{\gamma_i(k)} - D_i h, \\ s_i(k) + s_i(k+1) &= -k_i h[s_i(k)]^{\gamma_i(k)} + 2s_i(k) + \tilde{d}_i(k)h \\ &\geq -k_i h[s_i(k)]^{\gamma_i(k)} + 2s_i(k) - D_i h. \end{aligned}$$

If $s_i(k) > 1$, then $\frac{1+\beta}{2} < \gamma_i(k) < 1$ and $s_i(k) > k_i h[s_i(k)]^{\gamma_i(k)}$. Based on this, it can be further concluded that

$$\begin{aligned} s_i(k) - s_i(k+1) &> k_i h[s_i(k)]^{\frac{1+\beta}{2}} - D_i h \\ &> k_i h[\Psi(\beta)\Lambda_i]^{\frac{1+\beta}{2}} - D_i h = \omega_i, \\ s_i(k) + s_i(k+1) &> k_i h[s(k)]^{\gamma_i(k)} - D_i h \\ &> k_i h[\Psi(\beta)\Lambda_i]^{\frac{1+\beta}{2}} - D_i h = \omega_i. \end{aligned}$$

If $\zeta_i < s_i(k) \leq 1$, then $\beta < \gamma_i(k) \leq \frac{1+\beta}{2}$. By Lemma 7.4, we have $[s_i(k)]^{1-\gamma_i(k)} > [s_i(k)]^{1-\beta} > \Lambda_i^{1-\beta} \geq k_i h$ and $s_i(k) > k_i h[s_i(k)]^{\gamma_i(k)}$. After that, we can get

$$\begin{aligned} s_i(k) - s_i(k+1) &\geq k_i h[s_i(k)]^{\frac{1+\beta}{2}} - \tilde{d}_i(k)h \\ &> k_i h[\Psi(\beta)\Lambda_i]^{\frac{1+\beta}{2}} - D_i h = \omega_i, \\ s_i(k) + s_i(k+1) &> k_i h[s_i(k)]^{\gamma_i(k)} - D_i h \\ &> k_i h[\Psi(\beta)\Lambda_i]^{\frac{1+\beta}{2}} - D_i h = \omega_i. \end{aligned}$$

In addition, based on Lemma 7.5, we can get $\Psi(\beta) > 1$. By using Lemma 7.4 and considering the fact that $D_i \geq 0$. If $D_i > 0$, then $\omega_i \geq ([\Psi(\beta)]^{\frac{1+\beta}{2}} - 1)D_i h > 0$. If $D_i = 0$, then $\omega_i = k_i h[\Psi(\beta)\Lambda]^{\frac{1+\beta}{2}} > 0$, which means that $\omega_i > 0$ always holds.

Summarizing the results above, it can be finally obtained that

$$V(k+1) - V(k) \leq -\omega_i^2 < 0.$$

Moreover, based on the analysis above, we can summarize that $V_i(k+1) - V_i(k) = [s_i(k+1)]^2 - [s_i(k)]^2 < -\omega_i^2, k = 0, 1, 2, \ldots$. Then through mathematical induction, we can get that $[s_i(m)]^2 < [s_i(0)]^2 - m\omega_i^2, m \in \mathbb{N}^+$. Suppose that the sliding variable $s_i(k)$ converges into Z_i definitely after K_i steps, then we can give out the inequality

$$[s_i(K_i)]^2 < [s_i(0)]^2 - K_i\omega_i^2 \leq \zeta_i^2. \tag{7.29}$$

Based on the inequality above, it is apparent that (7.27) holds.Then the proof is completed. ■

Theorem 7.8 *For the n-dimensional discrete-time system (7.9) (7.10) under DFSMC (7.23), the sliding variable $s_i(k)$ maintains in the boundary layer Z_i after converging into it for the first time.*

Proof Supposing that $s_i(k)$ is in the boundary layer, i.e., $-\zeta_i \leq s_i(k) \leq \zeta_i$. Then we can get that $s_i(k) = \theta_i \zeta_i = \theta_i \Psi(\beta) \Lambda_i$, where θ_i is a constant satisfying $-1 \leq \theta_i \leq 1$.

On the one hand, we can get that $s_i(k+1) \leq \zeta_i$. The procedure of the proof can be divided into four cases, and they are as follows.

Case 1. $s_i(k) \geq 1 \Rightarrow [s(k)] \geq [s_i(k)]^{\gamma_i(k)} \geq [s_i(k)]^{\frac{1+\beta}{2}}$

1. Based on Lemma 7.4, it is obtained that

$$\begin{aligned} s_i(k+1) &= s_i(k) - k_i h [s_i(k)]^{\gamma_i(k)} + \tilde{d}_i(k) h \\ &\leq s_i(k) - k_i h [s_i(k)]^{\frac{1+\beta}{2}} + D_i h \\ &\leq s_i(k) + (1 - [\theta_i \Psi(\beta)]^{\frac{1+\beta}{2}}) D_i h. \end{aligned}$$

It is obvious that if $\theta_i \Psi(\beta) \geq 1$, then $s_i(k+1) \leq s_i(k)$. For $s_i(k) \leq \zeta_i$, we can obtain $s_i(k+1) \leq \zeta_i$. Nonetheless, if $\theta_i \Psi(\beta) < 1$, the proof becomes complicated. It is known that $s_i(k) \leq \theta_i \Psi(\beta) \Lambda_i$. Moreover, as $\theta_i \Psi(\beta) < [\theta_i \Psi(\beta)]^{\frac{1+\beta}{2}}$, considering Lemma 7.4, it is deduced that

$$\begin{aligned} s_i(k+1) &\leq s_i(k) + (1 - [\theta_i \Psi(\beta)]^{\frac{1+\beta}{2}}) D_i h \\ &< (\theta_i \Psi(\beta) + 1 - [\theta_i \Psi(\beta)]^{\frac{1+\beta}{2}}) \Lambda_i \\ &< \Lambda_i < \Psi(\beta) \Lambda_i = \zeta_i. \end{aligned}$$

2. Based on Lemma 7.4, we have

$$\begin{aligned} s_i(k+1) &= s_i(k) - k_i h [s_i(k)]^{\gamma_i(k)} + \tilde{d}_i(k) h \\ &\geq (1 - k_i h) s_i(k) - D_i h \\ &\geq -D_i h > -\Lambda_i > -\zeta_i. \end{aligned}$$

Case 2. $0 \leq s_i(k) < 1 \Rightarrow [s_i(k)]^{\beta} \geq [s_i(k)]^{\gamma_i(k)} \geq [s_i(k)]^{\frac{1+\beta}{2}}$

1. Similar to Case 1, we can get $s_i(k+1) < \zeta_i$ directly.
2. According to Lemma 7.4, we have

$$\begin{aligned} s_i(k+1) &\geq s_i(k) - k_i h [s_i(k)]^{\beta} - D_i h \\ &> \Lambda_i \left[\theta_i \Psi(\beta) - [\theta_i \Psi(\beta)]^{\beta} - 1 \right] \\ &\geq -\Psi(\beta) \Lambda_i = -\zeta_i. \end{aligned}$$

Case 3. $-1 < s_i(k) < 0 \Rightarrow |s_i(k)|^{\beta} > |s_i(k)|^{\gamma_i(k)} > |s_i(k)|^{\frac{1+\beta}{2}}$

1. According to Lemma 7.4, we can get

$$\begin{aligned} s_i(k+1) &= -|s_i(k)| + k_i h |s_i(k)|^{\gamma_i(k)} + \tilde{d}_i(k)h \\ &< -|s_i(k)| + k_i h |s_i(k)|^{\beta} + D_i h \\ &< -\Lambda_i(|\theta_i \Psi(\beta)| - |\theta_i \Psi(\beta)|^{\beta} - 1) \\ &< \Psi(\beta)\Lambda_i = \zeta_i. \end{aligned}$$

2. According to Lemma 7.4, we can get that

$$\begin{aligned} s_i(k+1) &= -|s_i(k)| + k_i h |s_i(k)|^{\gamma_i(k)} + \tilde{d}_i(k)h \\ &\geq -|s_i(k)| + k_i h |s_i(k)|^{\gamma_i(k)} - D_i h \\ &\geq -|\theta_i \Psi(\beta)\Lambda_i| + (|\theta_i \Psi(\beta)|^{\frac{1+\beta}{2}} - 1)D_i h. \end{aligned}$$

If $|\theta_i \Psi(\beta)| \geq 1$, then it is obvious that $|\theta_i \Psi(\beta)|^{\frac{1+\beta}{2}} \geq 1$, then, $s(k+1) \geq -|\theta_i \Psi(\beta)\Lambda_i| \geq -\zeta$.

If $|\theta_i \Psi(\beta)| < 1$, then $|\theta_i \Psi(\beta)| < |\theta_i \Psi(\beta)|^{\frac{1+\beta}{2}}$, based on Lemma 7.4, we have

$$\begin{aligned} s_i(k+1) &\geq -|\theta_i \Psi(\beta)\Lambda_i| - (1 - |\theta_i \Psi(\beta)|^{\frac{1+\beta}{2}})D_i h \\ &> -\Lambda_i \left[|\theta_i \Psi(\beta)| + 1 - |\theta_i \Psi(\beta)|^{\frac{1+\beta}{2}} \right] \\ &> -\Lambda_i > -\Psi(\beta)\Lambda_i = -\zeta_i. \end{aligned}$$

Case 4. $s_i(k) \leq -1 \Rightarrow |s_i(k)| > |s_i(k)|^{\gamma_i(k)} \geq |s_i(k)|^{\frac{1+\beta}{2}}$

1. According to Lemma 7.4, we can get

$$\begin{aligned} s_i(k+1) &= -|s_i(k)| + k_i h |s_i(k)|^{\gamma_i(k)} + \tilde{d}_i(k)h \\ &< -(1 - k_i h)|\theta \Psi(\beta)\Lambda_i| + D_i h \\ &< D_i h < \Lambda_i \\ &< \Psi(\beta)\Lambda_i = \zeta_i. \end{aligned}$$

2. Similar to Case 3, it can be obtained $s_i(k+1) > -\zeta_i$ directly.

Summarizing all the cases above, it is reasonable that if $s_i(k) \in Z_i$, then $s_i(k+1) \in Z_i$ always holds. Here is the proof. ■

7.3.3 Contouring Error Analysis

Theorems 7.7 and 7.8 describe a fact that the sliding variable $s_i(k)$ satisfies $|s_i(k)| \leq \zeta_i$ after the system becomes steady. Correspondingly, the contouring error satisfies $|\varepsilon_i(k)| \leq |\varepsilon_i|_s$, the theoretical range of which is given in the following theorem.

Theorem 7.9 *For the n-dimensional discrete-time system (7.9) (7.10) under DFSMC (7.23), the synthetical error* $|\check{p}_{ei}|_s$ *satisfies*

$$|\varepsilon_i|_s \leq \frac{\zeta_i}{\eta c_{1i} - \eta c_{31} h^{1-\lambda} \frac{\Gamma(l+2-\lambda)}{\Gamma(2-\lambda)\Gamma(l+1)}} + \frac{|p_{ei}|_s}{\eta}. \tag{7.30}$$

Proof First of all, by substituting (7.12) and related equations into the sliding surface (7.15), we can have

$$\check{p}_i(k+1) = (1 - c_{1i}h)\check{p}_i(k) - c_{3i}h\Delta^{\lambda-1}\check{p}_i(k) + hs_i(k).$$

By using inductive method, we can get that

$$\begin{aligned}\check{p}_i(k) =&(1 - c_{1i}h)^k \check{p}_i(0) + \sum_{j=0}^{k-1}(1 - c_{1i}h)^j hs_i(k-1-j)\\ &- c_{3i}h\sum_{j=0}^{k-1}(1 - c_{1i}h)^{k-1-j}\Delta^{\lambda-1}\check{p}_i(j).\end{aligned} \tag{7.31}$$

Assuming that after k_s and k_p steps, the stable states of $s(k)$ and $\check{p}_i$ are reached respectively. Apparently, the k_s and k_p are both bounded. Taking the fact

$$(-1)^j\binom{\alpha}{j} > 0,\ \alpha < 0, \tag{7.32}$$

into consideration [1], (7.33) and (7.34) can be obtained.

$$\begin{aligned}&c_{3i}h\sum_{j=0}^{k-1}(1 - c_{1i}h)^{k-1-j}|\Delta^{\lambda-1}\check{p}_i(j)|\\ &= c_{3i}h\sum_{j=0}^{k_p+l}(1 - c_{1i}h)^{k-1-j}|\Delta^{\lambda-1}\check{p}_i(j)|\\ &\quad + c_3h\sum_{j=k_p+l+1}^{k-1}(1 - c_{1i}h)^{k-1-j}|\Delta^{\lambda-1}\check{p}_i(j)|,\end{aligned} \tag{7.33}$$

$$\sum_{m=0}^{k-1}(1-c_{1i}h)^m h\,|s_i(k-1-m)| = \sum_{m=0}^{k-k_s}(1-c_{1i}h)^m h\,|s_i(k-1-m)| + \sum_{m=k-k_s+1}^{k-1}(1-c_{1i}h)^m h\,|s_i(k-1-m)|\,. \tag{7.34}$$

Because $1-c_{1i}h<1$, we can get

$$\lim_{k\to+\infty} c_{3i}h\sum_{j=0}^{k_p+l}(1-c_{1i}h)^{k-1-j}|\Delta^{\lambda-1}\check{p}_i(j)| = 0, \tag{7.35}$$

$$\lim_{k\to+\infty}\sum_{m=k-k_s+1}^{k-1}(1-c_{1i}h)^m h\,|s_i(k-1-m)| = 0, \tag{7.36}$$

$$\begin{aligned}
&\lim_{k\to+\infty} c_{3i}h\sum_{j=0}^{k-1}(1-c_{1i}h)^{k-1-j}|\Delta^{\lambda-1}\check{p}_i(j)| \\
=&\lim_{k\to+\infty} c_{3i}h\sum_{j=k_p+l+1}^{k-1}(1-c_{1i}h)^{k-1-j}\left|\frac{1}{h^{\lambda-1}}\sum_{m=0}^{l}(-1)^m\binom{\alpha}{m}\check{p}_i(j-m)\right| \\
\le&\lim_{k\to+\infty}\frac{c_{3i}}{h^{\lambda-2}}\sum_{j=k_p+l+1}^{k-1}(1-c_1h)^{k-1-i}\left(\frac{\Gamma(l+2-\lambda)}{\Gamma(2-\lambda)\Gamma(l+1)}|\check{p}_{ei}|_s\right) \\
=&\frac{c_{3i}}{h^{\lambda-2}}\frac{\Gamma(l+2-\lambda)}{\Gamma(2-\lambda)\Gamma(l+1)}|\check{p}_{ei}|_s\cdot\lim_{k\to+\infty}\sum_{j=k_p+l+1}^{k-1}(1-c_{1i}h)^{k-1-i} \\
=&\frac{c_{3i}}{c_{1i}h^{\lambda-1}}\frac{\Gamma(l+2-\lambda)}{\Gamma(2-\lambda)\Gamma(l+1)}|\check{p}_{ei}|_s.
\end{aligned} \tag{7.37}$$

Then, considering the definition of FO difference, the step when the steady states are achieved and the fact

$$\sum_{j=0}^{l}(-1)^j\binom{\alpha}{j} = \frac{\Gamma(l+1-\alpha)}{\Gamma(1-\alpha)\Gamma(l+1)}, \tag{7.38}$$

with the Gamma function $\Gamma(x)=\int_0^\infty e^{-t}t^{x-1}dt$ [1], (7.37) can be obtained. Based on (7.34), (7.36) and the fact that $|s(k)|\le\zeta_i$ if $k\ge k_{0s}$, we can get

$$\lim_{k\to+\infty}\sum_{m=0}^{k-1}(1-c_{1i}h)^m h\,|s_i(k-1-m)| \le \frac{\zeta_i}{c_{1i}}. \tag{7.39}$$

On the other hand, it is apparent that $|\check{p}_{ei}|_s = \lim_{k\to+\infty}|\check{p}_i(k)|$ Considering (7.31)–(7.36), (7.37) and the fact $1 - c_{1i}h < 1$, we can get

$$|\check{p}_{ei}|_s \le \frac{\zeta_i}{c_{1i} - c_{3i}h^{1-\lambda}\frac{\Gamma(l+2-\lambda)}{\Gamma(2-\lambda)\Gamma(l+1)}}. \tag{7.40}$$

Considering the fact $|\check{p}_{ei}|_s \ge \eta|\varepsilon_i|_s - |p_{ei}|$, (7.30) can be obtained. Here is the proof. ∎

Remark 7.10 As Theorem 7.9 shows, a properly larger value of α leads to a smaller steady contouring error $|\varepsilon_i|_s$, no matter $|p_{ei}|_s$ is known or not. This is helpful when we adjust parameters in the experiments.

Remark 7.11 The mathematical calculation shows that $\frac{\Gamma(l+2-\lambda)}{\Gamma(2-\lambda)\Gamma(l+1)}$ is a decreasing function of λ. This means that a larger λ results in a smaller $\frac{\Gamma(l+2-\lambda)}{\Gamma(2-\lambda)\Gamma(l+1)}$ and a smaller $|\varepsilon_i|_s$, according to Theorem 7.9. On the other hand, if $\lambda = 0$, then the controller is a classical integral controller. As a fractional controller has a larger λ, a higher theoretical precision can be obtained than an integral controller.

7.4 Experiments

7.4.1 Experimental Setup and Parameters

To validate the effectiveness of the simplified Newton-based CEE and n-dimensional DFSMC, two groups of experiments are carried out. Due to the contouring control is usually used in 2-dimensional systems in industrial applications, we conduct experiments on a 2-DOF linear-motor X-Y table. The structure of the experiments is depicted in Fig. 7.3.

As Fig. 4.1 shows, the linear motor table is made by Akribis company, which is mainly composed of a moving stage, a linear motor in the X-direction, and two independent linear motors in the Y-direction. The travel range of the X-direction linear motor is 700 mm, and that of the Y-direction linear motor is 650 mm. The position of the moving stage is measured by linear encoders with a resolution of 50 nm. The X-Y table is connected to computer by CANopen technology, and by programming in Twincat 3.0 PLC on the computer, the digital controller can be applied to the table.

In the experiments, one of the linear motors in the Y-direction is applied, and the other one is open-circuited. The sampling interval is set as 1 ms in the Twincat 3.0 PLC. The basic model of the X-Y table can be described as (7.9)

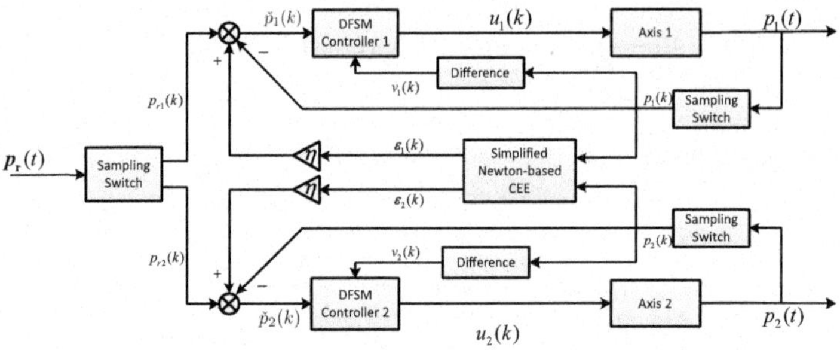

Fig. 7.3 Block diagram of the experiments

and (7.10) definitely [10], where the nominal parameters are $\bar{M} = diag\{\frac{\bar{m}_1}{k_{f1}}, \frac{\bar{m}_2}{k_{f2}}\}$, $\bar{B} = \{\frac{\bar{k}_{v1}}{k_{f1}}, \frac{\bar{k}_{v2}}{k_{f2}}\}$, $\bar{A} = \{\frac{\bar{k}_{v1}}{k_{f1}}, \frac{\bar{k}_{v2}}{k_{f2}}\}$ with m_1=11 kg, m_2=25 kg, $k_{f1} = k_{f2} = 80$. By referring to practical swarm optimization, parameters about fraction are determined as $\bar{k}_{v1} = 0.032$ N·s/mm, $\bar{k}_{v2} = 0.045$ N·s/mm, $\bar{k}_{c1} = 9.738$ N and $\bar{k}_{c2} = 11.6$ N. It is noticeable that the velocity of the moving stage cannot be collected directly, thus we use the difference of the position to estimate the velocity. As the sampling time is small enough and the velocity of the moving stage is not too large, the estimating error of velocity can be neglected.

For comparison, the integral sliding mode control (ISMC) with Newton-based CEE (NE) which is noted as NE-ISMC, the discrete-time FO terminal SMC (DFTSMC) with NE which is noted as NE-DFTSMC, and the model-free FO SMC (MFSMC) with simplified Newton-based CEE (SNE) which is noted as SNE-MFSMC, is also applied to the experiments [4, 14]. During the experiments, it was found that the parameters of NE-DFTSMC are hard to be tuned, thus the initial DFTSMC is modified with a position error term added to the sliding surface. Another part of the controller is not changed. We tried our best to adjust the parameters to guarantee these controllers with their corresponding estimators have achieved their best performance. It is apparent that the number of iterations affects the precision of CEE, but it has been proved that the precision is hard to improve with the increase of the number of iterations when the number is large enough [4]. In the experiments, the number of the iterations of the NE method is selected as 10, while that of the SNE method is 4.

7.4.2 Experimental Results

7.4.2.1 Slow Circular Motion

To investigate the low-speed performance of the SNE-DFSMC strategy, the reference contour is set as a circle with a radius of 10 mm and an angular velocity of 0.1π rad/s.

According to the definition of contouring error, the actual contouring errors of this group are calculated as

$$|\varepsilon^*| = \left|\left(p_1^2 + p_2^2\right)^{\frac{1}{2}} - 10\right|. \tag{7.41}$$

Figure 7.4 depicts the reference contour and the actual contour under the four strategies. Through the enlarged window, we can find that the contour under the SNE-DFSMC strategy is not only the closest to the reference contour but also the smoothest among these strategies.

The detailed results given in Fig. 7.5. (a) and (b) show the estimated contouring errors and the actual contouring errors respectively. According to (b), the contouring error under the SNE-DFSMC strategy is the smallest most of the time, accounting for only 1/3 of the error under NE-DFTSMC. By comparing (a) and (b), it can be found

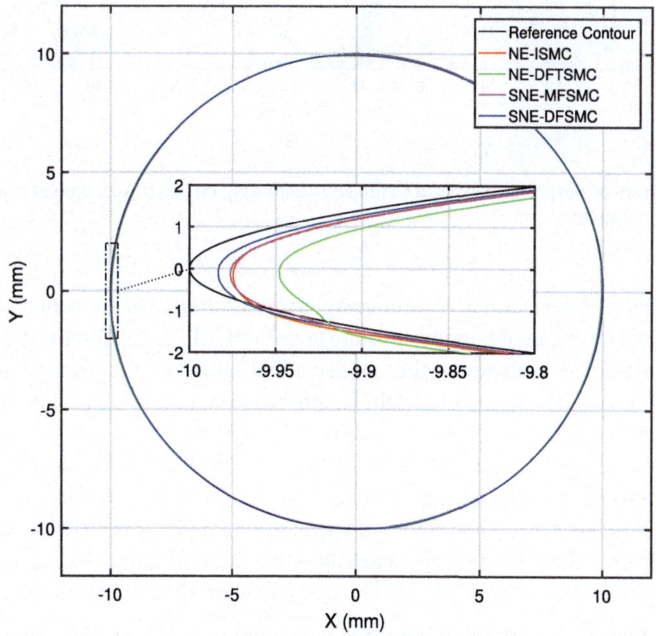

Fig. 7.4 Performance of tracking a slow circular contour

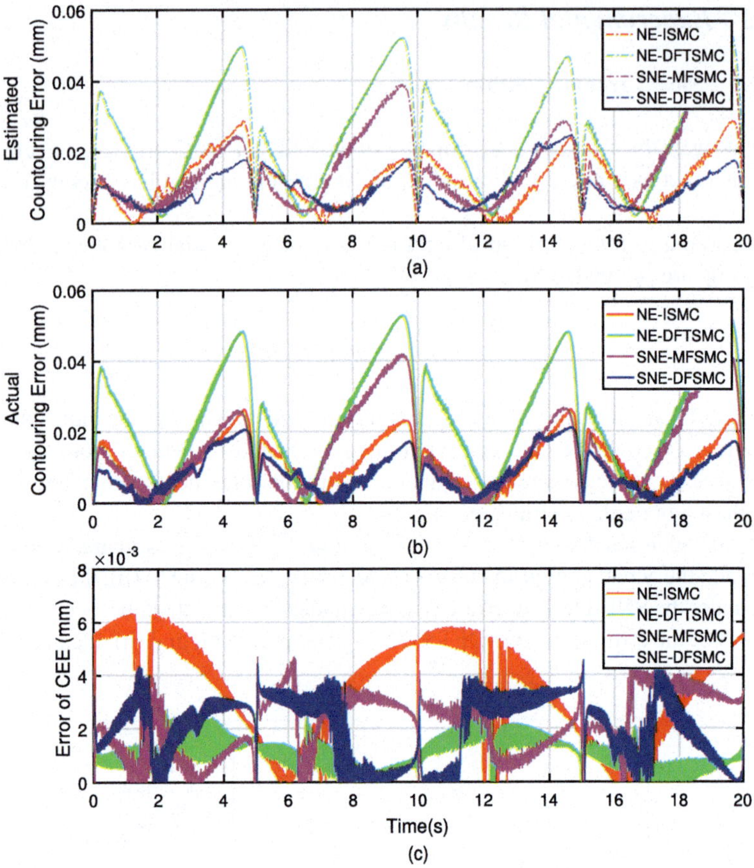

Fig. 7.5 Errors of the performances. **a** Estimated contouring errors; **b** Actual contouring errors; **c** Error of the estimators

that the curve of the estimated contouring errors is similar to the actual contouring errors. Further, we can obtain the absolute errors of CEE in (c). Obviously, (c) shows that both NE and SNE obtain a rather small estimating error. This proves that although the algorithm of SNE is simpler and the number of iterations is less, it can still achieve a high precision if the parameters are selected properly.

In industrial applications, varieties of uncertainties and disturbances exist [11], thus the robustness performance is also an important indicator to study the effectiveness of control strategy. Therefore, we attach a 3.3 kg payload (30% of the moving stage's mass) to the moving stage and apply the circular motion experiments again with all the parameters maintained. The results under all the circumstances are displayed in Fig. 7.6, where the RMS error is defined as $RMS\ error = \sqrt{\sum_{k=1}^{N} \frac{e^2(k)}{N}}$, where N is the sampling number and $e(k)$ is the sampled error at step k.

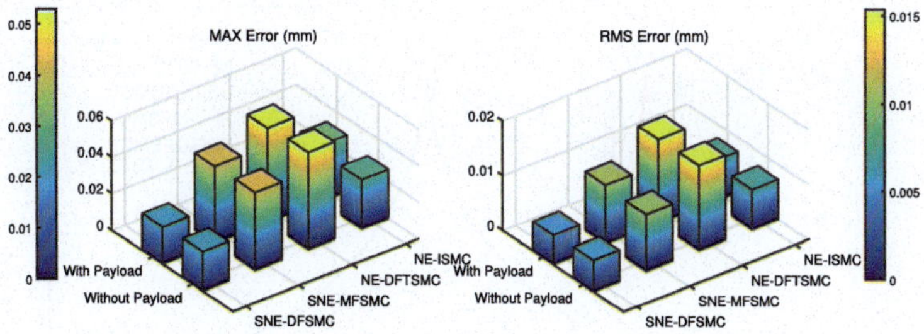

Fig. 7.6 Contouring error comparison among strategies

From Fig. 7.6, we can find that no matter whether a payload is attached to the moving stage or not, the SNE-DFSMC strategy has both the smallest MAX error and RMS error, which proves that the SNE-DFSMC has the highest precision. This is rational because compared with the SNE-MFSMC strategy, SNE-DFSMC is designed based on the discrete-time system, compared with the NE-DFTSMC strategy, SNE-DFSMC is designed with no "s^3" term in the control law which may cause chattering when s is not small enough, and compared with NE-ISMC strategy, SNE-DFSMC is designed with the fractional term which performs more steadily than the integral term. Moreover, the comparison between the cases with payload and without payload shows that all these four strategies have good robustness performances. This is to be expected, for these control methods are adapted from sliding mode control which has been proved robust in numerous literatures.

7.4.2.2 Fast Astroid Motion

To investigate the high-speed performance of the SNE-DFSMC strategy, the reference contour is set as a fast asteroid with sharp corners, which is composed of four symmetrical quarter circles with a radius of 10 mm and an angular velocity of 0.5π rad/s.

The actual contouring errors in this case are given as

$$|\varepsilon^*(k)| = \min\left\{|\varepsilon_1^*(k)|, |\varepsilon_2^*(k)|, |\varepsilon_3^*(k)|, |\varepsilon_4^*(k)|\right\}, \tag{7.42}$$

where

$$\begin{cases} \varepsilon_1^*(k) = \sqrt{(p_1(k) - 10)^2 + (p_2(k) - 10)^2} - 10, \\ \varepsilon_2^*(k) = \sqrt{(p_1(k) + 10)^2 + (p_2(k) - 10)^2} - 10, \\ \varepsilon_3^*(k) = \sqrt{(p_1(k) + 10)^2 + (p_2(k) + 10)^2} - 10, \\ \varepsilon_4^*(k) = \sqrt{(p_1(k) - 10)^2 + (p_2(k) + 10)^2} - 10. \end{cases} \tag{7.43}$$

The performance of tracking an astroid contour is displayed in Fig. 7.7. Intuitively, the actual contour of the SNE-DFSMC strategy fits the reference contour the

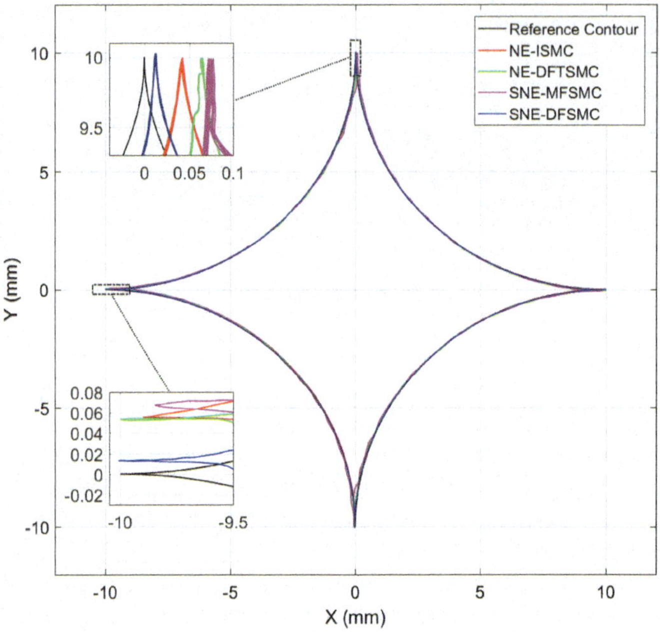

Fig. 7.7 Performance of tracking a fast astroid contour

best. To investigate the performance at highly curved corners, we give out the two enlarged windows. The upper window shows the sharp corners on the contour that exist when these four strategies are applied, which should be attributed to the excellent performance of NE and SNE at highly curved corners. SNE-DFSMC strategy and NE-ISMC strategy both have a good shape around the corner, and SNE-DFSMC is closer to the referenced corner. As for NE-DFTSMC and SNE-MFSMC, the contours are not smooth. Furthermore, through the lower window, we find that the actual contour of SNE-DFSMC is sharp enough and still the best.

The estimated contouring errors, the actual contouring errors, and the errors of CEE are demonstrated in Fig. 7.8. (a) and (b) present that the contouring errors of SNE-DFSMC are the smallest, and (c) presents the error of SNE and NE are rather small most of the time. However, we find that the curves in (a) are very different from those in (b) in shapes and values and there are many peaks in (c). This phenomenon is caused by the difference between the definition of actual contouring error and the method of CEE around the sharp corners, which is an interesting research direction and will be solved in future work. Despite all this, the SNE-DFSMC still has the highest precision.

Next, we will study the quantitative analysis of the errors and the robust performance of the proposed strategy. Similar to the situation before, the extra 3.3 kg payload is attached to the moving stage, and experiments under the same parameters are carried out. The statistics are shown as two bar charts in Fig. 7.9. The left bar

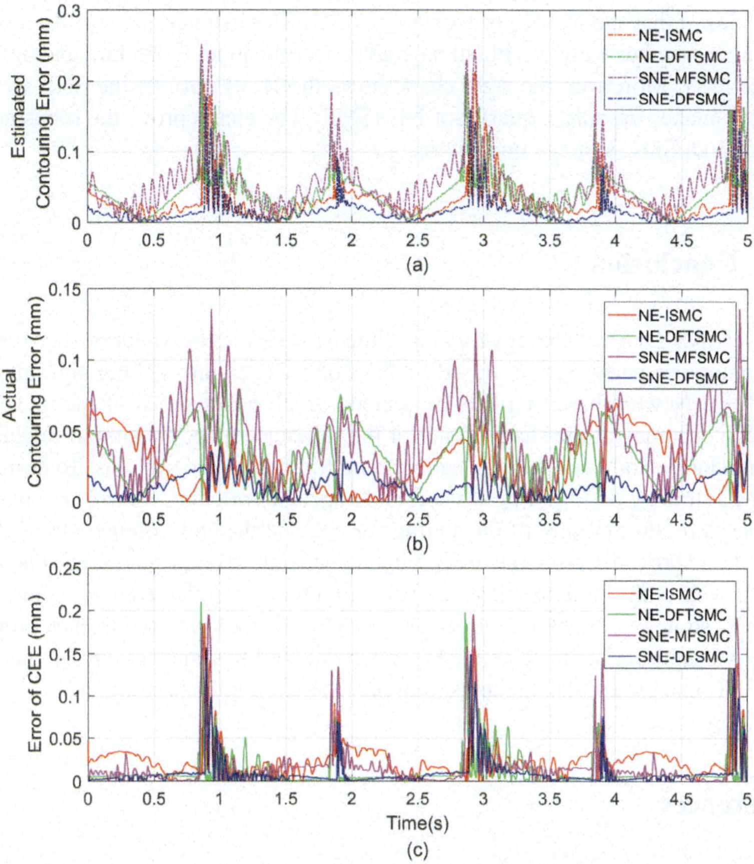

Fig. 7.8 Errors of the performances. **a** Estimated contouring errors; **b** Actual contouring errors; **c** Error of the estimators

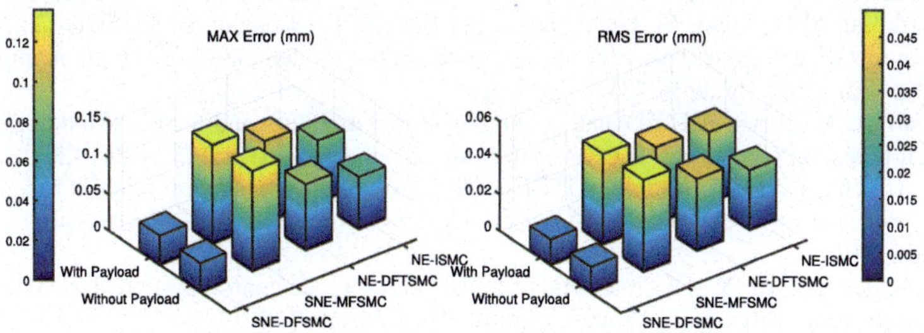

Fig. 7.9 Contouring error comparison among strategies

chart shows that the MAX error of the SNE-DFSMC strategy is still the smallest no matter whether with or without payload, covering about 0.025 mm and 0.01 mm respectively. Moreover, the right chart shows the RMS error of the SNE-DFSMC strategy makes up only a quarter of NE-ISMC. The charts prove the robustness of the SNE-DFSMC strategy once more.

7.5 Conclusion

A novel SNE-DFSMC strategy for multidimensional systems is addressed to improve the contouring performance. Based on Newton's extremum-seeking algorithm, the simplified Newton-based CEE is presented, which has a series of merits such as the high accuracy at the sharp corner of the reference map, the simple calculation, no singularity, and the easy application to multidimensional systems. To design the DFSMC strategy, a FO sliding surface, and a special switching control are given out. Furthermore, the stability of the sliding surface and the convergence of the sliding variable are proved. Comparative experiments are carried out, whose sufficient results accord well with the theoretical analysis and prove the effectiveness of the SNE-DFSMC strategy to track both low-speed and high-speed contouring. The proposed strategy is appropriate to be applied to industrial facilities, such as cutting machines and CNC machine tools, due to its high precision and good performance.

References

1. Bandyopadhyay, B., Kamal, S.: Stabilization and Control of Fractional Order Systems: a Sliding Mode Approach. Springer, Switzerland (2015)
2. Cheng, M.Y., Su, K.H., Wang, S.F.: Contour error reduction for free-form contour following tasks of biaxial motion control systems. Robot. Comput.-Integr. Manuf. **25**(2), 323–333 (2009)
3. Corradini, M.L., Fossi, V., Giantomassi, A., Ippoliti, G., Longhi, S., Orlando, G.: Minimal resource allocating networks for discrete time sliding mode control of robotic manipulators. IEEE Trans. Ind. Inf. **8**(4), 733–745 (2012)
4. Ghaffari, A., Ulsoy, A.G.: Dynamic contour error estimation and feedback modification for high-precision contouring. IEEE/ASME Trans. Mechatron. **21**(3), 1732–1741 (2015)
5. Hu, C., Zhu, Y., Hu, Z., Wang, Z.: GTCF based LARC contouring motion control of an industrial xy linear-motor-driven stage. In: American Control Conference (ACC), 2016, pp. 2162–2167. IEEE (2016)
6. Li, S., Du, H., Yu, X.: Discrete-time terminal sliding mode control systems based on Euler's discretization. IEEE Trans. Autom. Control **59**(2), 546–552 (2013)
7. Lin, Y., Shi, Y., Burton, R.: Modeling and robust discrete-time sliding-mode control design for a fluid power electrohydraulic actuator (EHA) system. IEEE/ASME Trans. Mechatron. **18**(1), 1–10 (2011)
8. Ortega, J.M., Rheinboldt, W.C.: Iterative solution of nonlinear equations in several variables. SIAM (2000)
9. Podlubny, I.: Fractional Differential Equations. Academic, New York, NY, USA (1999)
10. Qiu, L., Shi, Y., Pan, J., Xu, B., Li, H.: Robust control for a networked direct-drive linear motion control system: design and experiments. Inf. Sci. **370**, 725–742 (2016)

11. Shi, Y., Fang, H., Yan, M.: Kalman filter-based adaptive control for networked systems with unknown parameters and randomly missing outputs. Int. J. Robust Nonlinear Control: IFAC-Affiliated J. **19**(18), 1976–1992 (2009)
12. Shi, Y., Huang, J., Yu, B.: Robust tracking control of networked control systems: application to a networked DC motor. IEEE Trans. Ind. Electron. **60**(12), 5864–5874 (2013)
13. Su, W.C., Drakunov, S.V., Ozguner, U.: An O (T^2) boundary layer in sliding mode for sampled-data systems. IEEE Trans. Autom. Control **45**(3), 482–485 (2000)
14. Wang, H., Mustafa, G.I., Tian, Y.: Model-free fractional-order sliding mode control for an active vehicle suspension system. Adv. Eng. Softw. **115**, 452–461 (2018)
15. Yeh, S.S., Hsu, P.L.: Estimation of the contouring error vector for the cross-coupled control design. IEEE/ASME Trans. Mechatron. **7**(1), 44–51 (2002)

Part III
Fractional-Order SMC for the Deployment of Space Tethered System

Chapter 8
Fractional-Order Fuzzy SMC for the Deployment of TSS

8.1 Introduction

This chapter is dedicated to proposing a novel fractional-order fuzzy sliding mode control scheme for the deployment of TSS in the presence of input saturation. Firstly, the underactuated model of TSS is separated into two subsystems. A FO and a constrained integer-order sliding surfaces are constructed for the actuated and underactuated subsystems, respectively. Then, a novel hybrid sliding manifold can be readily delivered by coupling the two sub-sliding surfaces with a coupling coefficient. An adaptive fuzzy algorithm is employed to estimate the optimal coupling coefficient. Moreover, input saturation is also considered in this work. Finally, it is strictly proved that system trajectories can be steered on the hybrid sliding manifold and move to the equilibrium point along the surface by using the developed saturated control law. The main contributions of this work are as follows:

1. Compared with the analytical methods in [10], the proposed method is capable of dealing with both the nonlinear underactuated dynamics of TSS and external perturbations.
2. Compared with the conventional approaches in [4, 6] and the FOSMC approach in [3], in this chapter, the fractional-order calculus is introduced not merely into the sliding surface but also into control law. And owing to the existence of the additional tuning parameters in sliding mode surface and control law, better control performances are procured.
3. A new hybrid sliding manifold is presented to design FOSMC for an underactuated system for the first time. Furthermore, its asymptotic stability is theoretically demonstrated.
4. Compared with most existing literature about FOSMC [2, 3, 11], the saturation effect of control input is addressed and tackled during controller design.

G. Sun et al., *Fractional-Order Sliding Mode Control: Methodologies and Applications*, Studies in Systems, Decision and Control 532,
https://doi.org/10.1007/978-3-031-60847-6_8

8.2 System Description and Preliminaries

8.2.1 *Tethered Satellite System Description*

Figure 8.1 shows the illustration of the TSS to be considered. The system consists of two rigid satellites that are connected by a long and flexible tether and rotate around the Earth within a circular orbit. It is assumed that the space tether keeps straight, massless, and in-extensible during the process of deployment. Moreover, the ratio of tether length to satellite size is very large such that the two rigid satellites can be idealized as two lumped masses. θ represents the in-plane angle. We concentrate on the regulation of in-plane angle θ and ignore out-of-plan angle.

Then, by Lagrangian mechanics, the equations of motion of the considered TSS are stated as follows [10]

$$\begin{cases} \theta'' + 2(\theta' + \Omega)\dfrac{l'}{l} + 3\Omega^2 \sin\theta\cos\theta = \dfrac{F_\theta}{\overline{m}l^2}, \\ l'' - l[(\theta' + \Omega)^2 + \Omega^2(3\cos^2\theta - 1)] = -\dfrac{T_l}{\overline{m}}, \end{cases} \tag{8.1}$$

where $\overline{m} = \frac{m_1 m_2}{m_1+m_2}$ with m_1 and m_2 denoting the masses of main satellite and sub-satellite respectively. l is the instantaneous length of the tether between two satellites. l' (θ') refers to the derivative of l (θ) with respect to time t. T_l describes the tension of the tether. In light of the positive property of tensile force, it is assumed that $0 < \underline{T} \leq T_l \leq \overline{T}$ where $\underline{T}$ and $\overline{T}$ are the known lower and upper bounds of T_l respectively. Ω denotes the angular rate of orbit. F_θ represents the force in relation to θ and can be omitted in pure tension control.

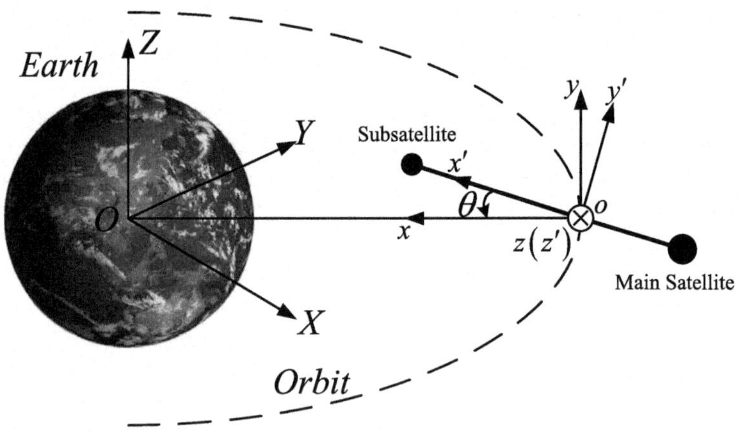

Fig. 8.1 The illustration of tethered satellite system

With the aid of the dimensionless transformations shown below

$$\delta \triangleq \frac{l}{L}, \quad u \triangleq \frac{-2T_l + \overline{T} + \underline{T}}{2\overline{m}L\Omega^2}, \quad \tau \triangleq \Omega t, \quad (\cdot)' \triangleq \Omega(\dot{\cdot}),$$

where L is the total length of the tether to be deployed, the dynamics of TSS (8.1) can be rewritten as the dimensionless formulation given by

$$\begin{cases} \ddot{\delta} - \delta[(1+\dot{\theta})^2 + 3\cos^2\theta - 1] = u + \omega, \\ \ddot{\theta} + 2\frac{\dot{\delta}}{\delta}(1+\dot{\theta}) + 3\sin\theta\cos\theta = 0, \end{cases} \tag{8.2}$$

where u is bounded within $[-\frac{\overline{T}-\underline{T}}{2\overline{m}L\Omega^2}, \frac{\overline{T}-\underline{T}}{2\overline{m}L\Omega^2}]$, and $\dot{l}$ ($\dot{\theta}$) indicates the derivative of l (θ) with respect to dimensionless time τ. ω refers to the lumped disturbance including modeling errors and external disturbances. Here, it is reasonable to assume that there exists a positive constant ω_m such that $|\omega| \leq \omega_m$.

To avoid the singularity, a new variable $\lambda \triangleq \delta - c$ where $c \in (0, 1]$ is introduced to describe tether length. Furthermore, by defining $x_1(\tau) \triangleq \lambda$, $x_2(\tau) \triangleq \dot{\lambda}$, $x_3(\tau) \triangleq \theta$, $x_4(\tau) \triangleq \dot{\theta}$, $x(\tau) \triangleq [x_1(\tau), x_2(\tau), x_3(\tau), x_4(\tau)]^T$, the state-space description of TSS (8.2) can be written as

$$\begin{cases} \dot{x}_1(\tau) = x_2(\tau), \\ \dot{x}_2(\tau) = (x_1(\tau) + c)\left[(x_4(\tau)+1)^2 - 1 + 3\cos^2 x_3(\tau)\right] + u(\tau) + \omega(\tau), \\ \dot{x}_3(\tau) = x_4(\tau), \\ \dot{x}_4(\tau) = -\dfrac{2x_2(\tau)}{x_1(\tau)+c}(1 + x_4(\tau)) - 3\cos x_3(\tau)\sin x_3(\tau). \end{cases} \tag{8.3}$$

8.2.2 *Control Objective*

The control objective of this chapter is to deploy the space tether of TSS from the initial condition $(\lambda, \dot{\lambda}, \theta, \dot{\theta}) = (\lambda_0, \dot{\lambda}_0, \theta_0, \dot{\theta}_0)$ to the final condition $(\lambda, \dot{\lambda}, \theta, \dot{\theta}) = (0, 0, 0, 0)$ in a fast and smooth behavior with coping with the saturated control input.

Before we proceed, we would like to stress the following lemma that will be used in the following sections.

Lemma 8.1 *Considering the FO autonomous system given as follows*

$$D^\beta x(t) = Ax(t), \quad x(0) = x_0,$$

where $\beta \in (0, 1)$ *denotes fractional operator,* $x(t) \in \mathbb{R}^n$ *is system vector, and* $A \in \mathbb{R}^{n \times n}$ *represents system matrix. This system is said to be asymptotically stable if and only if the following formula holds*

$$|agl(\varpi_i(A))| > \frac{\beta\pi}{2}, \quad (i = 1, 2, \cdots, n),$$

where $agl(\varpi_i(A))$ *represents the angle of the* i*th eigenvalue of matrix* A.

Remark 8.2 Lemma 8.1 tells us that a FO system possesses a larger stable field in comparison with an integer-order system (see Fig. 1.1). Clearly, the latter is a special case of the former by letting $\beta = 1$.

8.3 Control Strategy

In this section, the case without input saturation will be discussed first. Then, the achieved results will be extended to the case with input saturation.

8.3.1 *Fractional-Order Fuzzy Sliding Mode Control*

In this subsection, the control strategy for TSS without saturated input is constructed. First of all, the underactuated model of TSS (8.3) can be divided into two subsystems which are

$$\begin{cases} \dot{x}_1(\tau) = x_2(\tau), \\ \dot{x}_2(\tau) = f_1(x(\tau)) + u(\tau) + \omega(\tau), \end{cases} \tag{8.4}$$

$$\begin{cases} \dot{x}_3(\tau) = x_4(\tau), \\ \dot{x}_4(\tau) = f_2(x(\tau)), \end{cases} \tag{8.5}$$

where

$$f_1(x(\tau)) = (x_1(\tau) + c)\left[(x_4(\tau) + 1)^2 - 1 + 3\cos^2 x_3(\tau)\right],$$
$$f_2(x(\tau)) = -\frac{2x_2(\tau)}{x_1(\tau) + c}(1 + x_4(\tau)) - 3\cos x_3(\tau)\sin x_3(\tau).$$

Obviously, (8.4) is an actuated subsystem and (8.5) is an underactuated subsystem. Then, a fractional-order sliding surface $s_1(\tau)$ and an integer-order sliding surface $s_2(\tau)$ are designed for the actuated and underactuated subsystems respectively, which are

$$s_1(\tau) = c_1 x_1(\tau) + c_2 \mathrm{D}^{\beta} x_1(\tau) + c_3 x_2(\tau), \tag{8.6}$$
$$s_2(\tau) = c_4 x_3(\tau) + x_4(\tau), \tag{8.7}$$

where $\beta \in (0, 1)$ is the fractional order; c_i $(i = 1, 2, 3, 4)$ are the parameters to be designed to ensure control requirements.

Remark 8.3 From the actuated and underactuated subsystems shown in (8.4) and (8.5) respectively, it is seen that system states x_3 and x_4 are affected by system states x_1 and x_2 instead of the control input u. Clearly, system states x_1 and x_2 play a key role in controller design. Therefore, FOSMC is used in actuated subsystems to obtain better control performances.

The following constrained integer-order sliding surface $\sigma(\tau)$ is designed based on sub-sliding surface $s_2(\tau)$,

$$\sigma(\tau) = \frac{2\overline{\sigma}}{\pi} \arctan(s_2(\tau)), \quad \overline{\sigma} \in (0, 1). \tag{8.8}$$

The derivative of (8.8) is addressed as

$$\dot{\sigma}(\tau) = \frac{2\overline{\sigma}}{\pi(1 + s_2^2(\tau))} \dot{s}_2(\tau) \triangleq g(\tau)\dot{s}_2(\tau). \tag{8.9}$$

Now, a novel hybrid sliding manifold is achieved as follows:

$$s(\tau) = c_1 x_1(\tau) + c_2 \mathrm{D}^{\beta} x_1(\tau) + c_3 \left(x_2(\tau) - \frac{\phi^*}{c_3} \sigma(\tau) \right), \tag{8.10}$$

where $\phi^* > 0$ is a coupling coefficient. Taking (8.4), (8.5), (8.7), and (8.9) into consideration, the derivative of (8.10) is given by

$$\dot{s}(\tau) = c_1 x_2(\tau) + c_2 \mathrm{D}^{\beta} x_2(\tau) + c_3 \dot{x}_2(\tau) \tag{8.11}$$
$$-\phi^* g(\tau) \left[c_4 x_4(\tau) + \dot{x}_4(\tau) \right]. \tag{8.12}$$

It is worth mentioning that coupling coefficient ϕ^* affects the dynamic responses of closed-loop systems greatly. To approach this problem, in this case, adaptive fuzzy algorithm is made use of to procure optimal ϕ^*. The IF-THEN fuzzy rules are exhibited as

Fuzzy rule $\mathscr{R}^i$: IF h_1 is M_1^i and h_2 is M_2^i and $\cdots$ h_l is M_l^i,
THEN $\widehat{\phi}_f$ is N^i, $i = 1, 2, \cdots, r$,

where r is the number of IF-THEN fuzzy rules; $h = [h_1 \ h_2 \ \cdots \ h_l]^T$ denotes the vector of input variables; $\widehat{\phi}_f$ is the output variable used to estimate ϕ^*; M_j^i $(j =$

$1, 2, \cdots, l$) and N^i refer to the fuzzy sets related to the jth input variable and the output variable, respectively; $\mu_{M_j^i}(h_j)$ is defined as the membership function.

The output variable $\widehat{\phi}_f$ is figured out as

$$\widehat{\phi}_f = \frac{\sum_{i=1}^{r} \widehat{\eta}^i \Pi_{j=1}^{l} \mu_{M_j^i}(h_j)}{\sum_{i=1}^{r} \Pi_{j=1}^{l} \mu_{M_j^i}(h_j)} = \chi^T \widehat{\eta}, \tag{8.13}$$

where $\widehat{\eta} = \left[\widehat{\eta}^1 \ \widehat{\eta}^2 \ \cdots \ \widehat{\eta}^r\right]^T$ represents an adjustable parameter vector, and $\chi = \left[\chi^1 \ \chi^2 \ \cdots \ \chi^r\right]^T$ describes a fuzzy weighting vector with χ^i defined as

$$\chi^i = \frac{\Pi_{j=1}^{l} \mu_{M_j^i}(h_j)}{\sum_{i=1}^{r} \Pi_{j=1}^{l} \mu_{M_j^i}(h_j)}. \tag{8.14}$$

The optimal output ϕ_f^* is existent and determined by

$$\phi_f^* = \chi^T \eta^*, \tag{8.15}$$

where the constant vector η^* is chosen as

$$\eta^* = \arg \min_{\widehat{\eta} \in U_{\widehat{\eta}}} \left[\sup_{h_j \in U_{h_j}} \left| \widehat{\phi}_f - \phi^* \right| \right]. \tag{8.16}$$

In (8.16), we assume that $U_{\widehat{\eta}} = \left\{\widehat{\eta} : \|\widehat{\eta}\| \leq M_{\widehat{\eta}}\right\}$ and $U_{h_j} = \left\{h_j : \|h_j\| \leq M_{h_j}\right\}$ where $M_{\widehat{\eta}} > 0$ and $M_{h_j} > 0$. $\Delta = \phi_f^* - \phi^*$ is viewed as the minimum estimation error of ϕ^* and satisfies $|\Delta| \leq \Delta_m$ with $\Delta_m > 0$. Moreover, $\widetilde{\eta} = \widehat{\eta} - \eta^*$ is defined as the estimation error of η^*.

Now, we are ready to deliver the following theorem.

Theorem 8.4 *Consider the underactuated dynamics of TSS presented in (8.3), if the control law $u(\tau)$ is determined by*

$$\begin{aligned} u(\tau) = \frac{1}{c_3} \{ & -c_1 x_2(\tau) - c_2 D^\beta x_2(\tau) - c_3 f_1(x(\tau)) \\ & + \widehat{\phi}_f g(\tau) \left[c_4 x_4(\tau) + f_2(x(\tau))\right] - k_1 sgn\left(s(\tau)\right) \\ & - k_2 s(\tau) - k_3 sgn\left(s(\tau) s_1(\tau)\right) s_1(\tau) \\ & - \Delta_m sgn\left(\left[c_4 x_4(\tau) + f_2(x(\tau))\right] g(\tau) s(\tau)\right) \\ & \times \left[c_4 x_4(\tau) + f_2(x(\tau))\right] g(\tau) \}, \end{aligned} \tag{8.17}$$

where $k_i > 0$ ($i = 1, 2, 3$) are the parameters to be designed, and $u(\tau)$ is updated by the adaptive law as follows

$$\dot{\widehat{\eta}} = Proj_{\widehat{\eta}}(\zeta\chi) = \begin{cases} \zeta\chi, & \left(if\ \|\widehat{\eta}\| < M_{\widehat{\eta}} or\ \|\widehat{\eta}\| = M_{\widehat{\eta}}\ and\ \overline{\zeta}\chi \geq 0\right), \\ \zeta\chi - \frac{\widehat{\eta}\widehat{\eta}^T}{\|\widehat{\eta}\|^2}\zeta\chi, & \left(if\ \|\widehat{\eta}\| = M_{\widehat{\eta}} and\ \overline{\zeta}\chi < 0\right), \end{cases} \tag{8.18}$$

where

$$\zeta = -\gamma\left[c_4x_4(\tau) + f_2(x(\tau))\right]g(\tau)s(\tau),$$
$$\overline{\zeta} = \left[c_4x_4(\tau) + f_2(x(\tau))\right]g(\tau)s(\tau)\widehat{\eta}^T,$$

then the states of system (8.3) can be driven to the newly proposed hybrid sliding manifold (8.10) by $u(\tau)$.

Proof To prove this theorem, the Lyapunov function is chosen as

$$V(\tau) = \frac{1}{2}s^2(\tau) + \frac{1}{2\gamma}\widetilde{\eta}^T\widetilde{\eta}, \tag{8.19}$$

where γ is a positive constant.

By differentiating (8.19) with respect to dimensionless time τ and substituting (8.11) into the result, it can be known that

$$\begin{aligned} \dot{V}(\tau) &= s(\tau)\dot{s}(\tau) + \frac{1}{\gamma}\widetilde{\eta}^T\dot{\widetilde{\eta}} \\ &= s(\tau)\left\{c_1x_2(\tau) + c_2\mathrm{D}^\beta x_2(\tau) + c_3f_1(x(\tau)) + c_3u(\tau)\right. \\ &\quad \left. + c_3\omega(\tau) - \left(\widehat{\phi}_f - \Delta\right)g(\tau)\left[c_4x_4(\tau) + f_2(x(\tau))\right]\right\} \\ &\quad + \frac{1}{\gamma}\widetilde{\eta}^T\left\{\dot{\widehat{\eta}} + \gamma g(\tau)\left[c_4x_4(\tau) + f_2(x(\tau))\right]s(\tau)\chi\right\}. \end{aligned} \tag{8.20}$$

By further bringing (8.17) and (8.18) into (8.20), we obtain

$$\begin{aligned} \dot{V}(\tau) \leq &- (k_1 - c_3|\omega(\tau)|)|s(\tau)| - k_2s^2(\tau) - k_3|s(\tau)s_1(\tau)| \\ &- (\Delta_m - |\Delta|)\left|\left[c_4x_4(\tau) + f_2(x(\tau)\right]g(\tau)s(\tau)\right|. \end{aligned} \tag{8.21}$$

Owing to $|\omega(\tau)| \leq \omega_m$ and $|\Delta| \leq \Delta_m$, the following inequality is guaranteed if $k_1 \geq c_3\omega_m + \xi$ where $\xi > 0$,

$$\begin{aligned} \dot{V}(\tau) \leq &- k_2s^2(\tau) - (k_1 - c_3\omega_m)|s(\tau)| - k_3|s(\tau)s_1(\tau)| \\ &- (\Delta_m - |\Delta|)\left|\left[c_4x_4(\tau) + f_2(x(\tau)\right]g(\tau)s(\tau)\right| \\ \leq &- k_2s^2(\tau) - \xi|s(\tau)| \triangleq -Q(\tau) \leq 0. \end{aligned} \tag{8.22}$$

The integration of (8.22) from $\tau_0 = 0$ to ∞ leads to

$$\int_{\tau_0}^{\infty} Q(\tau)\mathrm{d}\tau \leq V(\tau_0) - V(\infty). \tag{8.23}$$

Since $V(\tau_0) - V(\infty)$ is bounded, it follows that $\int_{\tau_0}^{\infty} Q(\tau)\mathrm{d}\tau < \infty$. Moreover, noting the fact that $\dot{Q}(\tau)$ is also bounded, one can have $\lim_{t\to\infty} Q(\tau) = 0$ according to Barbalat's Lemma, which implies $\lim_{t\to\infty} s(\tau) = 0$. Therefore, it is proved that the state trajectories of the underactuated system (8.3) can be steered on the hybrid sliding manifold (8.10) by using the control law (8.17) updated by (8.18). The proof of this theorem is completed. ■

From Theorem 8.4, we have known that system states can reach the hybrid sliding manifold (8.10). In particular, if the hybrid sliding manifold (8.10) is asymptotically stable, system states will move to the equilibrium point $(\lambda, \dot{\lambda}, \theta, \dot{\theta}) = (0, 0, 0, 0)$ along manifold (8.10), which implies the control objective of this study is implemented. In what follows, the asymptotic stability of hybrid sliding manifold (8.10) will be demonstrated strictly.

Theorem 8.5 *Consider the underactuated dynamics of TSS presented in (8.3) and the hybrid sliding manifold shown in (8.10) where the fractional order β can be written as $\beta = \frac{q}{p}$ $(q, p \in N^+)$, the hybrid sliding manifold (8.10) is asymptotically stable if the following conditions are satisfied*

$$c_4 > 0, \tag{8.24}$$

$$|agl(\varpi_i(A))| > \frac{\pi}{2p}, \quad (i = 1, 2, \cdots, p), \tag{8.25}$$

where

$$A = \left[\begin{array}{c} 0\,1\,0 \cdots 0 \\ 0\,0\,1 \cdots 0 \\ \vdots\;\vdots\;\vdots\;\ddots\;\vdots \\ 0\,0\,0 \cdots 1 \\ \hline -\frac{c_1}{c_3}\;0 \cdots 0\; -\frac{c_2}{c_3}\;0 \cdots 0 \end{array}\right].$$

Proof Since $V(\tau) > 0$ and $\dot{V}(\tau) \le 0$ (see (8.19) and (8.22)), we have $V(\tau) = \frac{1}{2}s^2(\tau) + \frac{1}{2\gamma}\tilde{\eta}^T\tilde{\eta} \le V(\tau_0) < \infty$, therefore, $s(t) \in \mathcal{L}_\infty$. Furthermore, (8.8) implies $\sigma(\tau) \in (0, 1)$. Clearly, $\sigma(\tau) \in \mathcal{L}_\infty$. By noting the structure of (8.10), $s(\tau) \in \mathcal{L}_\infty$ and $\sigma(\tau) \in \mathcal{L}_\infty$ will result in $s_1(\tau) \in \mathcal{L}_\infty$.

Now, let us revisit (8.22), it can be derived that

$$\int_{\tau_0}^{\infty} k_2 s^2(\tau) + \xi|s(\tau)| + k_3|s(\tau)s_1(\tau)|\mathrm{d}\tau < \infty. \tag{8.26}$$

From (8.26), we have $s(\tau) \in \mathcal{L}_2$, $s(\tau) \in \mathcal{L}_1$, and

$$\int_{\tau_0}^{\infty} |s(\tau)s_1(\tau)|\mathrm{d}\tau < \infty. \tag{8.27}$$

In addition, it is obvious that

$$|s(\tau)| = |s_1(\tau) - \phi^*\sigma(\tau)| \geq |s_1(\tau)| - |\phi^*\sigma(\tau)|. \tag{8.28}$$

Multiplying the both sides of (8.28) by $|s_1(\tau)|$ gives rise to

$$|s(\tau)s_1(\tau)| \geq s_1^2(\tau) - |\phi^* s_1(\tau)\sigma(\tau)|. \tag{8.29}$$

Moreover, noting (8.27) and (8.29), one procures

$$\begin{aligned}&\int_{\tau_0}^{\infty} \left\{s_1^2(\tau) - 2|\phi^* s_1(\tau)\sigma(\tau)|\right\} \mathrm{d}\tau \leq \int_{\tau_0}^{\infty} \left\{s_1^2(\tau) - |\phi^* s_1(\tau)\sigma(\tau)|\right\} \mathrm{d}\tau \\ &\leq \int_{\tau_0}^{\infty} |s(\tau)s_1(\tau)| \mathrm{d}\tau < \infty.\end{aligned} \tag{8.30}$$

In view of $s(\tau) \in \mathcal{L}_2$, it is ensured that

$$\begin{aligned}&\int_{\tau_0}^{\infty} \left\{s_1^2(\tau) - 2\phi^* |s_1(\tau)\sigma(\tau)|\right\} \mathrm{d}\tau + \left(\phi^*\right)^2 \int_{\tau_0}^{\infty} \sigma^2(\tau)\mathrm{d}\tau \\ &= \int_{\tau_0}^{\infty} \left\{s_1^2(\tau) - 2\phi^* s_1(\tau)\sigma(\tau) + \left(\phi^*\sigma(\tau)\right)^2\right\} \mathrm{d}\tau \\ &= \int_{\tau_0}^{\infty} s^2(\tau)\mathrm{d}\tau < \infty.\end{aligned} \tag{8.31}$$

By combining (8.30) and (8.31), one finally obtains

$$\int_{\tau_0}^{\infty} \sigma^2(\tau)\mathrm{d}\tau < \infty, \tag{8.32}$$

which signifies $\sigma(\tau) \in \mathcal{L}_2$. Owing to $\dot{\sigma}(\tau) \in \mathcal{L}_\infty$, it is readily derived that $\lim_{\tau\to\infty} \sigma(\tau) = 0$ based on Barbalat's Lemma, which further results in $\lim_{\tau\to\infty} s_2(\tau) = 0$. Since Theorem 8.4 has concluded that $\lim_{\tau\to\infty} s(\tau) = 0$, $\lim_{\tau\to\infty} s_1(\tau) = 0$ is ensured as well.

The aforementioned steps have proved that the sub-sliding surfaces $s_1(\tau) = 0$ and $s_2(\tau) = 0$ can all be reached by using the proposed control law. Hence, if the asymptotic stabilities of $s_1(\tau)$ and $s_2(\tau)$ are satisfied, the asymptotic stability of hybrid sliding manifold (8.10) is guaranteed. According to Hurwitz's theorem, it is easy to know that $s_2(\tau)$ is asymptotically stable if (8.24) holds. In what follows, we will give the stability analysis of $s_1(\tau)$.

By recalling (8.6), $s_1(\tau) = 0$ is written as

$$s_1(\tau) = c_1 x_1(\tau) + c_2 \mathrm{D}^{\beta} x_1(\tau) + c_3 \dot{x}_1(\tau) = 0. \tag{8.33}$$

Due to $\beta = \frac{q}{p}$ and $q, p \in N^+$, (8.33) becomes

$$c_1 x_1(\tau) + c_2 \mathrm{D}^{\frac{q}{p}} x_1(\tau) + c_3 \mathrm{D}^{\frac{p}{p}} x_1(\tau) = 0. \tag{8.34}$$

Let us introduce a new set of state variables $v_i(\tau)$ $(i = 1, 2, \ldots, p)$ which are defined as follows

$$\begin{aligned} v_1(\tau) &= x_1(\tau), \\ v_2(\tau) &= \mathrm{D}^{\frac{1}{p}} v_1(\tau) = \mathrm{D}^{\frac{1}{p}} x_1(\tau), \\ &\vdots \\ v_p(\tau) &= \mathrm{D}^{\frac{1}{p}} v_{p-1}(\tau) = \mathrm{D}^{\frac{p-1}{p}} x_1(\tau). \end{aligned}$$

Consequently, the state-space description of (8.34) is addressed by

$$\mathrm{D}^{\frac{1}{p}} v(\tau) = A v(\tau), \tag{8.35}$$

where $v(\tau) = [v_1(\tau)\ v_2(\tau)\ \cdots v_p(\tau)]^T$. According to Lemma 8.1, it is proved that $s_1(\tau)$ is asymptotically stable if (8.25) holds. Therefore, the proof of this theorem is completed. ■

8.3.2 *Fractional-Order Fuzzy Sliding Mode Control in the Presence of Input Saturation*

In practice, any control action can not tend to infinity. In TSS, tensile force should be kept within a predefined range since the space tether should always be straight and can not be compressed. In this regard, the saturation effect of tension should be taken into account to attain an applicable controller for TSS.

With the saturated tension, the underactuated model of TSS can be expressed as

$$\begin{cases} \dot{x}_1(\tau) = x_2(\tau), \\ \dot{x}_2(\tau) = f_1(x(\tau)) + \mathrm{sat}\,(u(\tau)) + \omega(\tau), \\ \dot{x}_3(\tau) = x_4(\tau), \\ \dot{x}_4(\tau) = f_2(x(\tau)), \end{cases} \tag{8.36}$$

where the saturation function $\mathrm{sat}(u(\tau))$ is defined as

$$\mathrm{sat}(u(\tau)) = \mathrm{sgn}(u(\tau)) \min\{|u(\tau)|, u_{max}\}, \tag{8.37}$$

u_{max} is a known positive constant representing the bound of tension.

By virtue of Lyapunov function $V(\tau)$, the attraction domain of the origin of the closed-loop system (8.36) is estimated as [1]

$$\mathcal{E}(\rho) = \left\{x(\tau) \in \mathbb{R}^n \mid V(\tau) \leq \rho\right\}, \tag{8.38}$$

where the positive scalar ρ ensures $\max_{x(\tau)\in\mathcal{E}(\rho)} |z(\tau)| \leq u_{max}$ and $z(\tau)$ is to be defined in the following theorem.

Theorem 8.6 *Consider the underactuated dynamics of TSS described in (8.36). The saturated control law is designed as*

$$sat\,(u(\tau)) = sat\,(z(\tau) - \varphi s(\tau)), \tag{8.39}$$

where $z(\tau)$ is equivalent to the $u(\tau)$ defined in (8.17) and φ is a positive scalar. Then, with (8.39), the closed-loop system (8.36) is asymptotically stable, which means the control goal of this study is implemented.

Proof First, the control law (8.39) can be written as

$$\begin{aligned} \text{sat}\,(u(\tau)) &= \text{sat}\,(z(\tau) - \varphi s(\tau)) \\ &= z(\tau) - [\text{sat}\,(\varphi s(\tau) - z(\tau)) + z(\tau)]. \end{aligned} \tag{8.40}$$

Taking care of (8.37), one has

$$\begin{aligned} &\text{sat}\,(\varphi s(\tau) - z(\tau)) + z(\tau) \\ &= \begin{cases} \varphi s(\tau), & \text{if}\, |\varphi s(\tau) - z(t)| \leq u_{max}, \\ z(\tau) + u_{max}, & \text{if}\, (\varphi s(\tau) - z(\tau)) > u_{max}, \\ z(\tau) - u_{max}, & \text{if}\, (\varphi s(\tau) - z(\tau)) < -u_{max}, \end{cases} \end{aligned} \tag{8.41}$$

Here, we choose the Lyapunov function $V(\tau)$ which is the same as (8.19). Its derivative is stated as

$$\begin{aligned} \dot{V}(\tau) &= s(\tau)\dot{s}(\tau) + \frac{1}{\gamma}\widetilde{\eta}^T\widetilde{\eta} \\ &= s(\tau)\left\{c_1 x_2(\tau) + c_2 \mathrm{D}^{\beta} x_2(\tau) + c_3 f_1(x(\tau)) + c_3\omega(\tau)\right. \\ &\quad + c_3 z(\tau) - c_3\,[\text{sat}\,(\varphi s(t) - z(\tau)) + z(\tau)] \\ &\quad \left.- \left(\widehat{\phi}_f - \Delta\right) g(\tau)\,[c_4 x_4(\tau) + f_2(x(\tau))]\right\} + \frac{1}{\gamma}\widetilde{\eta}^T \\ &\quad \times \left\{\dot{\widehat{\eta}} + \gamma\,[c_4 x_4(t) + f_2(x(\tau))]\,g(\tau)s(\tau)\chi\right\}. \end{aligned} \tag{8.42}$$

By inserting (8.41) and $z(\tau)$ into (8.42), it is true that

$$\begin{aligned}
\dot{V}(\tau) =& -k_1|s(\tau)| - k_2 s^2(\tau) - k_3|s(\tau)s_1(\tau)| + c_3\omega(\tau)s(\tau) \\
& - \Delta_m \left|[c_4 x_4(\tau) + f_2(x(\tau))]\, g(\tau)s(\tau)\right| \\
& + \Delta\, [c_4 x_4(\tau) + f_2(x(\tau)]\, g(\tau)s(\tau) \\
& - c_3 s(\tau)\, [\mathrm{sat}\,(\varphi s(\tau) - z(\tau)) + z(\tau)] \\
\leq & - k_2 s^2(\tau) - (k_1 - c_3|\omega(\tau)|)|s(\tau)| - k_3|s(\tau)s_1(\tau)| \\
& - (\Delta_m - |\Delta|)\left|[c_4 x_4(\tau) + f_2(x(\tau)]\, g(\tau)s(\tau)\right| \\
& - c_3 s(\tau)\, [\mathrm{sat}\,(\varphi s(\tau) - z(\tau)) + z(\tau)] \\
\leq & - k_2 s^2(\tau) - \xi|s(\tau)| \\
& - \begin{cases} c_3\varphi s^2(\tau), & \text{if } |\varphi s(\tau) - z(\tau)| \leq u_{max}, \\ c_3 s(\tau)(z(\tau) + u_{max}), & \text{if } (\varphi s(\tau) - z(\tau)) > u_{max}, \\ c_3 s(\tau)(z(\tau) - u_{max}), & \text{if } (\varphi s(\tau) - z(\tau)) < -u_{max}, \end{cases}
\end{aligned} \tag{8.43}$$

Noting (8.41) and $|z(\tau)| \leq u_{max}$, we have $z(\tau) + u_{max} \geq 0$ and $z(\tau) - u_{max} \leq 0$. Moreover, $\varphi s(\tau) - z(\tau) > u_{max}$ and $\varphi s(\tau) - z(\tau) < -u_{max}$ lead to $s(\tau) > 0$ and $s(\tau) < 0$, respectively. Consequently, it can be inferred from (8.43) that

$$\dot{V}(\tau) < -k_2 s^2(\tau) - \xi|s(\tau)|.$$

According to Theorems 8.4 and 8.5, therefore, the state trajectories of system (8.36) can be driven to the hybrid manifold (8.10) and move to equilibrium point by employing (8.39). Hence, the proof of this theorem is completed. ■

8.4 Simulations

To validate the effectiveness and superiority of the proposed control strategy, detailed simulations, and comparisons are performed in this section. The physical parameters of TSS are defined as same as the ones employed in [3]. Therefore, the initial states of TSS are expressed as $(\lambda_0, \dot{\lambda}_0, \theta_0, \dot{\theta}_0) = (-0.99, 0.5, 0, 0)$ and the anticipated states are stated as $(\lambda, \dot{\lambda}, \theta, \dot{\theta}) = (0, 0, 0, 0)$. The parameters of sub-sliding surfaces $s_1(\tau)$ and $\sigma(\tau)$ are set to $c_1 = 3, c_2 = 0.65, c_3 = 1, c_4 = 2, \beta = 0.9$, and $\overline{\sigma} = 0.98$. For the adaptive fuzzy logic system used to estimate coupling coefficient ϕ^*, $s(\tau)$ and $\dot{s}(\tau)$ are selected as the input variables of fuzzy rules. The fuzzy membership functions $\mu_{M_j^i}(\cdot)$ are provided in Fig. 8.2. Additionally, the parameters of adaptive law are chosen as $\gamma = 0.05, \Delta_m = 0.05, M_{\widehat{\eta}} = 0.5$, and $\widehat{\eta}^i$ $(i = 1, 2, \cdots, r)$ are all initialized as 0.075. For the saturated control law (8.39), we define $k_1 = 2.1$, $k_2 = 1$, $k_3 = 1$, $u_{max} = 6$, and $\varphi = 120$. Furthermore, the external disturbance is assumed to be $\omega(\tau) = 0.1\sin(2\pi\tau)e^{-\tau}$ which implies $\omega_m = 0.1$.

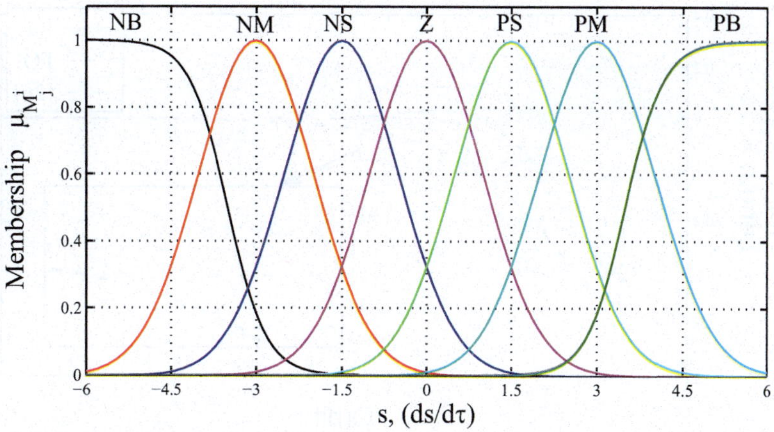

Fig. 8.2 Membership functions $\mu_{M_j^i}(\cdot)$

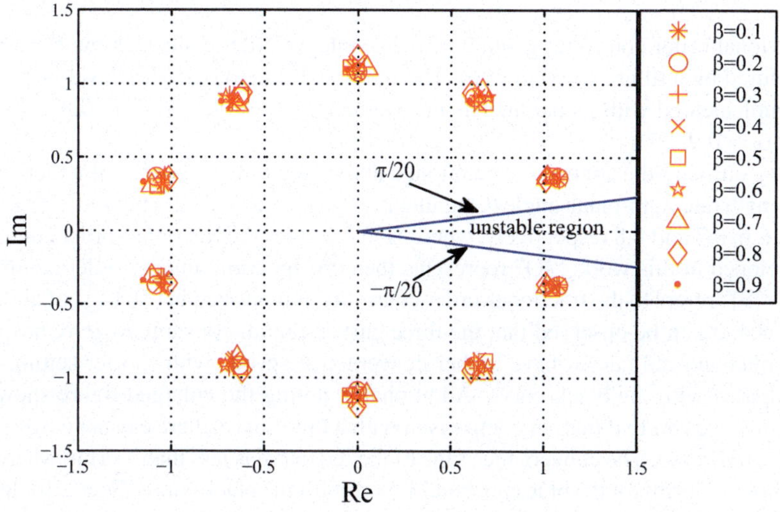

Fig. 8.3 Root distribution of sliding surface $s_1(\tau)$ with various β

According to Theorem 8.5, the values of c_i $(i = 1, 2, 3, 4)$ and β should satisfy the asymptotic stabilities of $s_1(\tau)$ and $s_2(\tau)$. For $s_2(\tau)$, it is clear that $c_4 = 2$ can ensure the asymptotic stability of $s_2(\tau)$. For $s_1(\tau)$, when specifying $c_1 = 3$, $c_2 = 0.65$, $c_3 = 1$, $p = 10$, and $q = 1, 2, \cdots, 9$, the root distribution of $s_1(\tau)$ with various β is displayed in Fig. 8.3. In this figure, it is easy to see that $\beta = 0.9$ can guarantee the asymptotic stability of $s_1(\tau)$.

What is more, to demonstrate the merits of this work, the proposed fractional-order controller will be compared with its integer-order counterpart and the existing

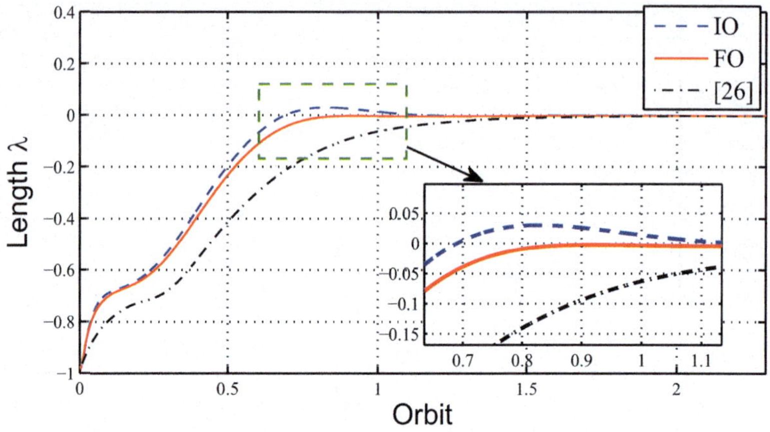

Fig. 8.4 Time responses of tether length λ

fractional-order controller given in [3]. The integer-order controller can be directly procured by letting $c_2 = 0$ in (8.6). The fractional-order controller developed in [3] is implemented with its parameters determined by $c_1 = c_2 = 1$, $c_3 = 0.8$, $\varepsilon = 0.01$ and $\alpha = 0.9$.

By utilizing the controllers mentioned above, the time responses of tether length λ, length rate $\dot{\lambda}$, in-plane angle θ, angular rate $\dot{\theta}$, and tension u are plotted in Figs. 8.4, 8.5, 8.6, 8.7 and 8.8 respectively, where "FO" denotes the fractional-order controller developed in this work, "IO" represents the corresponding integer-order controller, and "[3]" refers to the fractional-order controller synthesized in [3]. First of all, from Fig. 8.4, it can be observed that the three curves can all converge to zero, however, the blue and red curves have higher convergence speeds with shorter setting time compared with the black curve. Additionally, noting the enlarged figure shown in Fig. 8.4, we can find that, in comparison with a blue curve, there exists no overshoot in the red curve. The convergence time of the red curve is less than 1 orbit, whereas it is about 1.2 orbit for the blue curve and 1.5 orbit for the black curve. Hence, for length control, the proposed fractional-order algorithm attains better control performances than the other two algorithms. Similar phenomena can also be seen in the time responses of length rate $\dot{\lambda}$, in-plane angle θ and angular rate $\dot{\theta}$ (see Figs. 8.5, 8.6 and 8.7). What is more, the time responses of tension for the above three cases are depicted in Fig. 8.8. This figure states that the saturation effect of tension is considered and tackled by the proposed control scheme, which signifies the capability of the practical actuator is made full use of in our study. Therefore, based on previous discussions, we can conclude that the developed FO controller has gained the best control performances, indicating the fastest and most smooth tether deployment of TSS.

In fact, the FO control is an improvement of classical integer-order control. The FO operator can be regarded as an extra adjustable parameter for controller design. By

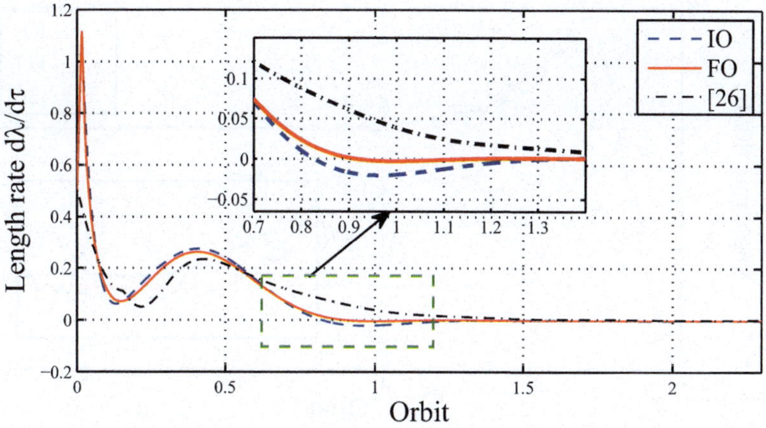

Fig. 8.5 Time responses of tether length rate $\dot{\lambda}$

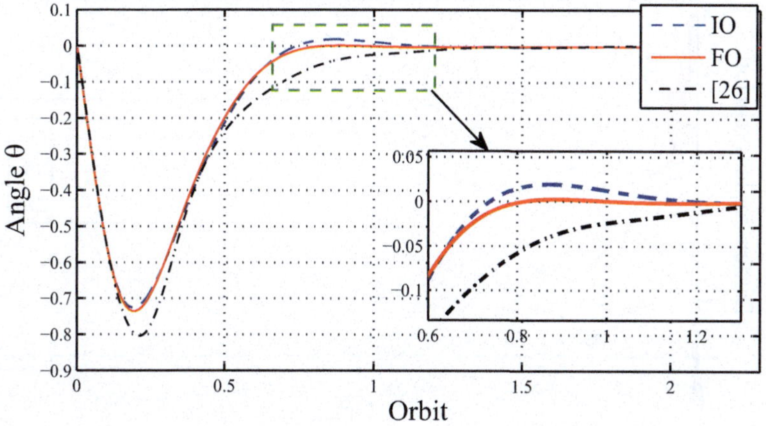

Fig. 8.6 Time responses of in-plane angle θ

tuning it appropriately, designers have the opportunity to obtain satisfactory transient responses, in the meantime, the merits of classical integer-order control can still be retained. This is why the FO control can achieve better control performances than its integer-order counterpart.

Remark 8.7 As mentioned in [4], the reel mechanism can be utilized to carry out the deployment or retrieval of space tether in practice. The real mechanism is composed of a servo motor, an optical encoder, and the reel where the space tether is stored. The real-time and fast tension control (such as the ones described in Fig. 8.8) can be performed by using a high-performance servo motor [4]. In addition, imposing

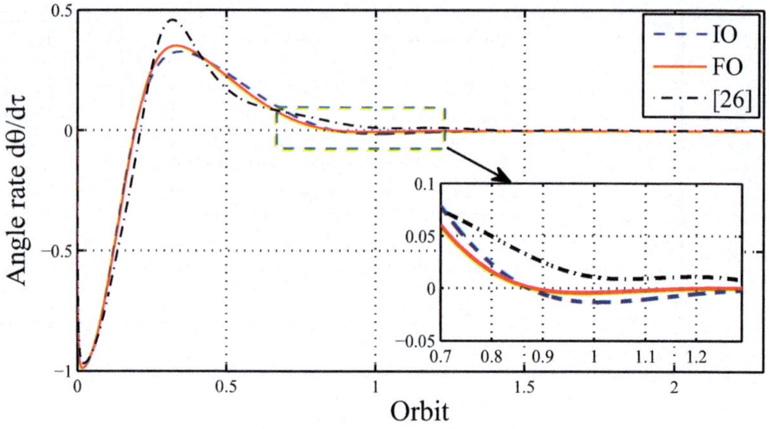

Fig. 8.7 Time responses of in-plane angular rate $\dot{\theta}$

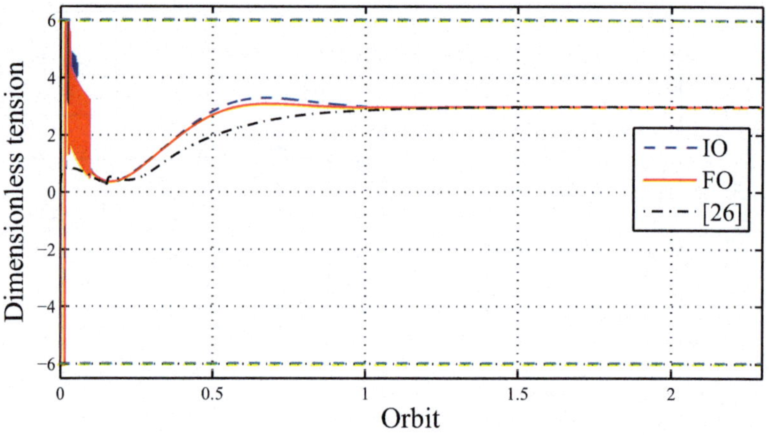

Fig. 8.8 Time responses of tension

some constraints on the rate of control input u when designing a controller is another effective way to further suppress the oscillation of tension force, which will be investigated in our future work.

8.5 Conclusion

This chapter is concerned with FO fuzzy sliding mode control for the deployment of TSS with input saturation. The underactuated dynamics of TSS are fully considered and tackled. The unique feature of this chapter is located in that a novel hybrid sliding manifold consisting of a FO sliding surface and a constrained integer-order sliding surface is proposed to design FOSMC for the first time. The coupling coefficient in a hybrid sliding manifold is approximated by the adaptive fuzzy technique to attain favorable performances. Moreover, the physical constraints on the tensile force are also dealt with. Finally, the effectiveness and superiority of the proposed control scheme are proved through convincing simulations. The future work for tether deployment can lie in how to combine the advantages of FO control with the advantages of observer-based control [9], backstepping control [7], hierarchical fuzzy control [5], or data-based control [8].

References

1. Bateman, A., Hull, J., Lin, Z.: A backstepping-based low-and-high gain design for marine vehicles. Int. J. Robust Nonlinear Control.: IFAC-Affil. J. **19**(4), 480–493 (2009)
2. Efe, M.Ö.: Fractional fuzzy adaptive sliding-mode control of a 2-DOF direct-drive robot arm. IEEE Trans. Syst. Man Cybern. Part B (Cybernetics) **38**(6), 1561–1570 (2008)
3. Kang, J., Zhu, Z.H., Wang, W., Li, A., Wang, C.: Fractional order sliding mode control for tethered satellite deployment with disturbances. Adv. Space Res. **59**(1), 263–273 (2017)
4. Meng, Z., Wang, B., Huang, P.: A space tethered towing method using tension and platform thrusts. Adv. Space Res. **59**(2), 656–669 (2017)
5. Sun, L., Huo, W.: Adaptive fuzzy control of spacecraft proximity operations using hierarchical fuzzy systems. IEEE/ASME Trans. Mechatron. **21**(3), 1629–1640 (2015)
6. Sun, L., Zhao, G., Huang, H.: Stability and control of tethered satellite with chemical propulsion in orbital plane. Nonlinear Dyn. **74**(4), 1113–1131 (2013)
7. Sun, L., Huo, W., Jiao, Z.: Adaptive backstepping control of spacecraft rendezvous and proximity operations with input saturation and full-state constraint. IEEE Trans. Ind. Electron. **64**(1), 480–492 (2016)
8. Tong, M., Pan, Y., Li, Z., Lin, W.: Valid data based normalized cross-correlation (VDNCC) for topography identification. Neurocomputing **308**, 184–193 (2018)
9. Wen, H., Zhu, Z.H., Jin, D., Hu, H.: Exponentially convergent velocity observer for an electrodynamic tether in an elliptical orbit. J. Guid. Control Dyn. **39**(5), 1113–1118 (2016)
10. Wen, H., Zhu, Z.H., Jin, D., Hu, H.: Space tether deployment control with explicit tension constraint and saturation function. J. Guid. Control. Dyn. **39**(4), 916–921 (2016)
11. Yin, C., Cheng, Y., Chen, Y., Stark, B., Zhong, S.: Adaptive fractional-order switching-type control method design for 3D fractional-order nonlinear systems. Nonlinear Dyn. **82**(1), 39–52 (2015)

Chapter 9
Fractional-Order Nonsingular Terminal SMC for the Deployment of STS

9.1 Introduction

Motivated by the remarkable merits of finite-time stability and perfect robustness from the nonsingular terminal sliding mode control scheme [5], this work proposes an integer-order nonsingular terminal sliding mode (IONTSM) controller at first, to obtain the finite-time stable deployment of TSS. Then, given the flexible and distributed structures of TSS, it is more suitable to involve the FO calculus in the tension control scheme, because of its unique historical memory effect, faster response, and smaller overshoot than the integer counterpart. Consequently, a novel FO nonsingular terminal sliding mode (FONTSM) controller is raised for the tether deployment, whose FO disturbance observer makes it more robust to external disturbances. Furthermore, considering the adverse impact of the saturated tension, this work presents a FO adaptive compensation law to ensure the uniform ultimate boundedness of the error between saturated input and ideal control.

9.2 Mathematical Description and Preliminaries

9.2.1 Mathematical Formulation

Given a typical TSS, we employ the two-body dumbbell model for the modeling and controlling of tether deployment. As shown in Fig. 8.1, the two satellites are connected by a long and flexible space tether. Assume that the space tether is non-extensible and massless, which remains straight in tether deployment with only length varies [8]. The working orbit of TSS will not be affected by the deployment status of the subsatellite, since the center of system mass is approximately located in the main satellite. In particular, when we focus on the research of control theory of

G. Sun et al., *Fractional-Order Sliding Mode Control: Methodologies and Applications*, Studies in Systems, Decision and Control 532,
https://doi.org/10.1007/978-3-031-60847-6_9

TSS, the in-plane motion will be more representative during the tether deployment, such that out-of-plane libration can be ignored [7]. Hence the motion of TSS can be comprised of the orbital motion of the main satellite, and the in-plane libration of the subsatellite. The deployment process can be described by the coordinate frames revealed in Fig. 8.1. Hence we can present the kinetic energy and the potential energy for the deployment of TSS as

$$\begin{aligned} K &= \frac{1}{2}m\Omega^2 R^2 + \frac{1}{2}\bar{m}l^2(\dot{\theta}+\Omega)^2 + \frac{1}{2}\bar{m}\dot{l}^2, \\ V &= -m\Omega^2 R^2 + \frac{1}{2}\bar{m}\Omega^2 l^2[1-3\cos^2\theta], \end{aligned} \tag{9.1}$$

where $\bar{m} = (m_1 m_2)/m$, $m = m_1 + m_2$, m_1 and m_2 denote the masses of the main satellite and subsatellite, respectively. θ and l express the state variables, namely, the in-plane angle and the instantaneous length of the space tether. $\dot{(\cdot)}$ indicates the derivative of $(\cdot)$ with respect to time t. Ω and R are the orbital angular velocity and the instantaneous distance OO_1 of TSS, respectively.

Based on the Lagrangian mechanics theory, define the Lagrangian function as $L = K - V$, the generalized coordinate vector as $q = (\theta, l)$, the differential equations for the in-plane libration of TSS can be established as [12]

$$\begin{aligned} &\ddot{\theta} + 2(\dot{l}/l)(\Omega+\dot{\theta}) + 3\Omega^2\cos\theta\sin\theta = \frac{\mathcal{F}_\theta}{\bar{m}l^2}, \\ &\ddot{l} - l\left[(\Omega+\dot{\theta})^2 + (3\cos^2\theta - 1)\Omega^2\right] = -\frac{\mathcal{F}_l}{\bar{m}}, \end{aligned} \tag{9.2}$$

where $\mathcal{F}_\theta$ and $\mathcal{F}_l$ represent the resultant forces in relation to θ and l, respectively. As for the pure-tension control scheme, it's remarkable that $\mathcal{F}_\theta$ can be regarded as 0 and $\mathcal{F}_l$ subjects to $0 < T_{min} < \mathcal{F}_l < T_{max}$, where T_{min} and T_{max} are the given lower and upper bounds of tether tension.

Due to the particularity of the space deployment mission, it is more appropriate to measure system time with the true anomaly τ [12]. Thus, the dimensionless transformation of TSS is given below

$$x_1 \triangleq \theta(\tau),\ x_2 \triangleq \frac{l - l_{max}}{l_{max}}, \tau \triangleq \Omega t,\ \frac{-\mathcal{F}_l}{\bar{m}\Omega^2 l_{max}} \triangleq -\hat{T} + d(t), \tag{9.3}$$

where $\hat{T}$ and l_{max} denote the dimensionless tension and the total length of the space tether. After this transformation, the system states will possess the same equilibrium point, i.e. the origin, and be bounded by the same upper bound 1. Meanwhile, based on this, we can transform the differential equations of TSS into the following dimensionless state space model

$$\begin{aligned} \ddot{x}_1 &= F(x_1, \dot{x}_1, x_2, \dot{x}_2), \\ \ddot{x}_2 &= G(x_1, \dot{x}_1, x_2, \dot{x}_2) + u(t) + d(t), \end{aligned} \tag{9.4}$$

where

$$
\begin{aligned}
F(x_1, \dot{x}_1, x_2, \dot{x}_2) &= -\frac{2\dot{x}_2(1+\dot{x}_1)}{1+x_2} - 3\cos(x_1)\sin(x_1), \\
G(x_1, \dot{x}_1, x_2, \dot{x}_2) &= (1+x_2)\left[(1+\dot{x}_1)^2 - 1 + 3\cos^2(x_1)\right],
\end{aligned} \tag{9.5}
$$

$u(t) = -\hat{T}$ and $d(t)$ express the control input from the proposed controller, and the disturbance input of external disturbances, respectively.

Remark 9.1 Considering the actual space environment of deployment mission [16], there exist some factors, such as gravity gradient, aerodynamic drag, solar radiation pressure, and so on, yielding the disturbance effect on the tether deployment. Based on [6], it can be concluded that the magnitudes of these disturbance torques are located in the range of 1×10^{-5} Nm. For convenience, we introduce the lumped disturbance $d(t)$ to represent all the external disturbances after dimensionless transformation.

According to the objective law of practical world, energy is continuous and bounded, not like the impulse function in mathematics, such that we assume $d(t)$ is bounded with $|d(t)| \le \mathcal{D}_1, |D^\alpha d(t)| \le \mathcal{D}_2$, where D^α denotes the fractional derivative operator. As for $D^\alpha d(t)$, it is a mathematical operation employed in this chapter, which is used for ensuring that the fractional derivative of disturbance is still bounded, just for mathematical derivation.

9.2.2 Preliminaries of Fractional-Order Calculus

Definition 9.2 (*Caputo fractional derivate* [11]) Given a continuously differentiable function $f(t) : \mathbb{R} \to \mathbb{R}$, the fractional derivate based on Caputo's definition is adopted with the following form

$$
{}_{t_0}D_t^\alpha f(t) = \begin{cases} \frac{d^n f(t)}{dt^n}, & \alpha = n, \\ \frac{1}{\Gamma(\alpha-n)} \int_{t_0}^t \frac{f^{(n)}(\tau)}{(t-\tau)^{\alpha-n+1}} d\tau, & n-1 < \alpha < n, \end{cases} \tag{9.6}
$$

where α denotes the order of fractional derivate, subject to $(n-1, n)$, and n is an integer. $\Gamma(\cdot)$ is the well-known Gamma function

$$
\Gamma(z) = \int_0^\infty \tau^{z-1} e^{-\tau} d\tau \tag{9.7}
$$

Definition 9.3 (*fractional integral* [11]) For the fractional integral, it has the typical unified expression as

$$
{}_{t_0}D_t^{-\alpha} f(t) = \frac{1}{\Gamma(\alpha)} \int_{t_0}^t \frac{f(\tau)}{(t-\tau)^{1-\alpha}} d\tau, \tag{9.8}
$$

where $0 < \alpha < 1$ indicates the fractional order.

Lemma 9.4 (Uniform ultimate boundedness [3]) *Consider a continuously differentiable system $\dot{x} = f(x, t)$, with its solution $x(t) \in \mathbb{R}^n, x(t_0) = x_0$, if there exists a set $\mathbb{W} \subset \mathbb{R}^n$ including the origin, which satisfies*

$$\|x(t_0)\| < \varepsilon \Rightarrow x(t) \in \mathbb{W}, \forall t > t_0 + t_w, \tag{9.9}$$

where $\varepsilon > 0$, $0 < t_w < \infty$, and $\mathbb{W}$ is not related to t_0. Then, the system can be judged to be uniformly ultimately bounded.

Remark 9.5 For some practical uncertain systems, the traditional asymptotic stability analysis method will not be applicable, due to the influence of the unknown disturbances. The concept of uniform ultimate boundedness is raised, to specifically analyze the stability of the uncertain systems.

Lemma 9.6 (Mittag-Leffler-based uniform ultimate boundedness) *Based on the Caputo definition, given a FO system ${}_{t_0}D_t^{\alpha} z(t) = \mathcal{F}(z(t)) + d(t)$, where $d(t)$ denotes the bounded disturbances and uncertainties. Its equilibrium point is $z(t_0) = 0$, and $\mathbb{V} \subset \mathbb{R}^n$ denotes the vicinity of origin. Suppose a Lyapunov function candidate $V(t, z(t)) : [0, \infty) \times \mathbb{V} \to \mathbb{R}$, is continuously differentiable and locally Lipschitz with respect to z such that*

$$\begin{aligned} &m_1\|z(t)\|^a \le V(t, z(t)) \le m_2\|z(t)\|^{ab} \\ &{}_{t_0}D_t^{\alpha} V(t, z(t)) \le -m_3\|z(t)\|^{ab} + \mathcal{G}(t), \end{aligned} \tag{9.10}$$

where $t \ge t_0$, $z(t) \in \mathbb{V}$, $\alpha \in (0, 1)$, $m_1, m_2, m_3 > 0$, $a, b > 0$. $\mathcal{G}(t)$ expresses a bounded term, whose bound may be state-related but must be time-independent. Then, there exist positive constants $\mathcal{D}$ and $t_1 < \infty$ satisfying that

$$\begin{aligned} &V(t, z(t)) < \mathcal{D}, \ \mathcal{D} > |\mathcal{G}(t)|, \\ &\|z(t)\| \le \left(\frac{\mathcal{D}}{m_1}\right)^{\frac{1}{a}}, \ t > t_1 \end{aligned} \tag{9.11}$$

then the Mittag-Leffler-based uniform ultimate boundedness of $V(t, z(t))$ and $z(t)$ can be ensured.

Remark 9.7 Lemma 9.6 is inspired by the concepts of Generalized Mittag–Leffler stability [9] and Lemma 9.4, which implies the theoretical stability of the FO uncertain system and is equivalent to the asymptotic stability. In practice, Mittag-Leffler-based uniform ultimate boundedness is an optimal stability that the FO uncertain system can achieve. Therefore, we employ this concept to analyze the theoretical stability of the FO control scheme.

9.3 Control Scheme

In this section, based on the concept of NTSM control, two pure-tension control schemes, i.e., the IONTSM controller and FONTSM controller, are proposed for the tether deployment of TSS. The former has finite-time stability and robustness to disturbances, while the latter is the improvement scheme with the aid of FO calculus, which successfully promotes dynamic performance and effectively deals with the adverse effect of external disturbances and tension saturation.

9.3.1 Integer-Order Nonsingular Terminal Sliding Mode Control

At first, thanks to the concept of NTSM control [5], a novel nonsingular terminal sliding equation is designed as

$$s_1 = -c_1 x_1 + c_2 x_2 + c_3 \operatorname{sig}(\dot{x}_2)^{p/q}, \tag{9.12}$$

where c_1, c_2, c_3 are positive constants, $\operatorname{sig}(\cdot) = |\cdot|^{\beta} \operatorname{sgn}(\cdot)$, and p, q express positive odd integers.

By the concept of equivalent control [14], we can obtain the equivalent control term of IONTSM controller via solving $\dot{s}_1 = 0$, in the absence of uncertainties and disturbances. Therefore, the following derivation can be constructed:

$$\begin{aligned} 0 &= -c_1 \dot{x}_1 + c_2 \dot{x}_2 + \frac{c_3 p}{q} |\dot{x}_2|^{p/q-1} \ddot{x}_2 \\ &= -c_1 \dot{x}_1 + c_2 \dot{x}_2 + \frac{c_3 p}{q} |\dot{x}_2|^{p/q-1} \left[G + u(t)\right] \\ &= \frac{c_3 p}{q} |\dot{x}_2|^{p/q-1} \left[\frac{-c_1 q \dot{x}_1}{c_3 p} |\dot{x}_2|^{1-p/q} + G + u(t) \right] \\ &\quad + \frac{c_3 p}{q} |\dot{x}_2|^{p/q-1} \left[\frac{c_2 q}{c_3 p} |\dot{x}_2|^{2-p/q} \operatorname{sgn}(\dot{x}_2) \right], \end{aligned} \tag{9.13}$$

which yields the equivalent control law as

$$u_{eq1} = \frac{c_1 q \dot{x}_1}{c_3 p} |\dot{x}_2|^{1-p/q} - \frac{c_2 q \operatorname{sgn}(\dot{x}_2)}{c_3 p} |\dot{x}_2|^{2-p/q} - G. \tag{9.14}$$

Hence the value of p/q should be limited within $(0, 1)$, to avoid the singularity problem of the controller.

Furthermore, to guarantee the finite-time convergence and the nonsingularity of the control scheme, a novel terminal-sliding-mode-type switching control law is proposed as

$$u_{sw1} = -k_0\left[(k_1|s_1|^{\gamma_1} + \mathcal{D}_1)\,\text{sgn}(s_1) + k_2 s_1^3\right], \tag{9.15}$$

where $k_0 = \frac{q}{c_3 p}(\dot{x}_2)^{1-p/q}$, $1 < \gamma_1 < 2$ and $k_1, k_2 > 0$.

Then, by combining the equivalent control term with switching control term, the control law of the IONTSM strategy can be expressed as

$$u_1(t) = u_{eq1} + u_{sw1}. \tag{9.16}$$

Lemma 9.8 ([2]) *Consider a continuous-time autonomous system $\dot{z}(t) = f(z(t))$ including the origin, and a positive definite and continuously differentiable function $V(z(t)) : \mathbb{U} \to \mathbb{R}$. If there exist constants $i > 0$, $0 < j < 1$ satisfying the following inequality, the given autonomous system is finite-time stable.*

$$\dot{V}(z(t)) + iV^j(z(t)) \leq 0,\ \ z(t) \in \mathbb{U}. \tag{9.17}$$

Theorem 9.9 *Given the underactuated TSS model shown in (9.4), the sliding mode can be driven to the presented sliding manifold (9.12) by the proposed IONTSM law (9.16) within the finite reaching time t_r*

$$t_r \leq \frac{1}{2k_1(2-\gamma_1)} V_1^{2-\gamma_1}(s(0)) \tag{9.18}$$

Proof Construct the Lyapunov function candidate as $V_1(t) = \frac{1}{2}s_1^2$. To eliminate the singular term, a nonsingular gain is involved in the input channel of disturbance, i.e. $k_0 d(t)$. Then, with the IONTSM control input $u_1(t)$, the derivative of $V_1(t)$ with respect to t can be derived as

$$\begin{aligned}
\dot{V}_1(t) &= -\left[(k_1|s_1|^{\gamma_1} + \mathcal{D}_1)\,\text{sgn}(s_1) + k_2 s_1^3 - d(t)\right] s_1 \\
&= -k_1|s_1|^{\gamma_1}|s_1| - k_2 s_1^4 - \mathcal{D}_1|s_1| + d(t)s_1 \\
&\leq -k_1|s_1|^{2+\gamma_1-1} - k_2 s_1^4 - [\mathcal{D}_1 - |d(t)|]\,|s_1| \\
&\leq -2k_1 V_1^{\gamma_1-1}(t) - 4k_2 V_1^2(t) \\
&\leq -2k_1 V_1^{\gamma_1-1}(t) < 0.
\end{aligned} \tag{9.19}$$

According to Lemma 9.8, it is clear that the derivative of $V_1(t)$ meets the finite-time stable condition, then the proposed IONTSM control scheme is finite-time stable. The finite reaching time t_r is related to the initial sliding mode $s(0)$, which can be given as

$$t_r \leq \left.\frac{1}{i(1-j)} V_1^{1-j}(s(0))\right|_{i=2k_1,\, j=\gamma_1-1} \leq \frac{1}{2k_1(2-\gamma_1)} V_1^{2-\gamma_1}(s(0)). \tag{9.20}$$

The proof is thus completed. ∎

Based on Theorem 9.9, we can get that under the proposed IONTSM control scheme, the sliding mode can be attracted to the sliding manifold within the finite convergence time t_r, which indicates that $s_1 = -c_1x_1 + c_2x_2 + c_3\,\mathrm{sig}(\dot{x}_2)^{p/q} = 0, t \geq t_r$. By introducing the variable substitutions $\boldsymbol{X}_1 = \left[x_1\ x_2\right]^T, \boldsymbol{X}_2 = \left[\dot{x}_1\ \dot{x}_2\right]^T$, the presented sliding manifold can be transformed as

$$\begin{aligned} s_1 &= \boldsymbol{C}_1\boldsymbol{X}_1 + \boldsymbol{C}_2\,\mathrm{sig}(\boldsymbol{X}_2)^{p/q} \\ &= \boldsymbol{C}_1\boldsymbol{X}_1 + \boldsymbol{C}_2\,\mathrm{sig}(\dot{\boldsymbol{X}}_1)^{p/q} = 0, \end{aligned} \tag{9.21}$$

where $\boldsymbol{C}_1 = \left[-c_1\ c_2\right]$ and $\boldsymbol{C}_2 = \left[0\ c_3\right]$.

Remark 9.10 Using the above differential equations, the time of system states converging to the origin along the sliding manifold, can be expressed as $t_s = \frac{p}{p-q}\boldsymbol{C}_3|\boldsymbol{X}_1(t_r)|^{\frac{p-q}{p}}$ [18], where $\boldsymbol{C}_3 = (\boldsymbol{C}_2\boldsymbol{C}_1^R)^{-q/p}$, and $\boldsymbol{C}_1^R = \boldsymbol{C}_1^T(\boldsymbol{C}_1\boldsymbol{C}_1^T)^{-1}$ denotes the pseudo-inverse matrix. Therefore, we can conclude that under the IONTSM controller, the system states of TSS are finite-time stable, and can asymptotically converge to the equilibrium point within the finite time $(t_r + t_s)$.

9.3.2 *Fractional-Order Nonsingular Terminal Sliding Mode*

Due to the unique historical memory effect, the FO calculus is more appropriate to be involved in the tension control of TSS. Thus, based on Definitions 9.2 and 9.3, an improved FO nonsingular terminal sliding mode is constructed as

$$s_2 = -c_4x_1 + c_5x_2 + c_6D^{1-\alpha}\dot{x}_2 + c_7D^{-\alpha}\,\mathrm{sig}(\hat{x}_2)^{\beta}, \tag{9.22}$$

where $\hat{x}_2 = x_2 + \epsilon\dot{x}_2$, $c_4, c_5, c_6, c_7, \epsilon > 0$, $0 < \alpha, \beta < 1$, especially, β is the ratio of two positive odd numbers.

Based on [13], it is clear that there doesn't exist a finite-time stable equilibrium for the FO system. And if there exists a finite-time t_p satisfying $D^{\alpha}s_2 = 0,\ t \geq t_p$, then $\lim_{t\to\infty} s_2 = 0$ holds [10]. Therefore, we propose the following theorem to verify the Mittag-Leffler-based uniform ultimate boundedness of the sliding manifold.

Theorem 9.11 *Consider the FO sliding mode dynamics (9.22). If the following equality holds, the system states are Mittag-Leffler-based uniformly ultimately bounded, and their dimensionless trajectories converge to the equilibrium point.*

$$D^{\alpha}\left(-c_4x_1 + c_5x_2\right) + c_6\ddot{x}_2 + c_7\,\mathrm{sig}(\hat{x}_2)^{\beta} = 0. \tag{9.23}$$

Proof Define the new system state $\boldsymbol{X} = \left[x_1\ x_2\right]^T$, then the given equation can be rewritten as

$$\boldsymbol{K}_1D^{\alpha}\boldsymbol{X} + c_6\boldsymbol{K}_2\ddot{\boldsymbol{X}} + c_7\,\mathrm{sig}(\hat{\boldsymbol{X}})^{\beta} = 0, \tag{9.24}$$

where $\boldsymbol{K}_1 = \left[-c_4\ c_5\right]$, $\boldsymbol{K}_2 = \left[0\ 1\right]$ and $\hat{\boldsymbol{X}} = \boldsymbol{K}_2\boldsymbol{X} + \epsilon\boldsymbol{K}_2\dot{\boldsymbol{X}}$. Subsequently, we design the Lyapunov function candidate as $V_2(t) = |\boldsymbol{X}(t)|$, then its α-order derivative can be expressed as

$$\begin{aligned} D^\alpha V_2(t) &= \mathrm{sgn}(\boldsymbol{X})D^\alpha \boldsymbol{X} \\ &= \mathrm{sgn}(\boldsymbol{X})\left[-\boldsymbol{K}_3\ddot{\boldsymbol{X}} - c_7\,\mathrm{sig}(\hat{\boldsymbol{X}}_1)^\beta\right], \end{aligned} \tag{9.25}$$

where $\boldsymbol{K}_3 = c_6\boldsymbol{K}_2\boldsymbol{K}_1^R$, $\hat{\boldsymbol{X}}_1 = \boldsymbol{K}_2\boldsymbol{K}_1^R\boldsymbol{X} + \epsilon\boldsymbol{K}_2\boldsymbol{K}_1^R\dot{\boldsymbol{X}}$ and $\boldsymbol{K}_1^R = \boldsymbol{K}_1^T(\boldsymbol{K}_1\boldsymbol{K}_1^T)^{-1}$ indicates the pseudo-inverse matrix. In the following, we will consider the two cases respectively.

Case 1. $\hat{\boldsymbol{X}}_1 \geq 0$

$$\begin{aligned} D^\alpha V_2(t) &= \mathrm{sgn}(\boldsymbol{X})D^\alpha \boldsymbol{X} \\ &= \mathrm{sgn}(\boldsymbol{X})\left[-\boldsymbol{K}_3\ddot{\boldsymbol{X}} - c_7(\hat{\boldsymbol{X}}_1)^\beta\right] \\ &= -\boldsymbol{K}_3\ddot{\boldsymbol{X}}\,\mathrm{sgn}(\boldsymbol{X}) - c_7\left[\hat{\boldsymbol{X}}_1\,\mathrm{sgn}(\boldsymbol{X})\right]^\beta, \end{aligned} \tag{9.26}$$

where $\hat{\boldsymbol{X}}_1\,\mathrm{sgn}(\boldsymbol{X}) = \boldsymbol{K}_2\boldsymbol{K}_1^R|\boldsymbol{X}| + \epsilon\boldsymbol{K}_2\boldsymbol{K}_1^R\dot{\boldsymbol{X}}\,\mathrm{sgn}(\boldsymbol{X})$, then $D^\alpha V_2(t)$ can be derived as the form of $-\boldsymbol{K}_4|\boldsymbol{X}|^\beta - \Psi(\boldsymbol{X}, \dot{\boldsymbol{X}}, \ddot{\boldsymbol{X}}) = -\boldsymbol{K}_4 V_2^\beta(t) - \Psi(\boldsymbol{X}, \dot{\boldsymbol{X}}, \ddot{\boldsymbol{X}})$, where $\boldsymbol{K}_4 > 0$, $\Psi(\boldsymbol{X}, \dot{\boldsymbol{X}}, \ddot{\boldsymbol{X}})$ denotes the polynomial which contains the system state and its derivatives. Based on the background of practical engineering, we can confirm that the system state and its derivatives are all bounded in the actual system [16]. Referred to Lemma 9.6, it can be concluded that $V_2(t)$ and $\boldsymbol{X}$ are Mittag-Leffler-based uniformly ultimately bounded, under the proposed sliding mode dynamics.

Case 2. $\hat{\boldsymbol{X}}_1 < 0$

$$\begin{aligned} D^\alpha V_2(t) &= \mathrm{sgn}(\boldsymbol{X})D^\alpha \boldsymbol{X} \\ &= \mathrm{sgn}(\boldsymbol{X})\left[-\boldsymbol{K}_3\ddot{\boldsymbol{X}} + c_7(-\hat{\boldsymbol{X}}_1)^\beta\right] \\ &= -\boldsymbol{K}_3\ddot{\boldsymbol{X}}\,\mathrm{sgn}(\boldsymbol{X}) + c_7\left[-\hat{\boldsymbol{X}}_1\,\mathrm{sgn}(\boldsymbol{X})\right]^\beta. \end{aligned} \tag{9.27}$$

Similarly, in this case, we can obtain the same form of $D^\alpha V_2(t)$, i.e., $-\boldsymbol{K}_4 V_2^\beta(t) - \Psi(\boldsymbol{X}, \dot{\boldsymbol{X}}, \ddot{\boldsymbol{X}})$. Hence the same conclusion of the Mittag-Leffler-based uniform ultimate boundedness for $V_2(t)$ and $\boldsymbol{X}$ can be received, based on Lemma 9.6. The proof is thus concluded. ■

Then, to ensure the sliding motion can converge to the presented sliding manifold, the disturbance observer-based FO terminal sliding mode tension control law is proposed below

$$
\begin{aligned}
u_2(t) = & -\frac{1}{c_6}\left[D^\alpha\left(-c_4 x_1 + c_5 x_2\right) + c_7 \operatorname{sig}(\hat{x}_2)^\beta\right] \\
& -\frac{1}{c_6}\left[k_3 \operatorname{sig}(s_2)^{\gamma_2} + k_4 s_2\right] - \eta(t) - G,
\end{aligned}
\tag{9.28}
$$

where $k_3, k_4 > 0$, $\eta(t)$ is a FO disturbance observer for the TSS model (9.4), which is designed to suppress the influence of the external disturbances on the controller's stability.

$$
\begin{aligned}
& \eta(t) = D^{1-\alpha} z + k_5 D^{1-\alpha} \dot{x}_2 \\
& \dot{z} = -k_5 D^{1-\alpha} z - k_5\left[G + u_2(t) + k_5 D^{1-\alpha} \dot{x}_2\right],
\end{aligned}
\tag{9.29}
$$

where $k_5 > 0$, z is an auxiliary variable to calculate the disturbance observer $\eta(t)$.

Lemma 9.12 (Young's inequality) *Suppose $a, b \in \mathbb{R}$, and subject to $a \geq, b \geq 0$, there will exist constants $p > 1, q > 1$ that satisfy the following inequality:*

$$
ab \leq \frac{a^p}{p} + \frac{b^q}{q}, \ s.t. \frac{1}{p} + \frac{1}{q} = 1. \tag{9.30}
$$

If and only if $a^p = b^q$, the equals sign holds.

Lemma 9.13 (Fractional derivative of Lyapunov function [1]) *Given a positive definite Lyapunov function $V(t) = z(t)^2$, its fractional derivative under the Leibniz rule of fractional differentiation can be expressed as:*

$$
\begin{aligned}
D^\alpha V(t) & = z(t) D^\alpha z(t) + \Sigma(t) \\
& \leq z(t) D^\alpha z(t) + |\Sigma(t)| \\
& \leq z(t) D^\alpha z(t) + \rho|z(t)|,
\end{aligned}
\tag{9.31}
$$

where ρ denotes a positive constant, and

$$
\Sigma(t) = \sum_{i=1}^{\infty} \frac{\Gamma(1+\alpha)}{\Gamma(1+i)\Gamma(1-i+\alpha)} D^i z(t) D^{\alpha-i} z(t). \tag{9.32}
$$

Theorem 9.14 *Consider the external disturbances of TSS (9.4), it can be effectively observed and estimated by the proposed FO disturbance observer (9.29). Furthermore, the disturbance observer $\eta(t)$ and its observation error $\hat{\eta}(t)$ can be proved as Mittag-Leffler-based uniformly ultimately bounded.*

Proof Firstly, define the observation error as $\hat{\eta}(t) = \eta(t) - d(t)$. According to (9.29), the α-order derivative of $\hat{\eta}(t)$ along t yields

$$
\begin{aligned}
D^{\alpha}\hat{\eta}(t) &= D^{\alpha}\eta(t) - D^{\alpha}d(t) \\
&= \dot{z} + k_5\ddot{x}_2 - D^{\alpha}d(t) \\
&= -k_5 D^{1-\alpha}z - k_5\left[G + u_2(t) + k_5 D^{1-\alpha}\dot{x}_2\right] \\
&\quad + k_5\left[G + u_2(t) + d(t)\right] - D^{\alpha}d(t) \\
&= -k_5\eta(t) + k_5 d(t) - D^{\alpha}d(t) \\
&= -k_5\hat{\eta}(t) - D^{\alpha}d(t).
\end{aligned} \tag{9.33}
$$

Then, construct the Lyapunov function candidate of the FO disturbance observer as $V_3(t) = \hat{\eta}(t)^2$, based on Lemmas 9.12 and 9.13, whose α-order derivative can be obtained as

$$
\begin{aligned}
D^{\alpha}V_3(t) &= \hat{\eta}(t)D^{\alpha}\hat{\eta}(t) + \sum_{i=1}^{\infty}\frac{\Gamma(1+\alpha)D^{i}\hat{\eta}(t)D^{\alpha-i}\hat{\eta}(t)}{\Gamma(1+i)\Gamma(1-i+\alpha)} \\
&\le \left[-k_5\hat{\eta}(t) - D^{\alpha}d(t)\right]\hat{\eta}(t) + \rho_1|\hat{\eta}(t)| \\
&\le -k_5\hat{\eta}(t)^2 - D^{\alpha}d(t)\hat{\eta}(t) + \rho_1|\hat{\eta}(t)| \\
&\le -k_5\hat{\eta}(t)^2 + \frac{1}{2}\left[D^{\alpha}d(t)\right]^2 + \hat{\eta}(t)^2 + \frac{1}{2}\rho_1^2 \\
&\le -(k_5 - 1)\,\hat{\eta}(t)^2 + \frac{1}{2}\left[D^{\alpha}d(t)\right]^2 + \frac{1}{2}\rho_1^2 \\
&\le -(k_5 - 1)\,V_3(t) + \frac{1}{2}\left[D^{\alpha}d(t)\right]^2 + \frac{1}{2}\rho_1^2.
\end{aligned} \tag{9.34}
$$

According to Lemma 9.6, we can conclude that $V_3(t)$ and $\hat{\eta}(t)$ are Mittag-Leffler-based uniformly ultimately bounded, due to that $D^{\alpha}d(t)$ is bounded with $\mathcal{D}_2$. Then, since $|d(t)| \le \mathcal{D}_1$ and $\eta(t) = \hat{\eta}(t) + d(t)$, the Mittag-Leffler-based uniform ultimate boundedness of the FO disturbance observer can also be ensured. Thus, by choosing the appropriate parameter $k_5 > 1$, there will exist positive constants $\mathcal{D}_3$ and $t_d < \infty$ fulfilling that

$$
\begin{aligned}
&V_3(t) < \mathcal{D}_3,\ |\hat{\eta}(t)| < \sqrt{\mathcal{D}_3}, \\
&\mathcal{D}_3 > \frac{1}{2}\left[D^{\alpha}d(t)\right]^2 + \frac{1}{2}\rho_1^2,\ \forall t > t_d.
\end{aligned} \tag{9.35}
$$

This completes the proof. ■

In the pure-tension control scheme, the unique control effect to govern the deployment mission is the tether tension generated by the reel type deployer [15]. It means that the pulling force along the space tether is the only driving force, which should be kept within a predefined range to meet the basic assumptions of TSS. On this condition, the tension saturation should be taken into account in the control law, to obtain a preferable deployment performance for TSS.

In this work, we employ an adaptive saturation compensator to compensate for the influence of tension saturation for the TSS, then the underactuated TSS model is redefined as

$$\begin{aligned} \ddot{x}_1 &= F(x_1, \dot{x}_1, x_2, \dot{x}_2), \\ \ddot{x}_2 &= G(x_1, \dot{x}_1, x_2, \dot{x}_2) + \mathrm{sat}(u_3(t)) + d(t). \end{aligned} \tag{9.36}$$

Due to $u(t) = -\hat{T}$, the input saturation of the tension control scheme can be expressed as follows

$$\mathrm{sat}(u_3(t)) = \begin{cases} u_{\max}, & u_3(t) \geq u_{\max}, \\ u_3(t), & u_{\min} < u_3(t) < u_{\max}, \\ u_{\min}, & u_3(t) \leq u_{\min}, \end{cases} \tag{9.37}$$

where $\mathrm{sat}(u_3(t))$ implies the saturated control effect, $u_3(t) = u_2(t) + u_a(t)$. $u_{\min} < u_{\max} < 0$ is designed to fulfill the predefined range of tether tension. $u_a(t) = k_7\xi(t)$ denotes the adaptive compensation law, and $\xi(t)$ is the adaptive variable:

$$D^\alpha \xi(t) = \begin{cases} 0, & |\xi| < \delta, \\ -k_6\xi - \frac{c_6 s_2}{\xi}\Delta u - \frac{1}{2\xi}\Delta u^2 + \Delta u, & |\xi| \geq \delta, \end{cases} \tag{9.38}$$

where $\Delta u = \mathrm{sat}(u_3(t)) - u_3(t)$ expresses the error between saturated control input and ideal control effect.

Theorem 9.15 *With the proposed FO adaptive law (9.38), its Mittag-Leffler-based uniform ultimate boundedness can be ensured, yielding that the error of saturated control input and ideal control effect is also Mittag-Leffler-based uniformly ultimately bounded.*

Proof Design the Lyapunov function candidate of the adaptive law as $V_4(t) = \xi^2(t)$. In the light of the Young's inequality, the following cases can be derived.

Case 1. $|\xi| < \delta \Rightarrow D^\alpha \xi = 0$

$$\begin{aligned} D^\alpha V_4(t) &= \xi D^\alpha \xi + \sum_{i=1}^{\infty} \frac{\Gamma(1+\alpha) D^i \xi D^{\alpha-i}\xi}{\Gamma(1+i)\Gamma(1-i+\alpha)} \\ &\leq \xi D^\alpha \xi + \rho_2 |\xi| \\ &\leq \rho_2 \delta. \end{aligned} \tag{9.39}$$

Case 2. $|\xi| \geq \delta \Rightarrow D^\alpha \xi = -k_6\xi - \frac{c_6 s_2}{\xi}\Delta u - \frac{1}{2\xi}\Delta u^2 + \Delta u$

$$
\begin{aligned}
D^{\alpha} V_4(t) &= \xi D^{\alpha}\xi + \sum_{i=1}^{\infty} \frac{\Gamma(1+\alpha) D^i \xi D^{\alpha-i}\xi}{\Gamma(1+i)\Gamma(1-i+\alpha)} \\
&\leq -k_6\xi^2 - c_6 s_2 \Delta u - \frac{1}{2}\Delta u^2 + \xi \Delta u + \rho_2|\xi| \\
&\leq (1-k_6)\xi^2 + \frac{1}{2}c_6^2 s_2^2 + \frac{1}{2}\Delta u^2 + \frac{1}{2}\rho_2^2 \\
&\leq (1-k_6)V_4(t) + \frac{1}{2}c_6^2 s_2^2 + \frac{1}{2}\Delta u^2 + \frac{1}{2}\rho_2^2 \\
\overset{\lim_{s_2\to 0}}{\Longrightarrow} &\leq (1-k_6)V_4(t) + \frac{1}{2}\Delta u^2 + \frac{1}{2}\rho_2^2.
\end{aligned} \tag{9.40}
$$

According to (9.37), the tether tension will eventually converge to u_{min} to maintain minimum tension keeping the space tether straight. Therefore, in Case 2, it is clear that $\frac{1}{2}\Delta u^2 + \frac{1}{2}\rho_2^2$ is bounded, such that V_4 and ξ are Mittag-Leffler-based uniformly ultimately bounded, referred to Lemma 9.6. Then, if $k_6 > 1$, there exists a positive constant $t_a < \infty$ satisfying that $|\xi| < \sqrt{\frac{1}{2}\Delta u^2 + \frac{1}{2}\rho_2^2}$, $t > t_a$.

And based on (9.38) and Case 1, we can get that $\xi(t)$ will converge into the terminal set $\Theta_\xi = \{\xi \,|0 < |\xi| < \delta\}$, which indicates that the convergence boundary of adaptive law subjects to $|\xi| < \sqrt{\frac{1}{2}\Delta u^2 + \frac{1}{2}\rho_2^2} \leq \delta$, yielding that $\frac{1}{2}\Delta u^2 < \delta \Rightarrow |\Delta u| < \sqrt{2\delta}$, $t > t_a$. This implies that the proposed adaptive law can ensure the Mittag-Leffler-based uniform ultimate boundedness of the error between saturated control input and ideal control effect. This concludes the proof. ■

Theorem 9.16 *Under the proposed adaptive compensation based FONTSM control law* sat$(u_3(t))$, *abbreviating as FOANTSM control law, sliding mode dynamics (9.22) can be verified as Mittag-Leffler-based uniformly ultimately bounded, with the finite convergence time.*

Proof Define the Lyapunov function candidate with the adaptive law as $V_5(t) = s_2^2(t) + \hat{\eta}(t)^2 + \xi^2(t)$. Under the control input $u_3(t)$, we can obtain that

$$
\begin{aligned}
s_2 D^{\alpha} s_2 &= s_2\left[D^{\alpha}\left(-c_4 x_1 + c_5 x_2\right) + c_6 \ddot{x}_2 + c_7 \,\mathrm{sig}(\hat{x}_2)^{\beta}\right] \\
&= s_2\left[D^{\alpha}\left(-c_4 x_1 + c_5 x_2\right) + c_7 \,\mathrm{sig}(\hat{x}_2)^{\beta}\right] \\
&\quad + s_2 c_6\left[u_2(t) + u_a(t) + \Delta u + d(t) + G\right] \\
&= s_2\left[-k_3 \,\mathrm{sig}(s_2)^{\gamma_2} - k_4 s_2 + c_6\Delta u - \kappa\xi - c_6\hat{\eta}(t)\right] \\
&= -k_3|s_2|^{\gamma_2+1} - k_4 s_2^2 - \kappa\xi s_2 - c_6\left(\hat{\eta}(t) - \Delta u\right)s_2,
\end{aligned} \tag{9.41}
$$

where $\kappa = c_6 k_7$. Then, in view of (9.34) and (9.38), the α-order derivative of $V_5(t)$ can be decomposed into two cases below.

Case 1. $|\xi| < \delta$

$$
\begin{aligned}
D^\alpha V_5(t) \leq\; & s_2 D^\alpha s_2 + \hat{\eta}(t) D^\alpha \hat{\eta}(t) + \xi D^\alpha \xi \\
& + \rho_3 |s_2| + \rho_1 |\hat{\eta}(t)| + \rho_2 |\xi| \\
\leq\; & -k_3 |s_2|^{\gamma_2+1} - k_4 s_2^2 - \kappa \xi s_2 - c_6 \left(\hat{\eta}(t) - \Delta u\right) s_2 \\
& + \rho_3 |s_2| - (k_5 - 1)\, \hat{\eta}(t)^2 + \frac{1}{2}\mathcal{D}_2^2 + \frac{1}{2}\rho_1^2 + \rho_2 \delta \\
\leq\; & -k_4 s_2^2 - \kappa \delta s_2 + \frac{1}{2}\hat{\eta}(t)^2 + c_6^2 s_2^2 + \frac{1}{2}\Delta u^2 + \frac{1}{2} s_2^2 \\
& - (k_5 - 1)\, \hat{\eta}(t)^2 + \frac{1}{2}\left(\rho_1^2 + \rho_2^2 + \rho_3^2 + \mathcal{D}_2^2 + \delta^2\right) \\
\leq\; & -\left(k_4 - 1 - c_6^2\right) s_2^2 - \left(k_5 - \frac{3}{2}\right) \hat{\eta}(t)^2 \\
& + \frac{1}{2}\left[\rho_1^2 + \rho_2^2 + \rho_3^2 + \mathcal{D}_2^2 + \Delta u^2 + \left(\kappa^2 + 1\right)\delta^2\right] \\
\leq\; & -\mu_1 V_5(t) + \Phi_1,
\end{aligned}
\tag{9.42}
$$

where

$$
\begin{aligned}
\mu_1 &= \min\left\{k_4 - 1 - c_6^2, k_5 - \frac{3}{2}\right\} > 0 \\
\Phi_1 &= \frac{1}{2}\left[\sum_{i=1}^{3} \rho_i^2 + \mathcal{D}_2^2 + \Delta u^2 + \left(2\mu_1 + \kappa^2 + 1\right)\delta^2\right].
\end{aligned}
\tag{9.43}
$$

Case 2. $|\xi| \geq \delta$

$$
\begin{aligned}
D^\alpha V_5(t) \leq\; & s_2 D^\alpha s_2 + \hat{\eta}(t) D^\alpha \hat{\eta}(t) + \xi D^\alpha \xi \\
& + \rho_3 |s_2| + \rho_1 |\hat{\eta}(t)| + \rho_2 |\xi| \\
\leq\; & -k_3 |s_2|^{\gamma_2+1} - k_4 s_2^2 - \kappa \xi s_2 - c_6 \left(\hat{\eta}(t) - \Delta u\right) s_2 \\
& + \rho_3 |s_2| - (k_5 - 1)\, \hat{\eta}(t)^2 + \frac{1}{2}\mathcal{D}_2^2 + \frac{1}{2}\rho_1^2 \\
& - k_6 \xi^2 - c_6 s_2 \Delta u - \frac{1}{2}\Delta u^2 + \xi \Delta u + \rho_2 |\xi| \\
\leq\; & -k_4 s_2^2 + \frac{\kappa^2}{2}\xi^2 + \frac{1}{2} s_2^2 + \frac{1}{2}\hat{\eta}(t)^2 + \frac{1}{2} c_6^2 s_2^2 + \frac{1}{2} s_2^2 \\
& - (k_5 - 1)\, \hat{\eta}(t)^2 + \frac{1}{2}\mathcal{D}_2^2 + \frac{1}{2}\rho_1^2 + \frac{1}{2}\rho_3^2 \\
& - k_6 \xi^2 - \frac{1}{2}\Delta u^2 + \xi^2 + \frac{1}{2}\Delta u^2 + \frac{1}{2}\rho_2^2 \\
\leq\; & -\left(k_4 - 1 - \frac{1}{2} c_6^2\right) s_2^2 - \left(k_5 - \frac{3}{2}\right) \hat{\eta}(t)^2 \\
& - \left(k_6 - \frac{\kappa^2 + 2}{2}\right) \xi^2 + \frac{1}{2}\left(\rho_1^2 + \rho_2^2 + \rho_3^2 + \mathcal{D}_2^2\right) \\
\leq\; & -\mu_2 V_5(t) + \Phi_2,
\end{aligned}
\tag{9.44}
$$

where

$$\mu_2 = \min\left\{k_4 - 1 - \frac{1}{2}c_6^2, k_5 - \frac{3}{2}, k_6 - \frac{\kappa^2+2}{2}\right\} > 0,$$
$$\Phi_2 = \frac{1}{2}\left(\rho_1^2 + \rho_2^2 + \rho_3^2 + \mathcal{D}_2^2\right). \tag{9.45}$$

Therefore, we can conclude that $D^\alpha V_5(t) \leq -\mu V_5(t) + \Phi$, where

$$\mu = \min\{\mu_1, \mu_2\} > 0,$$
$$\Phi = \max\{\Phi_1, \Phi_2\}. \tag{9.46}$$

Referred to Lemma 9.6, there will exist positive constants $\mathcal{D}_4$ and $t_f < \infty$, satisfying that

$$V_5(t) < \mathcal{D}_4, \ \mathcal{D}_4 > \Phi,$$
$$|s_2| < \sqrt{\mathcal{D}_4}, \ \forall t > t_f, \tag{9.47}$$

which implies that $V_5(t)$ and s_2 can be both ensured as Mittag-Leffler-based uniformly ultimately bounded with the finite convergence time t_f, under the FOANTSM control strategy with the disturbance observer. Thus, the proof is completed. ■

Remark 9.17 Consequently, with the proposed FOANTSM control scheme, it can be summarized that both external disturbances and input saturation are well considered and handled. Furthermore, the state variables of TSS can be verified as Mittag-Leffler-based uniformly ultimately bounded, by combining the conclusions of Theorems 9.11 and 9.16.

9.4 Simulations

To validate the effectiveness of the proposed controllers for the deployment of TSS, numerical simulations and comparisons are implemented in this section. Referred to the YES2 phase 1 deployment [16], the physical parameters of TSS are given as follows.

The TSS has the two-body structure shown in Fig. 8.1, with the masses of the main satellite and subsatellite denoting as $m_1 = 6530\,\text{kg}$ and $m_2 = 12\,\text{kg}$ respectively, whose simulation mission is to deploy a space tether with the total length $l_{max} = 3.5\,\text{km}$. The orbital altitude and the orbital angular velocity of TSS are indicated as $R = 6631\,\text{km}$ and $\Omega = 1.17 \times 10^{-3}\,\text{rad/s}$. To avoid model singularity of TSS, a certain length of space tether is to be deployed by the automatic ejection in advance, which is adopted as $l_0 = 3\,\text{m}$ in this section. Based on these parameters, the tether deployment of TSS can be described by the dimensionless system states $(x_1, \dot{x}_1, x_2, \dot{x}_2)$, i.e., via driving the initial states $(0.1, 0, 9 \times 10^{-4}, 0.7)$ to the desired

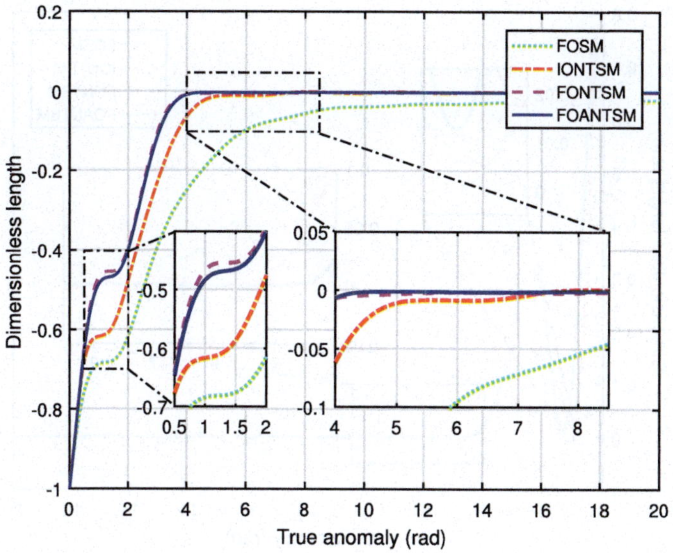

Fig. 9.1 Dimensionless length of tether

states (0, 0, 0, 0). It is noteworthy that in the nominal situation, the dimensionless tension of the TSS linearization model will converge to the static tension $\hat{T}_s = 3$. But when the time-variant disturbance is concerned, the dimensionless tension of the proposed controllers will be driven into the neighborhood of the static tension, in this work.

By the trial and error method, the parameters of the IONTSM manifold can be determined as $c_1 = 2.1$, $c_2 = 3.3$, $c_3 = 0.9$, $p/q = 7/9$, and $k_1 = 0.8$, $k_2 = 0.6$, $\gamma_1 = 1.2$ for the switching control law. For the FONTSM controller, according to the above Mittag-Leffler-based uniformly ultimately bounded analysis, the parameters of sliding mode can be selected as $c_4 = 2.7$, $c_5 = 2.8$, $c_6 = 1.35$, $c_7 = 1.8$, $\epsilon = 1.35$, $\alpha = 0.5$, $\beta = 79/99$, $k_3 = 5.4$, $k_4 = 2.83$, $\gamma_2 = 1.15$. Through the comparison of multi-group simulations, we choose the preferable parameters of the FO disturbance observer and the FO adaptive compensation law as $k_5 = 20$, $k_6 = 2.4$, $k_7 = 0.5$, $\delta = 0.01$, $\xi_0 = 10$. Additionally, the lumped disturbance is assumed as $d(t) = 5 \times 10^{-2} \sin(\tau)$.

Subsequently, the simulation results for the deployment of TSS are illustrated as Figs. 9.1, 9.2, 9.3, 9.4, 9.5 and 9.6, namely, the trajectories of tether length and its rate, in-plane angle, and its angular velocity, tether tension, and disturbance observer. In the simulation experiments, FOSM, IONTSM, FONTSM, and FOANTSM represent the control schemes of FO sliding mode, integer-order nonsingular terminal sliding mode, FO nonsingular terminal sliding mode, and FO nonsingular terminal sliding mode with adaptive compensation, respectively.

Figures 9.1 and 9.2 depict the dynamics trajectories of tether length for the deployment. The proposed NTSM controllers have significant advantages of deployment

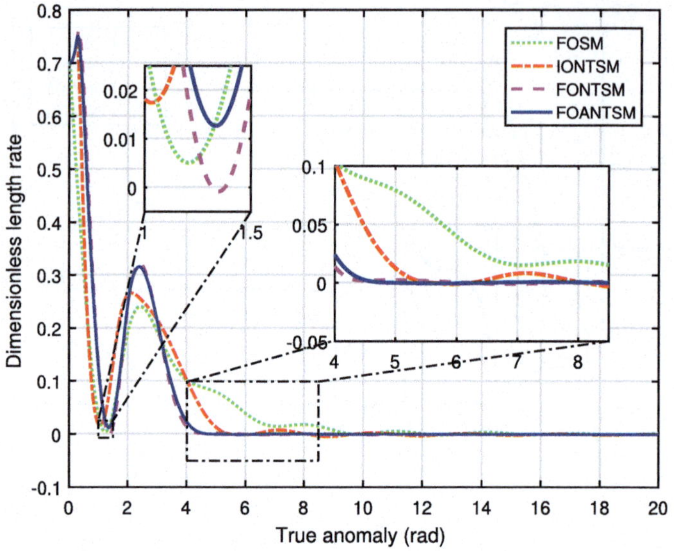

Fig. 9.2 Dimensionless length rate of tether

speed and convergence ability, compared with the FOSM controller [7]. In the steady phase, the length trajectory of the FOSM controller has a significant steady-state error, while proposed NTSM controllers can effectively decrease it and the FOANTSM controller has the best performance. With the help of the FO calculus, FONTSM and FOANTSM controllers accelerate the deployment process by about 3 rad, and eliminate the steady-state error of tether length (see Fig. 9.1). From the enlarged area in Fig. 9.2, we can find that the FOANTSM controller avoids the abnormal performance of negative tether length rate in the range of (1.31, 1.37) rad for the FONTSM controller, owing to the compensation effect of FO adaptive law, which can be also demonstrated by the enlarged area in Fig. 9.1. In conclusion, in terms of length dynamics, a faster and more stable tether length deployment can be achieved by the FOANTSM controller.

Figures 9.3 and 9.4 illustrate the dynamics trajectories of in-plane angle governed by the controllers of FOSM, IONTSM, FONTSM and FOANTSM. From the simulation results, the following conclusions can be drawn for the deployment of TSS. Firstly, thanks to the finite-time stability of terminal sliding mode [17], the proposed NTSM schemes possess better deployment performance of in-plane angle dynamics, compared to the longer convergence time of about 8.3 rad and more fluctuations in the steady phase of FOSM controller. Secondly, the proposed FO controllers effectively improve the dynamic and steady-state performances of TSS. And a faster convergence and more stable deployment can be guaranteed (see Fig. 9.3). Finally, in the light of FO adaptive compensation, the FOANTSM controller reduces the overshoots in the dynamic phase and stabilizes the fluctuations in the steady phase, contrasted to the FONTSM controller (see Fig. 9.4).

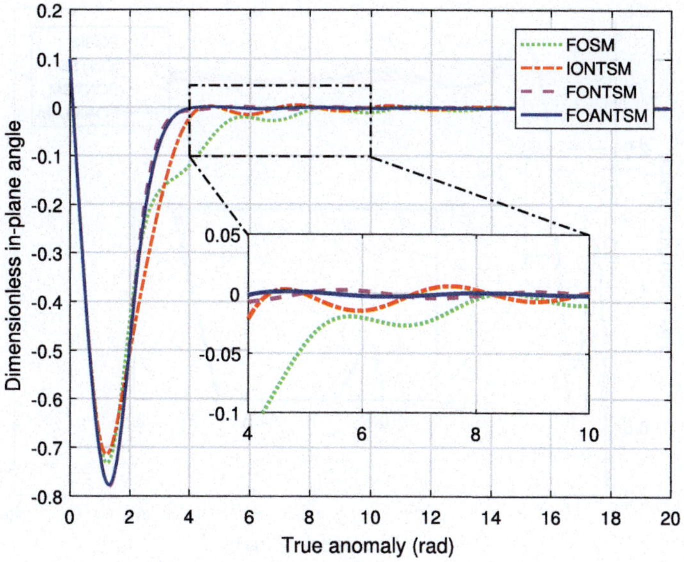

Fig. 9.3 Dimensionless angle of in-plane libration

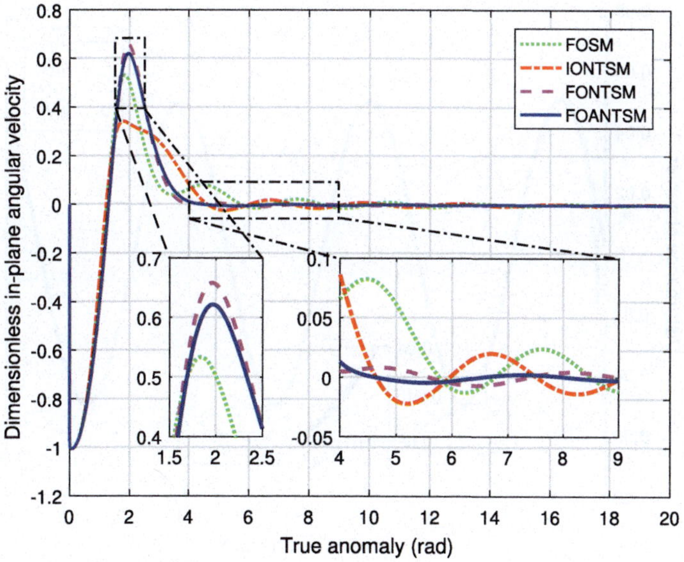

Fig. 9.4 Dimensionless angular velocity of in-plane libration

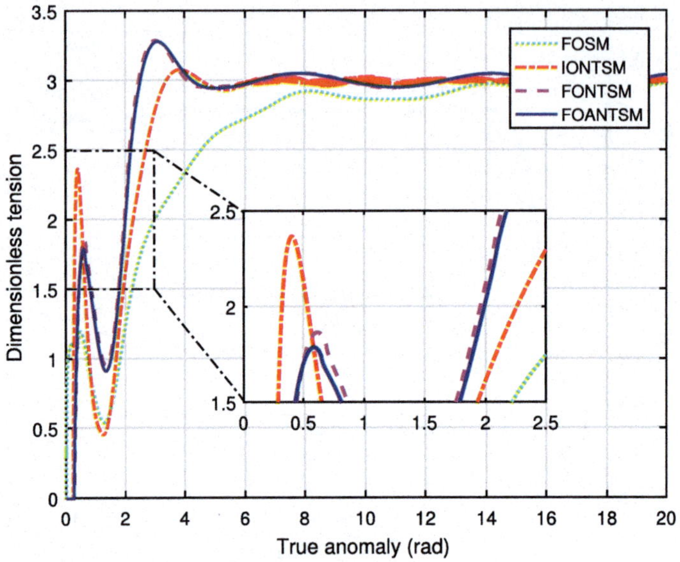

Fig. 9.5 Dimensionless tension in tether

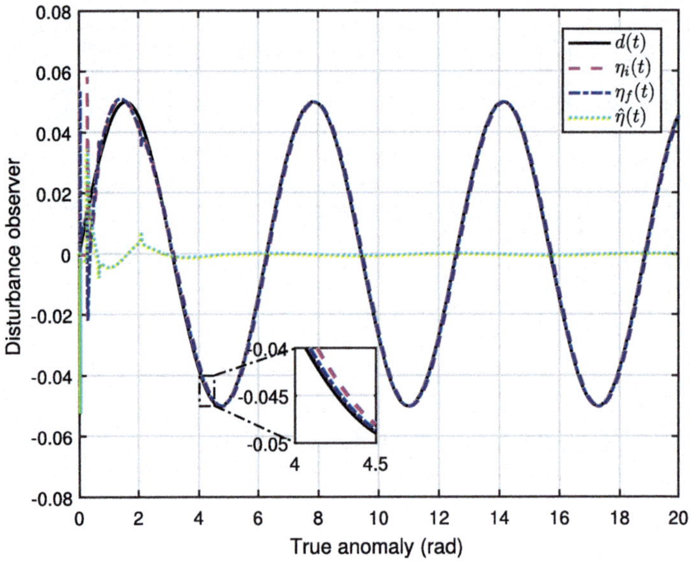

Fig. 9.6 Disturbance and FO disturbance observer

Figure 9.5 presents the trajectories of dimensionless tether tension for the deployment under the various control schemes above. It is shown by Fig. 9.5 that the tension trajectory of the FOSM controller has a longer regulation time of about 14 rad, while the proposed NTSM controllers possess an obvious improvement effect on the tension regulation, about 10 rad ahead of schedule. According to the enlarged area in Fig. 9.5, we can see that in the dynamic phase, the FONTSM controller effectively reduces the overshoots of tension compared with the IONTSM controller, which fully implies the effect of the FO control law and corresponds to the above conclusions in tether length and in-plane angle. Similarly, the FO adaptive compensation law is also helpful in regulating the tether tension with smaller overshoots in the dynamic phase (same conclusions see Figs. 9.2 and 9.4). To further analyze the effectiveness of the proposed FO control strategy, the trajectories of disturbance and FO disturbance observer are plotted in the Fig. 9.6, where $d(t)$, $\eta_i(t)$, $\eta_f(t)$, $\hat{\eta}(t)$ denote the lumped disturbance, the integer-order observer [4], the proposed FO observer and its observation error, respectively. Based on the enlarged area shown in Fig. 9.6, it is clear that the proposed FO observer has a better tracking effect on the time-varying disturbance, by comparison with the integer-order observer. From the trajectory of the $\hat{\eta}(t)$, it can be concluded that after 4.5 rad, the FO observer's error in observation will be in a preferable neighborhood of the desired value, which accords with the stability analysis and verifies the effectiveness of the proposed control scheme.

9.5 Conclusion

This chapter has proposed a novel pure-tension control scheme for the deployment of TSS, via a FO nonsingular terminal sliding mode controller. The underactuated dynamics, external disturbances, and input saturation of TSS are fully considered and handled. The key contribution of this work is the innovative design of a FO nonsingular terminal sliding manifold for tether deployment, which avoids the singularity problem of terminal sliding mode and improves the dynamic performance of the integer-order method. Furthermore, we develop a FO disturbance observer to effectively deal with the time-variant disturbance, whose superiority to the conventional observer is verified by the simulations. Meanwhile, a FO adaptive law is raised to compensate for the influence of the saturated tension, which tackles the abnormal length rate of the proposed FO controller. Finally, by the convincing simulations for the deployment of TSS, the effectiveness and superiority of the proposed control schemes are demonstrated.

References

1. Aghababa, M.P.: A Lyapunov-based control scheme for robust stabilization of fractional chaotic systems. Nonlinear Dyn. **78**(3), 2129–2140 (2014)
2. Bhat, S.P., Bernstein, D.S.: Finite-time stability of continuous autonomous systems. SIAM J. Control. Optim. **38**(3), 751–766 (2000)
3. Corless, M., Leitmann, G.: Continuous state feedback guaranteeing uniform ultimate boundedness for uncertain dynamic systems. IEEE Trans. Autom. Control. **26**(5), 1139–1144 (1981)
4. Du, J., Hu, X., Krstić, M., Sun, Y.: Robust dynamic positioning of ships with disturbances under input saturation. Automatica **73**, 207–214 (2016)
5. Feng, Y., Yu, X., Man, Z.: Non-singular terminal sliding mode control of rigid manipulators. Automatica **38**(12), 2159–2167 (2002)
6. Inamori, T., Kawashima, R., Saisutjarit, P., Sako, N., Ohsaki, H.: Magnetic plasma deorbit system for nano-and micro-satellites using magnetic torquer interference with space plasma in low earth orbit. Acta Astronaut. **112**, 192–199 (2015)
7. Kang, J., Zhu, Z.H., Wang, W., Li, A., Wang, C.: Fractional order sliding mode control for tethered satellite deployment with disturbances. Adv. Space Res. **59**(1), 263–273 (2017)
8. Li, X., Sun, G., Shao, X.: Discrete-time pure-tension sliding mode predictive control for the deployment of space tethered satellite with input saturation. Acta Astronaut. **170**, 521–529 (2020)
9. Li, Y., Chen, Y., Podlubny, I.: Mittag-Leffler stability of fractional order nonlinear dynamic systems. Automatica **45**(8), 1965–1969 (2009)
10. Pisano, A., Rapaić, M.R., Jeličić, Z.D., Usai, E.: Sliding mode control approaches to the robust regulation of linear multivariable fractional-order dynamics. Int. J. Robust Nonlinear Control. **20**(18), 2045–2056 (2010)
11. Podlubny, I.: Fractional Differential Equations. Academic, New York, NY, USA (1999)
12. Pradeep, S.: A new tension control law for deployment of tethered satellites. Mech. Res. Commun. **24**(3), 247–254 (1997)
13. Shen, J., Lam, J.: Non-existence of finite-time stable equilibria in fractional-order nonlinear systems. Automatica **50**(2), 547–551 (2014)
14. Utkin, V.I.: Sliding mode control design principles and applications to electric drives. IEEE Trans. Ind. Electron. **40**(1), 23–36 (1993)
15. Williams, P.: A review of space tether technology. Recent. Pat. Space Technol. **2**(1), 22–36 (2012)
16. Williams, P., Hyslop, A., Stelzer, M., Kruijff, M.: YES2 optimal trajectories in presence of eccentricity and aerodynamic drag. Acta Astronaut. **64**(7–8), 745–769 (2009)
17. Yu, S., Yu, X., Shirinzadeh, B., Man, Z.: Continuous finite-time control for robotic manipulators with terminal sliding mode. Automatica **41**(11), 1957–1964 (2005)
18. Zheng, J., Wang, H., Man, Z., Jin, J., Fu, M.: Robust motion control of a linear motor positioner using fast nonsingular terminal sliding mode. IEEE/ASME Trans. Mechatron. **20**(4), 1743–1752 (2014)

Chapter 10
Fractional-Order SMC for the Deployment of STS

10.1 Introduction

The integral operator can suppress the steady-state error and has enabled a wide scope of application areas including control in robotics, electric drives, etc. [10]. In addition, FO calculus can be introduced to enhance the flexibility of controller design to make their constructs more accurate. Due to the nature of FO calculus, the FO sliding surface specified for integer-order systems will lead to some residual errors in stability analysis and the boundedness of FO systems with uncertainties has not been well analyzed.

Motivated by the advantages of FO calculus and based on previous works, to deal with the underactuated dynamics and constrained tension, a FO SMC for the deployment of tethered spacecraft systems is proposed. For more accurate dynamics, a novel nonlinear fractional integral sliding surface is presented to establish a reduced stable FO system due to FO uniform ultimate boundedness that is discussed in this chapter for the first time, and the corresponding disturbance observers are designed by using FO calculus to eliminate the external disturbances while the FO adaptive law works to remove the adverse effect in stability analysis. The novelty of this work is to allow an adjustable fractional order as a parameter to tune in the control system for seeking a better dynamic and steady-state performance. Moreover, the FO sliding mode controller is rigorous nonlinear, consisting of fractional derivate and integral.

10.2 Dynamics and Control

In the light of Lagrangian mechanics, the deployment dynamics of TSS can be described by [12]

G. Sun et al., *Fractional-Order Sliding Mode Control: Methodologies and Applications*, Studies in Systems, Decision and Control 532,
https://doi.org/10.1007/978-3-031-60847-6_10

$$
\begin{aligned}
&\theta'' + 2\left(\theta' + \Omega\right) l'/l + 3\Omega^2 \sin\theta \cos\theta = Q_\theta/(\bar{m}l^2), \\
&l'' - l\left[\left(\theta' + \Omega\right)^2 + \Omega^2\left(3\cos^2\theta - 1\right)\right] = -\tau/\bar{m} + d,
\end{aligned}
\tag{10.1}
$$

where τ denotes the tension acting along the space tether, and the generalized force Q_θ is related to the phase of θ and is regarded as zero as the tension is the only force to regulate the in-plane dynamics. To maintain the space tether taut, the condition $0 < \tau_{min} \leq \tau \leq \tau_{max} < \infty$ should be ensured during the entire liberation, that is, τ_{max} averts the breakage of space tether and $\tau_{min} > 0$ precludes a slack tether. The derivative with respect to time is represented by $()'$, and d represents the disturbance. $\bar{m}$ is the relative system mass related to the mother spacecraft and the subspacecraft, and Ω means the orbital angular speed. l and θ are referred to as the space tether length and the in-plane angle, respectively, and these two states together depict the in-plane dynamics [11].

Introducing the dimensionless conversion (10.2) into (10.1) obtains the normalized formulation of deployment dynamics (10.3).

$$
\begin{aligned}
&\lambda = l/L, \\
&d(\cdot)/dt = \Omega d(\cdot)/dv, \\
&\hat{\tau} = \tau/(\bar{m}\Omega^2 L), \\
&d_2 = d/(\bar{m}\Omega^2 L),
\end{aligned}
\tag{10.2}
$$

where L is the desired length of space tether, and $v = \Omega t$ means the dimensionless time.

$$
\begin{aligned}
&\ddot{\theta} + 2\left(\dot{\lambda}/\lambda\right)\left(1 + \dot{\theta}\right) + 3\cos\theta \sin\theta = 0, \\
&\ddot{\lambda} - \lambda\left[\left(1 + \dot{\theta}\right)^2 - 1 + 3\cos^2\theta\right] = -\hat{T} + d_2.
\end{aligned}
\tag{10.3}
$$

It is noted that the dimensionless space tether length is positive. It is worth pointing out that the system (10.3) is singular before the beginning of the deployment since $\lambda = 0$, and to make the deployment model physically meaningful, the dimensionless space tether length λ is artificially initialized with a small positive scalar. In reality, there indeed exists a certain initial distance between the two spacecraft to avert colliding.

10.2.1 Integral Sliding Mode Controller Design

To complete the deployment of subspacecraft, it is required that the space tether be unreeled as the desired length along the local vertical direction, and the desired condition can hence be determined: $\theta_d = \dot{\theta}_d = 0$, $\lambda_d = 1$ and $\dot{\lambda}_d = 0$. For the sake of controller design, it is essential to turn the system (10.3) into an error formulation by

$$e_1 = \theta - \theta_d, \quad e_2 = \dot{\theta} - \dot{\theta}_d, \quad e_3 = \lambda - \lambda_d, \quad e_4 = \dot{\lambda} - \dot{\lambda}_d, \tag{10.4}$$

and substituting (10.4) into system (10.3) yields

$$\begin{aligned} \dot{e}_1 &= e_2, \\ \dot{e}_2 &= f_1(e_1, e_2, e_3, e_4) + d_1, \\ \dot{e}_3 &= e_4, \\ \dot{e}_4 &= f_2(e_1, e_3, e_3, e_4) + u + d_2, , \end{aligned} \tag{10.5}$$

where

$$\begin{aligned} f_1(e_1, e_2, e_3, e_4) &= -\frac{2e_4(1+e_2)}{e_3+1} - 3\cos e_1 \sin e_1 - \varepsilon e_3, \\ f_2(e_1, e_3, e_3, e_4) &= (e_3+1)\left[(1+e_2)^2 - 1 + 3\cos^2 e_1\right], \end{aligned} \tag{10.6}$$

and $d_1 = \varepsilon e_3$ is constructed as a virtual disturbance. Owing to the presence of the virtual disturbance, it is clear to prove that $f_1(0, 0, e_3, e_4)$ is specified to be asymptotically stable by combing with the third row of (10.5). According to the physical feature of the tension, the input in the system (10.5) is strictly constrained and is expressed by

$$u_c = u - \Delta u, \tag{10.7}$$

where u_c is the control command generated by the controller, while u is the actual input given out by the actuator. For the subsequent entire controller design, we have the following assumptions and lemmas for the system (10.5).

Assumption 10.1 To achieve a smooth deployment, the angular rate with $\dot{\theta} + 1 > 0$ should be satisfied. In this regard, $\partial f_1 / \partial e_4$ is invertible, while this assumption precludes the singularity of the subsequent controller design. The disturbances are assumed to be bounded with $\left|\dot{d}_1\right| \le \rho_1$, $\left|\dot{d}_2\right| \le \rho_2$, $|d_3| \le \rho_3$, $|d_4| \le \rho_4$ and $\rho_i > 0, i = 1, 2, 3, 4$.

Lemma 10.1 ([3]) *Consider a second-order system:*

$$\begin{aligned} \dot{x}_1 &= x_2, \\ \dot{x}_2 &= f(x_1, x_2, u) + d, \end{aligned} \tag{10.8}$$

where $f(x_1, x_2, u)$ is a known function, and d is bounded with $\left|\dot{d}\right| < C_d$. The construction of disturbance observer can achieve that the estimator error $\tilde{d} = (\hat{d} - d)$ exponentially converges to a ball centered at the origin, and the ball can be made arbitrarily small by a suitable selection of the parameters.

$$\begin{aligned} \hat{d} &= q + kx_2, \\ \dot{q} &= -kq - k(f(x_1, x_2, u) + kx_2), \end{aligned} \tag{10.9}$$

where k is a positive scalar.

Lemma 10.2 *Based on the above lemma and system (10.5), we can design the following observers to compensate for the negative effect of disturbances.*

$$\begin{aligned} \dot{q}_1 &= -k_1 q_1 - k_1(f_1 + k_1 e_2), \\ \hat{d}_1 &= q_1 + k_1 e_2, \end{aligned} \tag{10.10}$$

$$\begin{aligned} \dot{q}_2 &= -k_2 q_2 - k_2(f_2 + u + k_2 e_4), \\ \hat{d}_2 &= q_2 + k_2 e_4, \end{aligned} \tag{10.11}$$

where $k_1 > 0$, $k_2 > 0$, q_1 *and* q_2 *are auxiliary variables of the disturbance estimators* $\hat{d}_1$ *and* $\hat{d}_2$, *respectively.*

Lemma 10.3 ([2]) *To cope with the input constraint, the auxiliary dynamic system (10.12) is employed to guarantee a valid stability analysis.*

$$\dot{\eta} = \begin{cases} -k_3\eta - \dfrac{(\partial f_1/\partial e_4)s_i\Delta u + \frac{1}{2}(\Delta u)^2}{\eta^2}\eta + \Delta u, & |\eta| \geq \sigma \\ 0, & |\eta| < \sigma \end{cases}, \tag{10.12}$$

where $k_3 > 0$ *and* σ *is a small positive scalar.*

Regarding (10.7) and the external disturbance, a trivial integral sliding mode surface can be designed as

$$\begin{aligned} s &= c_1 e_1 + c_2 e_2 + f_1, \\ s_i &= s + \chi, \\ \dot{\chi} &= s. \end{aligned} \tag{10.13}$$

Based on (10.10), (10.11), (10.12) and (10.13), the following controller can be constructed to ensure that the system state errors are all uniformly ultimately bounded.

$$\begin{aligned} u_c = &-\left(\frac{\partial f_1}{\partial e_4}\right)^{-1}\left(s + c_1 e_2 + c_2 f_1 + \frac{\partial f_1}{\partial e_1}e_2 + \frac{\partial f_1}{\partial e_2}f_1 + \frac{\partial f_1}{\partial e_3}e_4 + \frac{\partial f_1}{\partial e_4}f_2\right) \\ &-\left(\frac{\partial f_1}{\partial e_4}\right)^{-1}k_4 s_i - \hat{d}_2 - \left(\frac{\partial f_1}{\partial e_4}\right)^{-1}\left(c_2 + \frac{\partial f_1}{\partial e_2}\right)\hat{d}_1 - k_5\left(\frac{\partial f_1}{\partial e_4}\right)^{-1}\eta. \end{aligned} \tag{10.14}$$

Define the Lyapunov function candidate as

$$V_i = \frac{1}{2}s_i^2 + \frac{1}{2}\tilde{d}_1^2 + \frac{1}{2}\tilde{d}_2^2 + \frac{1}{2}\eta^2. \tag{10.15}$$

The derivative of V_i is

$$\dot{V}_i = s_i(\dot{s} + s) + \tilde{d}_1\dot{\tilde{d}}_1 + \tilde{d}_2\dot{\tilde{d}}_2 + \eta\dot{\eta}. \tag{10.16}$$

In the light of (10.13), (10.14) and Young's inequality, one can obtain

$$\begin{aligned} s_i\dot{s}_i &= s_i\Bigg[-k_4 s_i - \frac{\partial f_1}{\partial e_4}\hat{d}_2 + \frac{\partial f_1}{\partial e_4}d_2 - \left(c_2 + \frac{\partial f_1}{\partial e_2}\right)\hat{d}_1 \\ &\quad + \left(c_2 + \frac{\partial f_1}{\partial e_2}\right)d_1 + \frac{\partial f_1}{\partial e_4}\Delta u - k_5\eta\Bigg] \\ &= -k_4 s_i^2 + \frac{\partial f_1}{\partial e_4}s_i\tilde{d}_2 + \left(c_2 + \frac{\partial f_1}{\partial e_2}\right)s_i\tilde{d}_1 + \frac{\partial f_1}{\partial e_4}s_i\Delta u - k_5\eta s_i \\ &\le -k_4 s_i^2 + \frac{1}{2}\left|\frac{\partial f_1}{\partial e_4}\right|s_i^2 + \frac{1}{2}\left|\frac{\partial f_1}{\partial e_4}\right|\tilde{d}_2^2 + \frac{1}{2}\left|c_2 + \frac{\partial f_1}{\partial e_2}\right|s_i^2 \\ &\quad + \frac{1}{2}\left|c_2 + \frac{\partial f_1}{\partial e_2}\right|\tilde{d}_1^2 + \frac{1}{2}s_i^2 + \frac{1}{2}k_5^2\eta^2 + \frac{\partial f_1}{\partial e_4}s_i\Delta u \\ &\le -\left(k_4 - \frac{1}{2}\left|\frac{\partial f_1}{\partial e_4}\right| - \frac{1}{2}\left|c_2 + \frac{\partial f_1}{\partial e_2}\right| - \frac{1}{2}\right)s_i^2 + \frac{1}{2}\left|\frac{\partial f_1}{\partial e_4}\right|\tilde{d}_2^2 \\ &\quad + \frac{1}{2}\left|c_2 + \frac{\partial f_1}{\partial e_2}\right|\tilde{d}_1^2 + \frac{1}{2}k_5^2\eta^2 + \frac{\partial f_1}{\partial e_4}s_i\Delta u. \end{aligned} \tag{10.17}$$

Using (10.10), (10.11) and Young's inequality results in

$$\begin{aligned} \tilde{d}_1\dot{\tilde{d}}_1 + \tilde{d}_2\dot{\tilde{d}}_2 &= -k_1\tilde{d}_1^2 - \dot{d}_1\tilde{d}_1 - k_2\tilde{d}_2^2 - \dot{d}_2\tilde{d}_1 \\ &\le -\left(k_1 - \frac{1}{2}\right)\tilde{d}_1^2 + \frac{1}{2}\dot{d}_1^2 - \left(k_2 - \frac{1}{2}\right)\tilde{d}_2^2 + \frac{1}{2}\dot{d}_2^2 \\ &\le -\left(k_1 - \frac{1}{2}\right)\tilde{d}_1^2 - \left(k_2 - \frac{1}{2}\right)\tilde{d}_2^2 + \frac{1}{2}\rho_1^2 + \frac{1}{2}\rho_2^2. \end{aligned} \tag{10.18}$$

One can proceed with the stability analysis according to the different ranges in which η is located.

Case 1. When $|\eta| \ge \sigma$, introducing (10.12), (10.17) and (10.18) into (10.16) obtains

$$\begin{aligned}\dot{V}_i \leq & -\left(k_4 - \frac{1}{2}\left|\frac{\partial f_1}{\partial e_4}\right| - \frac{1}{2}\left|c_2 + \frac{\partial f_1}{\partial e_2}\right| - \frac{1}{2}\right)s_i^2 - \left(k_1 - \frac{1}{2} - \frac{1}{2}\left|c_2 + \frac{\partial f_1}{\partial e_2}\right|\right)\tilde{d}_1^2 \\ & - \left(k_2 - \frac{1}{2} - \frac{1}{2}\left|\frac{\partial f_1}{\partial e_4}\right|\right)\tilde{d}_2^2 - \left(k_3 - \frac{1}{2}k_5^2 - \frac{1}{2}\right)\eta^2 + \frac{1}{2}\rho_1^2 + \frac{1}{2}\rho_2^2 \\ \leq & -2\mu_1 V + \Phi_1,\end{aligned} \tag{10.19}$$

where

$$\begin{aligned}\mu_1 =& \min\left\{k_4 - \frac{1}{2}\left|\frac{\partial f_1}{\partial e_4}\right| - \frac{1}{2}\left|c_2 + \frac{\partial f_1}{\partial e_2}\right| - \frac{1}{2}, k_1 - \frac{1}{2} - \frac{1}{2}\left|c_2 + \frac{\partial f_1}{\partial e_2}\right|,\right. \\ & \left. k_2 - \frac{1}{2} - \frac{1}{2}\left|\frac{\partial f_1}{\partial e_4}\right|, k_3 - \frac{1}{2}k_5^2 - \frac{1}{2}\right\} > 0, \\ \Phi_1 =& \frac{1}{2}\rho_1^2 + \frac{1}{2}\rho_2^2.\end{aligned}$$

Case 2. When $|\eta| < \sigma$, introducing (10.12), (10.17) and (10.18) into (10.16) obtains

$$\begin{aligned}\dot{V}_i \leq & -\left[k_4 - \frac{1}{2}\left|\frac{\partial f_1}{\partial e_4}\right| - \frac{1}{2}\left|c_2 + \frac{\partial f_1}{\partial e_2}\right| - \frac{1}{2}\left(\frac{\partial f_1}{\partial e_4}\right)^2 - \frac{1}{2}\right]s_i^2 \\ & - \left(k_1 - \frac{1}{2} - \frac{1}{2}\left|c_2 + \frac{\partial f_1}{\partial e_2}\right|\right)\tilde{d}_1^2 \\ & - \left(k_2 - \frac{1}{2} - \frac{1}{2}\left|\frac{\partial f_1}{\partial e_4}\right|\right)\tilde{d}_2^2 + \frac{1}{2}k_5^2\sigma^2 + \frac{1}{2}(\Delta u)^2 + \frac{1}{2}\rho_1^2 + \frac{1}{2}\rho_2^2 \\ \leq & -2\mu_2 V + \Phi_2,\end{aligned} \tag{10.20}$$

where

$$\begin{aligned}\mu_2 =& \min\left\{k_4 - \frac{1}{2}\left|\frac{\partial f_1}{\partial e_4}\right| - \frac{1}{2}\left|c_2 + \frac{\partial f_1}{\partial e_2}\right| - \frac{1}{2}\left(\frac{\partial f_1}{\partial e_4}\right)^2 - \frac{1}{2},\right. \\ & \left. k_1 - \frac{1}{2} - \frac{1}{2}\left|c_2 + \frac{\partial f_1}{\partial e_2}\right|, k_2 - \frac{1}{2} - \frac{1}{2}\left|\frac{\partial f_1}{\partial e_4}\right|\right\} > 0, \\ \Phi_2 =& \frac{1}{2}\rho_1^2 + \frac{1}{2}\rho_2^2 + \frac{1}{2}k_5^2\sigma^2 + \frac{1}{2}(\Delta u)^2.\end{aligned} \tag{10.21}$$

Therefore, one has

$$\dot{V}_i \leq -2\mu V_i + \Phi, \tag{10.22}$$

with $\mu = \min\{\mu_1, \mu_2\}$ and $\Phi = \max\{\Phi_1, \Phi_2\}$, and one can solve (10.22) to get

$$0 \leq V_i(t) \leq \frac{\Phi}{2\mu} + \left[V_i(0) - \frac{\Phi}{2\mu} \right] e^{-2\mu t}. \tag{10.23}$$

With $\mu > 0$, the above expression implies

$$0 \leq |s_i(t)| \leq \sqrt{\frac{\Phi}{\mu} + 2\left[V_i(0) - \frac{\Phi}{2\mu} \right] e^{-2\mu t}}. \tag{10.24}$$

One can select the Lyapunov function candidate $V_1 = \chi^2$, and along the line of deriving $|s_i(t)|$, it is easy to obtain

$$0 \leq |\chi| \leq \sqrt{s_i^2(t) + 2\left[V_1(0) - \frac{s_i^2(t)}{2} \right] e^{-t}}, \tag{10.25}$$

which implies that $\lim_{t \to \infty} \chi$ exists and is bounded. Therefore, one has $\lim_{t \to \infty} s = 0$ according to Barbalat's lemma. Hence the existence of sliding surfaces has been proved.

Remark 10.4 Utilizing the compensation methods mentioned above to deal with the disturbance and input constraint, compared to the trivial sliding mode control, the integral sliding mode control is capable of reaching the desired sliding surface instead of the boundary layer with a certain thickness.

Once the sliding surface is established, the overall system dynamics are handled by the reduced system

$$\begin{aligned} \dot{e}_1 &= e_2, \\ \dot{e}_2 &= -c_1 e_1 - c_2 e_2 + d_1. \end{aligned} \tag{10.26}$$

For investigating the convergence region of e_1, select the Lyapunov function candidate

$$V_2 = \frac{1}{2}(c_1 - c_2 + 1)e_1^2 + \frac{1}{2}(e_1 + e_2)^2, \tag{10.27}$$

where $c_1 > c_2$, and using Young's inequality, the derivate of V_3 along the trajectory of (10.26) is given by

$$\begin{aligned} \dot{V}_2 &= -(c_1 - c_2 + 1)e_1^2 - \left(c_2 - \frac{3}{2} \right)(e_1 + e_2)^2 + \frac{\rho_3^2}{2} \\ &\leq -2V_2 + \frac{\rho_3^2}{2}, \end{aligned} \tag{10.28}$$

where $c_2 > \frac{5}{2}$. Along the line from (10.22) to (10.25), e_1 can be proved to be uniformly ultimately bounded by

$$0 \le |e_1(t)| \le \sqrt{\frac{\rho_3^2}{2} + 2\left[V_3(0) - \frac{\rho_3^2}{4}\right] e^{-2t}}, \tag{10.29}$$

and for convenience of notation, we denote $|e_1(t)| \le \xi_1$ with $t \to \infty$. Hence, e_1 is uniformly ultimately bounded and the corresponding convergence precisions are also given out. Moreover, based on Barbalat's lemma, $\lim_{t\to\infty} e_1(t)$ exists and is bounded, $e_2 = \dot{e}_1$ is asymptotically stable. By the analyses above, with $t \to \infty$, one can proceed to investigate the convergence precision of e_3.

Consider the following Lyapunov function candidate.

$$V_3 = \frac{1}{2} e_3^2. \tag{10.30}$$

Employing (10.13), $e_2 + 1 > 0$ and $\lambda = e_3 + 1 > 0$, the derivate of V_3 can be expressed by

$$\begin{aligned} \dot{V}_3 &= -\frac{\varepsilon e_3^2(e_3+1)}{2(e_2+1)} - \frac{e_3(e_3+1)(3\cos e_1 \sin e_1 - c_1 e_1 - c_2 e_2)}{2(e_2+1)} \\ &\le -\left(\frac{\varepsilon}{2(e_2+1)} - \frac{|3\cos e_1 \sin e_1 - c_1 e_1 - c_2 e_2|}{|2(e_2+1)||e_3|}\right) e_3^2(e_3+1), \end{aligned}$$

which, according to LaSalle's Lagrange stability theorem [4], implies that e_3 can converge onto the bound of the compact set Σ_2 with $t \to \infty$

$$\Sigma_2 = \left\{ e_3 \in R \,\middle|\, |e_3| \le \frac{(2c_1+3)\xi_1}{2\varepsilon} \right\}, \tag{10.31}$$

which can be arbitrarily small with a suitable selection of parameters. Therefore, based on Barbalat's lemma, $\lim_{t\to\infty} e_3(t)$ exists and is bounded, and hence $e_4 = \dot{e}_3$ is asymptotically stable.

10.2.2 Fractional-Order Sliding Mode Controller Design

At this point, the proof of integral sliding mode control for deployment of tethered spacecraft system has been completed. One can use FO calculus to modify the integral controller, that is, the order of the sliding surface is designed to be a fraction. To develop the FO controller, some fundamental definitions and lemmas are presented previously.

Definition 10.5 (*Fractional integral* [9]) The fractional integral of function $f(t)$ with order ν is

$$ {}_{t_0}D_t^{-\nu} f(t) = \frac{1}{\Gamma(\nu)} \int_{t_0}^{t} (t-\tau)^{\nu-1} f(\tau) d\tau, \tag{10.32} $$

where $0 < \nu < 1$ and $\Gamma(x) = \int_0^\infty e^{-\tau} \tau^{x-1} d\tau$.

Definition 10.6 (*Riemann–Liouville fractional derivate* [9]) For any integrable function $f(t)$, ν-th derivate operator can be defined as

$$ {}_{t_0}^{RL}D_t^{\nu} f(t) = \frac{1}{\Gamma(1-\nu)} \frac{d}{dt} \int_{t_0}^{t} (t-\tau)^{-\nu} f(\tau) d\tau, \tag{10.33} $$

where $0 < \nu < 1$.

Definition 10.7 (*Caputo fractional derivate* [9]) For continuously differentiable $f(t) : \mathbb{R} \to \mathbb{R}$, its FO derivative under Caputo definition is

$$ {}_{t_0}^{C}D_t^{\nu} f(t) = \frac{1}{\Gamma(1-\nu)} \int_{t_0}^{t} \frac{f'(\tau)}{(t-\tau)^{\nu}} d\tau, \tag{10.34} $$

where $0 < \nu < 1$.

Remark 10.8 For the convenience of notation, the upper and lower limits of Riemann-Liouville FO calculus are omitted during the derivation, and hence ${}_{t_0}^{RL}D_t^{\nu}$ $f(t)$ equals $D^{\nu} f(t)$ if any confusion doesn't occur. It has been proved that $D^{\nu}\left(D^{\beta} f(t)\right) = D^{\nu+\beta} f(t)$ [9].

Lemma 10.9 (Generalized Mittag–Leffler stability [5, 6]) *Suppose $V(t, x(t)) : [0, \infty) \times D \to R$ with $D \subset R^n$ containing the origin to be a continuously differentiable function and locally Lipschitz with respect to x such that*

$$ \alpha_1 \|x\|^a \le V(t, x(t)) \le \alpha_2 \|x\|^{ab}, \tag{10.35} $$

$$ D^{\nu} V(t, x(t)) \le -\alpha_3 \|x\|^{ab}, \tag{10.36} $$

where $t \le 0$ and $x \in D$. $0 < \nu < 1$, α_1, α_2, α_3, a and b are positive scalars. Hence, one can say $x = 0$ is generalized Mittag–Leffler stable.

Remark 10.10 Generalized Mittag–Leffler stability implies asymptotic stability [6]. (10.36) can be written as $D^{\beta} V(t, x(t)) \le -(\nu_3/\nu_2)V$ due to $V(t) \ge 0$.

Definition 10.11 ([9]) Mittag–Leffler function is given by

$$ E_{\alpha}(z) = \int_{k=0}^{\infty} \frac{z^k}{\Gamma(k\alpha + 1)}, \tag{10.37} $$

where $\alpha > 0$.

Mittag–Leffler function with two parameters is defined as:

$$E_{\alpha,\beta}(z) = \int_{k=0}^{\infty} \frac{z^k}{\Gamma(k\alpha+\beta)}, \tag{10.38}$$

where $\alpha > 0$ and $\beta > 0$. If $\beta = 1$, then $E_\alpha(z) = E_{\alpha,\beta}(z)$, and $E_{\alpha=1,\beta=1}(z) = e^z$.

Lemma 10.12 ([1]) *If $V(t, x(t)) = x^2(t)$ is a continuously differentiable and positive definite function, the FO derivative of $V(t, x(t))$ can be given out by*

$$\begin{aligned} D^\nu V(t, x(t)) &= x(t)D^\nu x(t) + \textstyle\sum_{i=1}^{\infty} \frac{\Gamma(1+\nu)}{\Gamma(1+i)\Gamma(1-i+\nu)} D^i x(t) D^{\nu-i} x(t) \\ &\le x(t)D^\nu x(t) + \vartheta |x(t)|, \end{aligned}$$

where ϑ is a positive constant.

Assumption 10.2 The disturbances d_1 and d_2 are assumed to be bounded under the definition of fractional order such that $|D^\nu d_1| < \rho_5$ and $|D^\nu d_2| < \rho_6$.

Based on the aforementioned definitions and lemmas, the FO calculus can be synthesized into the integer-order sliding mode controller (10.14) to construct a fractional integral sliding mode controller, which can guarantee that all the system state errors are uniformly ultimately bounded.

Before the FO controller design and stability analyses, the uniform ultimate boundedness of a FO system is presented first.

Theorem 10.13 *Considering a nonlinear system $\dot{\boldsymbol{x}} = \boldsymbol{f}(\boldsymbol{x}) + \boldsymbol{b}(t, \boldsymbol{x})$ with disturbance and uncertainties, $\boldsymbol{x} = 0$ is an equilibrium and $\mathbb{D} \subset \mathbb{R}^n$ is the vicinity of $\boldsymbol{x} = 0$. Selecting a Lyapunov function candidate $V(t, \boldsymbol{x}(t)) : [0, \infty) \times \mathbb{D} \to \mathbb{R}$ with $\boldsymbol{x}$ satisfying Lipschitz condition, if one can obtain*

$$\begin{aligned} &\alpha_1 \|\boldsymbol{x}\|^a \le V(t, \boldsymbol{x}(t)) \le \alpha_2 \|\boldsymbol{x}\|^{ab}, \\ &{}_{t_0}^{C}D_t^\nu V(t, \boldsymbol{x}(t)) \le -\alpha_3 \|\boldsymbol{x}\|^{ab} + g(t), \end{aligned} \tag{10.39}$$

where $t \ge 0, \boldsymbol{x} \in \mathbb{D}$, $\beta \in (0, 1)$, $\alpha_1 > 0$ $\alpha_2 > 0$, $\alpha_3 > 0$, $a > 0$ and $b > 0$. $g(t) : [0, \infty) \to \mathcal{D}$ is the term meaning the bounded disturbance and uncertainties. $V(t, \boldsymbol{x}(t))$ is uniformly ultimately bounded by ξ, satisfying

$$\xi > |g(t)|, t \to \infty.$$

Therefore, $\boldsymbol{x}$ is uniformly ultimately bounded by

$$\|\boldsymbol{x}\| \le \left(\frac{\xi}{\alpha_1}\right)^{\frac{1}{a}}, t \to \infty.$$

Proof In the light of (10.39), one has

$$ {}_{t_0}^{C}D_t^{\nu} V(t, \boldsymbol{x}(t)) \le -\frac{\alpha_3}{\alpha_2} V(t, \boldsymbol{x}(t)) + g(t). \tag{10.40} $$

There must exist a function $M(t) > 0$ to satisfy

$$ {}_{t_0}^{C}D_t^{\nu} V(t, \boldsymbol{x}(t)) + M(t) = -\frac{\alpha_3}{\alpha_2} V(t, \boldsymbol{x}(t)) + g(t). \tag{10.41} $$

The Laplace of (10.41) can be transformed into

$$ s^{\nu} V(s) - V(0)s^{\nu-1} + M(s) = -\frac{\alpha_3}{\alpha_2} V(s) + g(s), \tag{10.42} $$

where $V(0) = V(0, \boldsymbol{x}(0))$, $V(s) = \mathcal{L}\{V(t, \boldsymbol{x}(t))\}$. The above equation can be rewritten as

$$ V(s) = \frac{V(0)s^{\nu-1} - M(s) + g(s)}{s^{\nu} + \frac{\alpha_3}{\alpha_2}}. \tag{10.43} $$

When $\boldsymbol{x} = 0$, it is obvious that $V(0) = 0$, while $\boldsymbol{x} \neq 0$ leads to $V(0) > 0$. Due to the existence of solution of FO derivative equation [9], the solution of $V(t, \boldsymbol{x}(t))$ can be expressed by

$$ \begin{aligned} V(t) =& V(0)E_{\nu}\left(-\frac{\alpha_3}{\alpha_2}t^{\nu}\right) - M(t)\left[t^{\nu-1}E_{\nu,\nu}\left(-\frac{\alpha_3}{\alpha_2}t^{\nu}\right)\right] \\ &+ g(t)\left[t^{\nu-1}E_{\nu,\nu}\left(-\frac{\alpha_3}{\alpha_2}t^{\nu}\right)\right]. \end{aligned} $$

Because $t^{\nu-1}$ and $E_{\nu,\nu}\left(-\frac{\alpha_3}{\alpha_2}t^{\nu}\right)$ are both nonnegative [8], the above expressions can be rewritten as

$$ V(t) \le V(0)E_{\nu}\left(-\frac{\alpha_3}{\alpha_2}t^{\nu}\right) + |g(t)|\left[t^{\nu-1}E_{\nu,\nu}\left(-\frac{\alpha_3}{\alpha_2}t^{\nu}\right)\right]. \tag{10.44} $$

Moreover, the monotonicity of exponent-like Mittag–Leffler function results in the boundedness of (10.44), and trajectories of functions $E_{\nu,\nu}\left(-t^{\nu}\right)$ and $t^{\nu-1}E_{\nu,\nu}\left(-t^{\nu}\right)$ are illustrated in Fig. 10.1. It indicates that the second term on the right side $|g(t)|\left[t^{\nu-1}E_{\nu,\nu}\left(-\frac{\alpha_3}{\alpha_2}t^{\nu}\right)\right]$ will converge to a small vicinity of origin with $t \to \infty$. Due to bounded $g(t)$ and Mittag–Leffler function, it is easy to find $T_g < \infty$ satisfying $V(t) < \xi$ when $t > T_g$, where $\xi > |g(t)|$.

Based on (10.39) and $V(t) < \xi, t > T_g$, it follows

$$ \alpha_1 \|\boldsymbol{x}\|^a < \xi,\ t > T_g \Rightarrow \|\boldsymbol{x}\| < \left(\frac{\xi}{\alpha_1}\right)^{\frac{1}{a}},\ t > T_g, $$

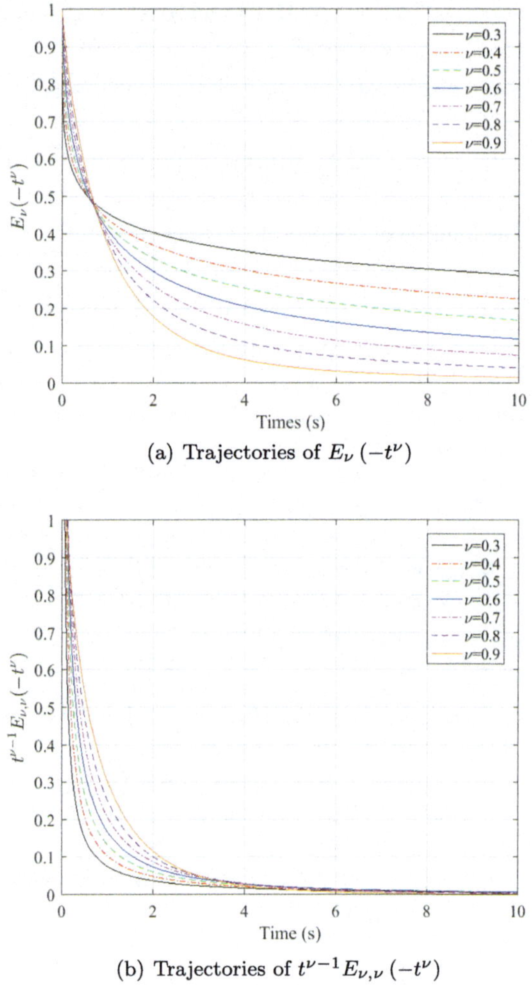

(a) Trajectories of $E_\nu(-t^\nu)$

(b) Trajectories of $t^{\nu-1}E_{\nu,\nu}(-t^\nu)$

Fig. 10.1 Mittag-Leffler function

for $t > T_g$, one has $\|\boldsymbol{x}\| < \left(\frac{\xi}{\alpha_1}\right)^{\frac{1}{a}}$, implying $\boldsymbol{x}$ is uniformly ultimately bounded. It completes the proof. ■

Remark 10.14 It is noted that Theorem 10.13 utilizes Caputo FO calculus to indicate the uniform ultimate boundedness of FO systems, and Riemann-Liouville FO calculus can be synthesized with Theorem 10.13 by the following lemma.

Lemma 10.15 ([5]) *Considering* $\nu \in (0, 1)$ *and* $M(t)$ *with* $M(0) \geq 0$, *one has the relation between Caputo and Riemann-Liouville FOr calculus* ${}_{t_0}^{C}D_t^{\nu}M(t) \leq {}_{t_0}^{RL}D_t^{\nu}M(t)$.

Lemma 10.16 *If a nonlinear system meets the condition in Theorem 10.13, it is capable of replacing Caputo FO operator* ${}^{C}D$ *with Riemann-Liouville FO operator* D *to obtain the same result.*

Proof Consider $\boldsymbol{x}$ and the following Lyapunov function candidate $V(t, \boldsymbol{x})$.

$$\begin{aligned} \alpha_1 \|\boldsymbol{x}\|^a \le V(t, \boldsymbol{x}(t)) \le \alpha_2 \|\boldsymbol{x}\|^{ab}, \\ D^{\nu} V(t, \boldsymbol{x}(t)) \le -\alpha_3 \|\boldsymbol{x}\|^{ab} + g(t). \end{aligned} \tag{10.45}$$

If the following inequality exists

$$D^{\nu} V(t, \boldsymbol{x}(t)) \le -\frac{\alpha_3}{\alpha_2} V(t, \boldsymbol{x}(t)) + g(t),$$

due to $V(0) \ge 0$ and Lemma 10.15, one has

$${}_{t_0}^{C}D_t^{\nu} V(t, \boldsymbol{x}(t)) \le D^{\nu} V(t, \boldsymbol{x}(t)) \le -\frac{\alpha_3}{\alpha_2} V(t, \boldsymbol{x}(t)) + g(t).$$

Referred to the analyses in Theorem 10.13, $V(t, \boldsymbol{x})$ and $\boldsymbol{x}$ are uniformly ultimately bounded. This completes the proof. □

The FO sliding surface contains fractional derivate and integral of integer-order sliding surface as follows:

$$s_{fo} = D^{1-\nu} s + D^{-\nu} s, \tag{10.46}$$

and FO disturbance observers can be described by

$$\begin{aligned} \hat{d}_3 &= D^{1-\nu} q_3 + k_6 D^{1-\nu} e_2, \\ \dot{q}_3 &= -k_6 D^{1-\nu} q_3 - k_6 (f_1 + k_6 D^{1-\nu} e_2), \end{aligned} \tag{10.47}$$

$$\begin{aligned} \hat{d}_4 &= D^{1-\nu} q_2 + k_7 D^{1-\nu} e_4, \\ \dot{q}_4 &= -k_7 D^{1-\nu} q_2 - k_7 (f_2 + u + k_7 D^{1-\nu} e_4). \end{aligned} \tag{10.48}$$

To deal with the input limitation in (10.5), a novel FO auxiliary variable is presented as follows:

$$\eta_{fo} = \begin{cases} -D^{-\nu}\left(k_9 \eta_{fo} + \dfrac{(\partial f_1/\partial e_4) s_{fo} \Delta u + \frac{1}{2}\Delta u^2}{\eta_{fo}^2} \eta_{fo} - \Delta u\right), & |\eta_{fo}| \ge \sigma, \\ 0, & |\eta_{fo}| < \sigma. \end{cases} \tag{10.49}$$

Remark 10.17 Lemma 10.3 presents an integer-order adaptive law which is modified to a FO version. The FO adaptive law is a part of the control command to deal with input limitation in tension control for the deployment of TSS.

Hence the FO sliding mode controller can be designed according to the following theorem.

Theorem 10.18 *Considering system (10.5) representing the deployment dynamics of TSS with input constraint and external disturbance, in the light of* $\eta_{fo} = D^{1-\nu}\eta$, *(10.47) and (10.48), the following FO SMC law can be constructed to ensure all the system state errors are uniformly ultimately bounded.*

$$\begin{aligned} u_c =& -\left(\frac{\partial f_1}{\partial e_4}\right)^{-1}\left(s + c_1 e_2 + c_2 f_1 + \frac{\partial f_1}{\partial e_1} e_2 + \frac{\partial f_1}{\partial e_2} f_1 + \frac{\partial f_1}{\partial e_3} e_4 + \frac{\partial f_1}{\partial e_4} f_2\right) \\ & - \left(\frac{\partial f_1}{\partial e_4}\right)^{-1} k_8 s_{fo} - \hat{d}_4 - \left(\frac{\partial f_1}{\partial e_4}\right)^{-1}\left(c_2 + \frac{\partial f_1}{\partial e_2}\right)\hat{d}_3 - k_{10}\left(\frac{\partial f_1}{\partial e_4}\right)^{-1}\eta_{fo}. \end{aligned} \tag{10.50}$$

Proof The Lyapunov function candidate is as follows.

$$V_{fo} = s_{fo}^2 + \tilde{d}_3^2 + \tilde{d}_4^2 + \eta_{fo}^2. \tag{10.51}$$

Using Lemma 10.12, the ν-th derivative of V_{fo} results in

$$\begin{aligned} D^\nu V_{fo} \leq & s_{fo} D^\nu s_{fo} + \tilde{d}_3 D^\nu \tilde{d}_3 + \tilde{d}_4 D^\nu \tilde{d}_4 + \eta_{fo} D^\nu \eta_{fo} \\ & + \vartheta_1 |s_{fo}| + \vartheta_2 |\tilde{d}_3| + \vartheta_3 |\tilde{d}_4| + \vartheta_4 |\eta_{fo}|. \end{aligned} \tag{10.52}$$

In the view of (10.46)–(10.50), one has

$$\begin{aligned} s_{fo} D^\nu s_{fo} =& s_{fo}(\dot{s} + s) \\ =& s_{fo}\left[-k_8 s_{fo} - \left(\frac{\partial f_1}{\partial e_4}\right)\hat{d}_4 + \left(\frac{\partial f_1}{\partial e_4}\right) d_2 - \left(c_2 + \frac{\partial f_1}{\partial e_2}\right)\hat{d}_3 \right. \\ & \left. + \left(c_2 + \frac{\partial f_1}{\partial e_2}\right) d_1 + \frac{\partial f_1}{\partial e_4}\Delta u - k_{10}\eta_{fo}\right] \\ =& -k_8 s_{fo}^2 + \left(\frac{\partial f_1}{\partial e_4}\right) s_{fo}\tilde{d}_4 + \left(c_2 + \frac{\partial f_1}{\partial e_2}\right) s_{fo}\tilde{d}_3 \\ & + \frac{\partial f_1}{\partial e_4} s_{fo}\Delta u - k_{10}\eta_{fo} s_{fo} \\ \leq & -k_8 s_{fo}^2 + \frac{1}{2}\left(\frac{\partial f_1}{\partial e_4}\right)^2 s_{fo}^2 + \frac{1}{2}\tilde{d}_4^2 + \frac{1}{2}\left(c_2 + \frac{\partial f_1}{\partial e_2}\right)^2 s_{fo}^2 \\ & + \frac{1}{2}\tilde{d}_3^2 + \frac{1}{2}s_{fo}^2 + \frac{1}{2}k_{10}^2\eta_{fo}^2 + \frac{\partial f_1}{\partial e_4} s_{fo}\Delta u \\ \leq & -\left[k_8 - \frac{1}{2} - \frac{1}{2}\left(c_2 + \frac{\partial f_1}{\partial e_2}\right)^2 - \frac{1}{2}\left(\frac{\partial f_1}{\partial e_4}\right)^2\right] s_{fo}^2 + \frac{1}{2}\tilde{d}_4^2 \\ & + \frac{1}{2}\tilde{d}_3^2 + \frac{1}{2}k_{10}^2\eta_{fo}^2 + \frac{\partial f_1}{\partial e_4} s_{fo}\Delta u. \end{aligned} \tag{10.53}$$

According to $\tilde{d}_3 = \hat{d}_3 - d_1$, (10.47), Assumption 10.2 and Young's inequality, one has

$$\begin{aligned}
\tilde{d}_3 D^{\nu}\tilde{d}_3 &= \tilde{d}_3\left(D^{\nu}\hat{d}_3 - D^{\nu}d_1\right)\\
&= \tilde{d}_3\left(\dot{q}_3 + k_6\dot{e}_2 - D^{\nu}d_1\right)\\
&= \tilde{d}_3\left[-k_6D^{1-\nu}q_3 - k_6(f_1 + k_6D^{1-\nu}e_2) + k_6(f_1 + d_1) - D^{\nu}d_1\right]\\
&= \tilde{d}_3\left[-k_6(D^{1-\nu}q_3 + k_6D^{1-\nu}e_2 - d_1) - D^{\nu}d_1\right]\\
&= -k_6\tilde{d}_3^2 - \tilde{d}_3D^{\nu}d_1\\
&\le -\left(k_6 - \frac{1}{2}\right)\tilde{d}_3^2 + \frac{1}{2}\rho_5^2,
\end{aligned} \tag{10.54}$$

and along the same line, one can also obtain

$$\tilde{d}_4 D^{\nu}\tilde{d}_4 \le -\left(k_7 - \frac{1}{2}\right)\tilde{d}_4^2 + \frac{1}{2}\rho_6^2. \tag{10.55}$$

Similar to the integral sliding mode controller, the overall stability analysis will be divided into two cases based on the region in which η_{fo} is located.
Case 1. When $|\eta_{fo}| \ge \sigma$, introducing (10.49), (10.53), (10.54) and (10.55) into (10.52) obtains

$$\begin{aligned}
D^{\nu}V_{fo} \le &-\left[k_8 - \frac{1}{2} - \frac{1}{2}\left(c_2 + \frac{\partial f_1}{\partial e_2}\right)^2 - \frac{1}{2}\left(\frac{\partial f_1}{\partial e_4}\right)^2\right]s_{fo}^2\\
&- (k_6 - 1)\,\tilde{d}_3^2 - (k_7 - 1)\,\tilde{d}_4^2\\
&- \left(k_9 - \frac{1}{2}k_{10}^2 - \frac{1}{2}\right)\eta_{fo}^2 + \frac{1}{2}\rho_5^2 + \frac{1}{2}\rho_6^2 + \vartheta_1|s_{fo}|\\
&+ \vartheta_2|\tilde{d}_3| + \vartheta_3|\tilde{d}_4| + \vartheta_4|\eta_{fo}|\\
\le &-\left[k_8 - 1 - \frac{1}{2}\left(c_2 + \frac{\partial f_1}{\partial e_2}\right)^2 - \frac{1}{2}\left(\frac{\partial f_1}{\partial e_4}\right)^2\right]s_{fo}^2\\
&- \left(k_6 - \frac{3}{2}\right)\tilde{d}_3^2 - \left(k_7 - \frac{3}{2}\right)\tilde{d}_4^2 - \left(k_9 - \frac{1}{2}k_{10}^2 - 1\right)\eta_{fo}^2\\
&+ \frac{1}{2}\rho_5^2 + \frac{1}{2}\rho_6^2 + \frac{1}{2}\vartheta_1^2 + \frac{1}{2}\vartheta_2^2 + \frac{1}{2}\vartheta_3^2 + \frac{1}{2}\vartheta_4^2\\
\le &-\xi_1 V + \Psi_1,
\end{aligned} \tag{10.56}$$

where

$$
\begin{aligned}
\xi_1 = \min \Bigg\{ & k_8 - 1 - \frac{1}{2}\left(c_2 + \frac{\partial f_1}{\partial e_2}\right)^2 - \frac{1}{2}\left(\frac{\partial f_1}{\partial e_4}\right)^2, \\
& k_6 - \frac{3}{2}, k_7 - \frac{3}{2}, k_9 - \frac{1}{2}k_{10}^2 - 1 \Bigg\} > 0, \\
\Psi_1 = & \frac{1}{2}\rho_5^2 + \frac{1}{2}\rho_6^2 + \frac{1}{2}\vartheta_1^2 + \frac{1}{2}\vartheta_2^2 + \frac{1}{2}\vartheta_3^2 + \frac{1}{2}\vartheta_4^2.
\end{aligned}
\tag{10.57}
$$

Case 2. When $|\eta_{fo}| < \sigma$, introducing (10.49), (10.53), (10.55) and (10.54) into (10.52) obtains

$$
\begin{aligned}
D^{\nu} V_{fo} \leq & - \left[k_8 - \frac{1}{2} - \frac{1}{2}\left(c_2 + \frac{\partial f_1}{\partial e_2}\right)^2 - \left(\frac{\partial f_1}{\partial e_4}\right)^2 \right] s_{fo}^2 \\
& - (k_6 - 1)\,\tilde{d}_3^2 - (k_7 - 1)\,\tilde{d}_4^2 - \frac{1}{2}k_{10}^2\eta_{fo}^2 + \frac{1}{2}\Delta u^2 \\
& + \frac{1}{2}\rho_5^2 + \frac{1}{2}\rho_6^2 + \vartheta_1|s_{fo}| + \vartheta_2|\tilde{d}_3| + \vartheta_3|\tilde{d}_4| + \vartheta_4|\eta_{fo}| \\
\leq & - \left[k_8 - 1 - \frac{1}{2}\left(c_2 + \frac{\partial f_1}{\partial e_2}\right)^2 - \left(\frac{\partial f_1}{\partial e_4}\right)^2 \right] s_{fo}^2 \\
& - \left(k_6 - \frac{3}{2}\right)\tilde{d}_3^2 - \left(k_7 - \frac{3}{2}\right)\tilde{d}_4^2 - (\frac{1}{2}k_{10}^2 - \frac{1}{2})\eta_{fo}^2 \\
& + \frac{1}{2}\sigma^2 + \frac{1}{2}\rho_5^2 + \frac{1}{2}\rho_6^2 + \frac{1}{2}\vartheta_1^2 + \frac{1}{2}\vartheta_2^2 + \frac{1}{2}\vartheta_3^2 + \frac{1}{2}\vartheta_4^2 \\
\leq & - \xi_2 V + \Psi_2,
\end{aligned}
\tag{10.58}
$$

where

$$
\begin{aligned}
\xi_2 = \min \Bigg\{ & k_8 - 1 - \frac{1}{2}\left(c_2 + \frac{\partial f_1}{\partial e_2}\right)^2 - \left(\frac{\partial f_1}{\partial e_4}\right)^2, \\
& k_6 - \frac{3}{2}, k_7 - \frac{3}{2}, \frac{1}{2}k_{10}^2 - \frac{1}{2} \Bigg\} > 0, \\
\Psi_2 = & \frac{1}{2}\sigma^2 + \frac{1}{2}\rho_5^2 + \frac{1}{2}\rho_6^2 + \frac{1}{2}\vartheta_1^2 + \frac{1}{2}\vartheta_2^2 + \frac{1}{2}\vartheta_3^2 + \frac{1}{2}\vartheta_4^2.
\end{aligned}
$$

Therefore, one has

$$
D^{\nu} V_{fo} \leq -\xi V_{fo} + \Psi.
$$

According to Theorem 10.16, V_{fo} and s_{fo} are uniformly ultimately bounded. Therefore, one can proceed to prove that all the system state errors are uniformly ultimately bounded by referring to the deduction from Eqs. (10.24) to (10.31), which is omitted here. This completes the proof. ■

10.3 Simulations

To verify the effectiveness of the proposed nonlinear FO sliding mode controller for deployment dynamics of TSS, numerical simulations are carried out in this section.

Firstly, The simulation parameters including the orbital parameters and physical properties of the space tether are presented, and we select the conditions of YES2 phase 1 deployment task as the reference, where the mother spacecraft mass is 6530 kg and subspacecraft mass is 12 kg. Accordingly, this work will deploy 3.5 km of space tether, and the space tether should remain as close to the local vertical as possible during the task. The orbital altitude of the TSS running is approximately 260 km, and it is reasonable to assume that the true anomaly is 0 rad when ejecting the space tether and subspacecraft from the mother spacecraft. Therefore, the initial length of the space tether between the subspacecraft and the mother spacecraft is set as 3 m. The ejection speed of the subspacecraft is 1.25 m/s, and the initial in-plane angle is described by the following conditions: $\theta_0 = 0.01$ rad and $\dot{\theta}_0 = 0$ rad/s. Without loss of generality external disturbance can be expressed by a periodical function $d_2 = 1 \times 10^{-2} \sin(\nu)$ N. Based on these conditions, the initial state of the entire dimensionless system can be expressed approximatively: $\lambda_0 = 9 \times 10^{-4}$, $\dot{\lambda}_0 = 0.3$, $\theta_0 = 0.01$, $\dot{\theta}_0 = 0$, and the desired state follows: $\lambda_d = 1$, $\dot{\lambda}_d = 0$, $\theta_d = 0$, $\dot{\theta}_d = 0$, indicating the deployment has been finished.

It is well known that the space tether can not be compressed like a stick while the tension beyond the maximum tolerance will lead to fragmentation. Physically, the tension along the space tether should be positive to ensure the space tether is tightened, and the constrained input of the system can be calculated with (10.2). Selecting a very small $\tau_{min} = 10^{-6}\ N$ to guarantee the space tether taut and straight during the task, and the assumed dimensionless tension: $10^{-3} \leq \hat{\tau} \leq 10$. The linearization result exhibits a relevant explicit form of the system input: $u = -\hat{\tau}$, thus one can obtain input limitation: $-10 \leq u \leq -10^{-3}$, making the system meet the realistic situation.

According to the proposed integer-order controller design, to ensure the existence of a sliding surface, time-varying parameters of the controller and observer can be designed as follows:

$$
\begin{aligned}
k_1 &= \max\left\{\frac{1}{2} + \frac{1}{2}\left|c_2 + \frac{\partial f_1}{\partial \dot{e}_1}\right| + \varrho/100, \varrho\right\}, \\
k_2 &= \max\left\{\frac{1}{2} + \frac{1}{2}\left|\frac{\partial f_1}{\partial \dot{e}_2}\right| + \varrho/100, \varrho\right\}, \\
k_3 &= \max\left\{\frac{1}{2} + \frac{1}{2}k_5^2 + \varrho/100, \varrho\right\}, \quad k_5 = 10, \\
k_4 &= \max\left\{\frac{1}{2}\left|c_2 + \frac{\partial f_1}{\partial \dot{e}_1}\right| + \frac{1}{2}\left|\frac{\partial f_1}{\partial \dot{e}_2}\right| + \frac{1}{2}\left(\frac{\partial f_1}{\partial \dot{e}_2}\right)^2 + \frac{1}{2} + \varrho/100,\right. \\
&\quad \left.\frac{1}{2}\left|c_2 + \frac{\partial f_1}{\partial \dot{e}_1}\right| + \frac{1}{2}\left|\frac{\partial f_1}{\partial \dot{e}_2}\right| + \frac{1}{2} + \varrho/100, \varrho\right\},
\end{aligned}
$$

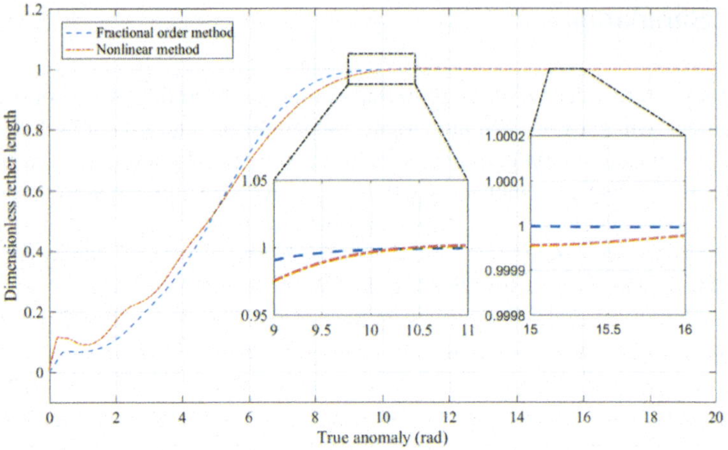

Fig. 10.2 Trajectories of dimensionless tether length compared with nonlinear SMC

where ϱ is a constant, without loss of generality, one can select $\varrho = 100$, $\eta(0) = 100$ and $\sigma = 0.01$ to obtain satisfactory performance. Similarly, the parameters of the FO controller can be selected as $k_6 = 10$, $k_7 = 10$, $k_9 = 5$, $k_{10} = 2$ and

$$k_8 = \max\left\{1 + \frac{1}{2}\left(c_2 + \frac{\partial f_1}{\partial \dot{e}_1}\right)^2 + \frac{1}{2}\left(\frac{\partial f_1}{\partial \dot{e}_2}\right)^2 + \varrho/100,\right.$$
$$\left. 1 + \frac{1}{2}\left(c_2 + \frac{\partial f_1}{\partial \dot{e}_1}\right)^2 + \left(\frac{\partial f_1}{\partial \dot{e}_2}\right)^2 + \varrho/100, \varrho\right\}.$$

Parameters of sliding surface can be chosen as $c_1 = 75$, $c_2 = 3$, and $\varepsilon = 35$ to guarantee that the reduced system is uniformly ultimately bounded.

Figures 10.2 and 10.3 illustrate the comparison between nonlinear sliding mode control for deployment of TSS in [7] and integral sliding mode control which is proposed. In the dynamic phase, the space tether length trajectory of the integral method rises more smoothly than the nonlinear method and reaches the desired value earlier without any overshoot. In the steady phase, the integral method has smaller steady errors, and contrarily, the trajectory of the nonlinear method fluctuates around the desired value. Figure 10.3 shows that the integral method performs a greater swing of the space tether and completes a faster deployment with smaller steady errors than the nonlinear method.

The integral SMC can be viewed as a special FO SMC with order $\nu = 1$. Its length and in-plane angle trajectories are also drawn in Figs. 10.4 and 10.6, and simulation results indicate that the controllers with fractional order $0 < \nu < 1$ perform better. Furthermore, it concludes that the proposed methods in this chapter work more effectively than the method in [7].

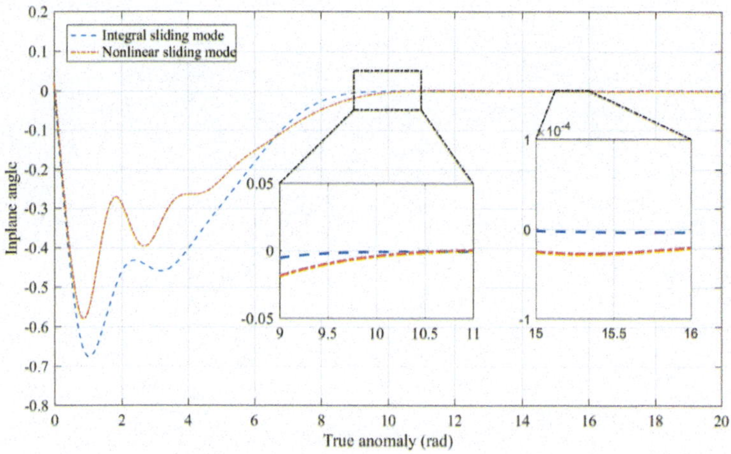

Fig. 10.3 Trajectories of in-plane angle compared with nonlinear SMC

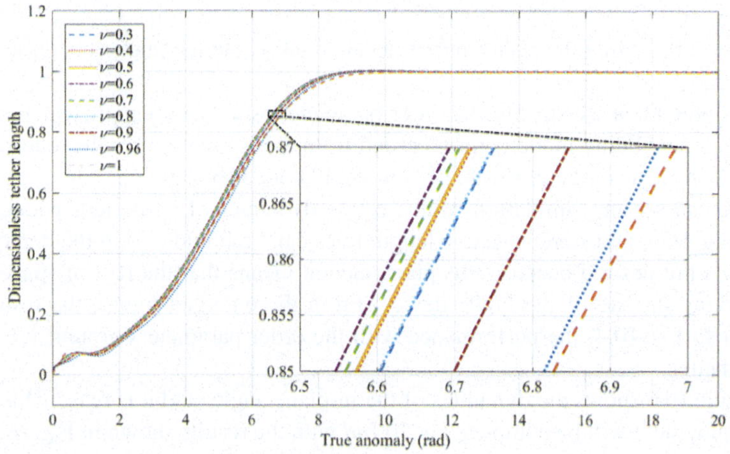

Fig. 10.4 Trajectories of dimensionless tether length

Figure 10.4 shows the dynamics of the length during the deployment, and it is obvious from the simulation results that there exists a cluster of trajectories expressing the deployment dynamics with different FO parameters. It is hard to find the overshoot from this cluster of trajectories while the proposed FO sliding mode controllers can guarantee a smooth and fast deployment subjected to pure tension.

The difference in FO parameters will result in gradual varying performance that can be seen in the enlarged figures. Because the FO sliding mode controller consists of fractional derivate and integral operators, the interaction is reflected in the simulation results, and one can find $\nu = 0.5$, $\nu = 0.6$ and $\nu = 0.7$ produce better performance than other order parameters, and the performance improves as the parameter increases

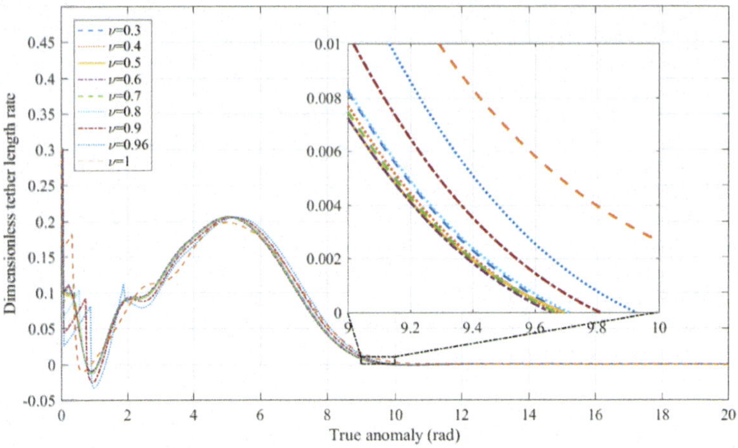

Fig. 10.5 Trajectories of dimensionless tether length rate

when $\nu \leq 0.5$ while the performance degenerates as the parameter decreases when $\nu > 0.5$.

It is noted that we specify the order parameter as $\nu = 0.96$ approximating to an integer-order sliding mode controller, and from the simulation results, one can find that they have a similar performance but the FO one is better.

One can see the simulation results of the dynamics of space tether length rate from Fig. 10.5, and some negative length rates can be found during the deployment which is not desired because this phenomenon means that the reel of space tether is working instead of the brake device. From the enlarged figure, the results are similar to Fig. 10.4, the performance with the order parameters around $\nu = 0.5$ is much better.

Figure 10.6 shows the dynamics of the in-plane angle, and it is reasonable to say the deployment will be completed at 10 *rad* with the results shown in Fig. 10.4. The varying regions of the in-plane angles are feasible which will not lead to a great swing of the space tether. The results also indicate the derivation caused by the gradual order parameters. The best performances appear when the order parameters are selected around $\nu = 0.5$, and all the results with fractional order are better than the integer-order one.

Figure 10.7 shows the dynamics of the in-plane angle rate, and it is worth mentioning that all the values satisfy the assumption about the controller design, which will not cause any singularity in the control laws based on the nonlinear FO sliding surface. One can see a fine-tuning after 10 *rad* at the level of 10^{-3} when the deployment is finishing.

The trajectories of tension are illustrated in Fig. 10.8, from which it is obvious that there is not any negative value occurring during the deployment, indicating that all the inputs during the deployment are feasible, and the final stable values are equivalent

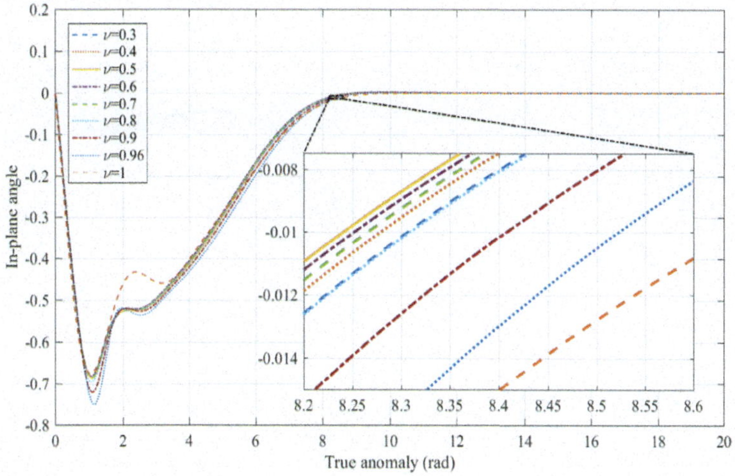

Fig. 10.6 Trajectories of in-plane angle

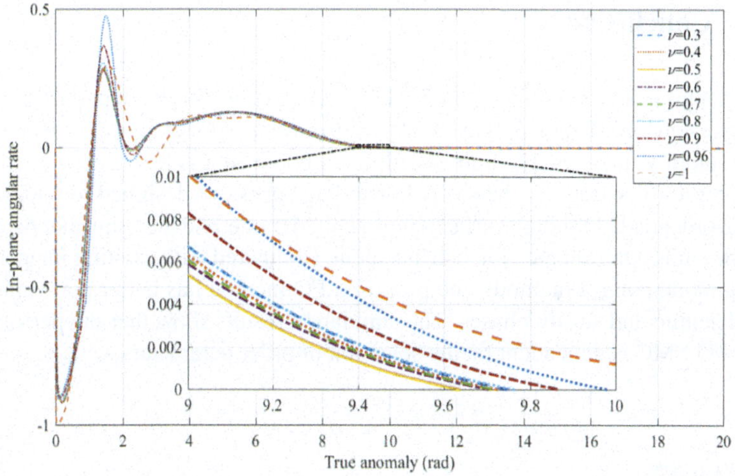

Fig. 10.7 Trajectories of in-plane angular rate

to 3 that is coincident with the reality. There do not exist obvious fluctuations in the stable states of all these trajectories because of the disturbance eliminating external disturbances.

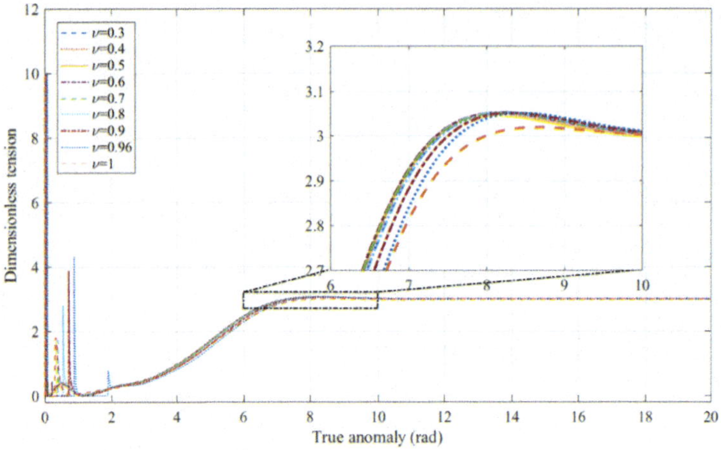

Fig. 10.8 Trajectories of dimensionless tension

10.4 Conclusion

This chapter has proposed a nonlinear FO SMC scheme for the deployment of TSS with pure tension. A FO adaptive law has been developed to remove the adverse effect in the stability analysis. FO uniform ultimate boundedness has been proved based on Mittag-Leffler stability theory. A FO sliding surface is designed to integrate the integer-order and FO sliding surfaces into a united frame, and the controller parameter guidance aids control precision enhancement. Compared with existing integer-order sliding mode control methods, the proposed FO method has better performance in both dynamic and steady phases. The simulation results show that the performance of the FO SMC is dependent on the selection of order parameters.

References

1. Aghababa, M.P.: A Lyapunov-based control scheme for robust stabilization of fractional chaotic systems. Nonlinear Dyn. **78**(3), 2129–2140 (2014)
2. Chen, M., Ge, S.S., Ren, B.: Adaptive tracking control of uncertain MIMO nonlinear systems with input constraints. Automatica **47**(3), 452–465 (2011)
3. Do, K.D.: Practical control of underactuated ships. Ocean. Eng. **37**(13), 1111–1119 (2010)
4. LaSalle, J.: Some extensions of Liapunov's second method. IRE Trans. Circuit Theory **7**(4), 520–527 (1960)
5. Li, Y., Chen, Y., Podlubny, I.: Mittag-Leffler stability of fractional order nonlinear dynamic systems. Automatica **45**(8), 1965–1969 (2009)
6. Li, Y., Chen, Y., Podlubny, I.: Stability of fractional-order nonlinear dynamic systems: Lyapunov direct method and generalized Mittag-Leffler stability. Comput. Math. Appl. **59**(5), 1810–1821 (2010)

7. Ma, Z., Sun, G., Cheng, Z., Li, Z.: Pure-tension non-linear sliding mode control for deployment of tethered satellite system. Proc. Inst. Mech. Eng. Part G J. Aerosp. Eng. **232**(13), 2541–2551 (2018)
8. Miller, K.S., Samko, S.G.: Completely monotonic functions. Integr. Transform. Spec. Funct. **12**(4), 389–402 (2001)
9. Podlubny, I.: Fractional Differential Equations. Academic, New York, NY, USA (1999)
10. Utkin, V., Shi, J.: Integral sliding mode in systems operating under uncertainty conditions. In: Proceedings of 35th IEEE Conference on Decision and Control, vol. 4, pp. 4591–4596. IEEE (1996)
11. Williams, P.: Deployment/retrieval optimization for flexible tethered satellite systems. Nonlinear Dyn. **52**(1–2), 159–179 (2008)
12. Williams, P., Hyslop, A., Stelzer, M., Kruijff, M.: YES2 optimal trajectories in presence of eccentricity and aerodynamic drag. Acta Astronaut. **64**(7–8), 745–769 (2009)

Chapter 11
Fractional-Order Adaptive SMC for the Deployment of STS

11.1 Introduction

In this chapter, a novel FO SMC scheme based on a nonlinear mathematical model in elliptical orbits is presented. A new dimensionless transform based on coordinate translation is constructed to convert input limitation into the form of saturation. With the help of this transform and the Euler-Lagrange mechanical equation, a dynamic model of STS deployment in elliptical orbits is established. To tackle tether tension limitation and thruster saturation, we introduce a quasi-saturated function and a bounded function, with their properties discussed completely. Furthermore, a FO sliding surface is constructed by bringing in a FO derivative operator, reaching higher robustness, smaller overshoots, and shorter settling times. Moreover, we design the controller according to the nonlinear dynamic model, which is a progress than linearized one. Simulation results are presented to evaluate the performance of the proposed methods. To the best of our knowledge, it is the first time in the literature that tether tension limitation and thruster saturation have been considered and handled for nonlinear models in elliptical orbits using a FO sliding mode controller simultaneously.

11.2 Mathematical Model

As shown in Fig. 11.1, the considered STS is composed of a mother satellite, a subsatellite, and a tether. To establish the dynamic equations of the deployment process, two coordinate systems are introduced at first. O is located at the earth center. Coordinate O_1xyz stands for the orbital coordinate system, with the origin at the centroid of STS. The directions of axes are specified as follows: O_1x and O_1z point to the moving forward direction of the orbit and the core of the earth downward, respectively. According to the right-hand rule, one can define the axis

G. Sun et al., *Fractional-Order Sliding Mode Control: Methodologies and Applications*, Studies in Systems, Decision and Control 532,
https://doi.org/10.1007/978-3-031-60847-6_11

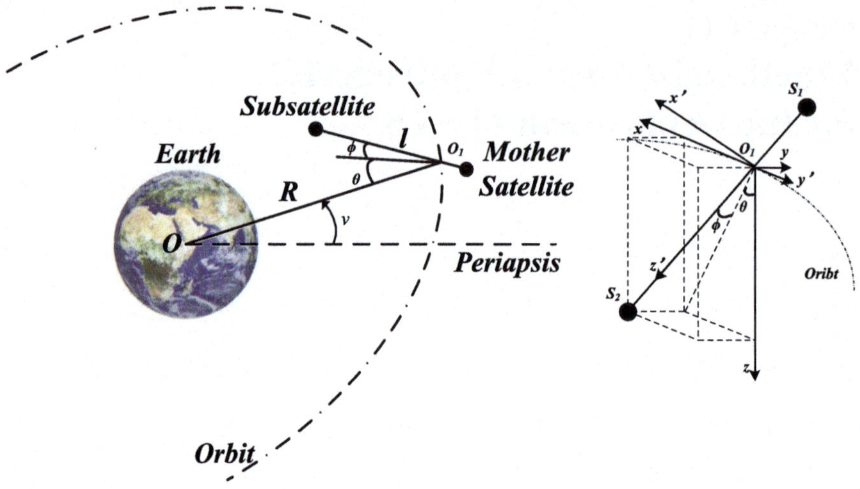

Fig. 11.1 STS geometric representation

O_1y. The other coordinate system $O_1x'y'z'$ is the body frame of the STS, which means that the origin of $O_1x'y'z'$ is the same as that of O_1xyz. O_1z' is collinear with the tether direction, pointing to the subsatellite. $O_1x'y'z'$ can be obtained via coordinate rotation of O_1xyz around the fixed origin O_1, that is, O_1x' can be obtained by rotating O_1x along O_1y by angle θ and O_1y' can be acquired by rotating O_1y along O_1x by angle ϕ.

The following assumptions are made, which can help reduce the complexity of derivation. Furthermore, these assumptions are reasonable, without losing key features and generality of the STS.

Assumption 11.1 The satellites of the STS are taken as non-volumetric particles without consideration of their attitude variations because during the deployment the tether is practically long enough. Take the earth as a perfect sphere so that the centroid of the earth coincides with the center of it.

Assumption 11.2 The orbit of the STS is elliptical.

Assumption 11.3 The mass of the mother satellite is considerably larger compared to that of the subsatellite. When the mass ratio of the mother satellite to the subsatellite is not less than 100, we think that the mother satellite can keep the nominal orbit and it is not affected by the subsatellite during the deployment task.

Assumption 11.4 ([17]) The tether is springless and massless. Furthermore, it is considered a rigid rod during the deployment, which is referred to as the dumbbell model with three degrees of freedom.

Remark 11.1 In Assumption 11.4, the tether is regarded as a rod. However, a rod can provide both normal force and tension while tension is the only force a tether can offer. Therefore, one knows that the tether must be stretched during the deployment mission, which means that the tether tension should fulfill equation $\tau_t > 0$. $\tau_t = 0$ implies that the tether is loose and contradictory with the dumbbell model. Hence, a small positive number τ_{tmin} is given to denote the minimum tether tension to assure that the tether is stretched. On the other hand, the maximum tension provided by the tether is limited, which can be denoted by τ_{tmax}. Once $\tau_t > \tau_{tmax}$, the tether would break which will make the control of the STS fail. In summary, τ_t is bounded by two positive values, whose lower bound and upper bound are τ_{tmin} and τ_{tmax}, respectively.

Before giving out a deduction of dynamics of deployment, some preliminary notations are exhibited in advance. R represents the instantaneous distance of OO_1. The coefficient of the earth's gravity field and the true anomaly measured from perigee are denoted as μ_e and f, respectively. Define the symbols m_1 and m_2 as the mass of the mother satellite and the subsatellite, respectively. As a result, $m = m_1 + m_2$ corresponds to the total mass of the STS. The remaining symbols are the state variables of the STS, with l representing the tether length between the two satellites, θ being the in-plane angle, and ϕ indicating the out-of-plane angle, respectively. L stands for the total length of the tether. In this chapter, we use notation $\dot{(\cdot)}$ to represent the time derivative of the first order. Based on the above assumptions and notations, one has the equations of kinetic energy and potential energy during the deployment established as follows [12, 13].

$$T=\frac{1}{2}m(\dot{R}^2+R^2\dot{f}^2)+\frac{1}{2}\bar{m}l^2\left[\dot{\phi}^2+(\dot{\theta}+\dot{f})^2cos^2\phi\right]+\frac{1}{2}\bar{m}\dot{l}^2, \tag{11.1}$$

$$V = -\frac{\mu_e m}{R} + \frac{\mu_e \bar{m} l^2}{2R^3}(1 - 3\cos^2\theta\cos^2\phi), \tag{11.2}$$

where $\bar{m} = m_1 m_2 / m$.

Then applying the Lagrangian mechanics theory, we can get (11.3)

$$\begin{aligned}
\bar{m}\ddot{l} - \tfrac{2e\sin f}{k}\dot{l}f - \bar{m}l\left[\dot{\phi}^2 + (\dot{\theta}+\dot{f})^2\cos^2\phi\right]&\\
+\tfrac{\bar{m}l\dot{f}^2}{k}(1 - 3\cos^2\phi\cos^2\theta) = -\tau_t + \tau_{td},&\\
\bar{m}l^2\cos^2\phi\ddot{\theta} - \tfrac{2e\sin f}{k}\dot{\theta}fl^2\cos^2\phi - \tfrac{2e\sin f}{k}\theta\dot{f}^2l^2\cos^2\phi + 2\bar{m}(\dot{\theta}+\dot{f})l^2&\\
\times\cos^2\phi(\tfrac{\dot{l}}{l} - \dot{\phi}\tan\phi) + \tfrac{3}{k}\bar{m}\dot{f}^2l^2\sin\theta\cos\theta\cos^2\phi = \tau_\theta + \tau_{\theta d},&\\
\bar{m}l^2\ddot{\phi} - \tfrac{2e\sin f}{k}\dot{\phi}fl^2 + 2\bar{m}\dot{\phi}\dot{l}l&\\
+\bar{m}l^2\sin\phi\cos\phi\left[(\dot{\theta}+\dot{f})^2 + 3\bar{m}\dot{f}\cos^2\theta\right] = \tau_\phi + \tau_{\phi d},&
\end{aligned} \tag{11.3}$$

where $k = 1 + ecosf$, e is the eccentricity of the Keplerian orbit, τ_t means the tether tension while τ_θ and τ_ϕ represent thruster torques affecting in-plane angle and out-of-plane angle, respectively. τ_{td}, $\tau_{\theta d}$ and $\tau_{\phi d}$ are the external disturbances and they are assumed to be bounded. Notably, tether tension τ_t is positive whereas thruster torques τ_θ and τ_ϕ have no such limitation, increasing the complexity of the controller design.

To reduce the difficulty of designing a saturated controller, a novel dimensionless conversion is presented as follows

$$\begin{aligned}
&\lambda = l/L \\
&d(\cdot)/dt = \dot{f} d(\cdot)/df, \\
&\hat{\tau}_t = -\tau_t/(\bar{m}\dot{f}^2 L) + \rho, \\
&\hat{\tau}_\theta = \tau_\theta/(\bar{m}\dot{f}^2 L^2), \\
&\hat{\tau}_\phi = \tau_\phi/(\bar{m}\dot{f}^2 L^2), \\
&\hat{\tau}_{td} = \tau_{td}/(\bar{m}\dot{f}^2 L), \\
&\hat{\tau}_{\theta d} = \tau_{\theta d}/(\bar{m}\dot{f}^2 L^2), \\
&\hat{\tau}_{\phi d} = \tau_{\phi d}/(\bar{m}\dot{f}^2 L^2),
\end{aligned} \tag{11.4}$$

in which ρ is a positive number. According to (11.4), proper ρ can be selected such that $\hat{\tau}_{tmin} < 0$ and $\hat{\tau}_{tmax} > 0$. Thus, $\hat{\tau}_t$, $\hat{\tau}_\theta$ and $\hat{\tau}_\phi$ are transformed into the unified form. Based on transformation in (11.4), one can convert (11.3) into dimensionless Euler-Lagrange dynamical equation shown below:

$$H(q)\ddot{q} + C(q,\dot{q})\dot{q} + B_0\dot{q} + G(q) = \tau + \tau_d, \tag{11.5}$$

where

$$\begin{aligned}
q &= \left(\lambda\ \theta\ \phi\right)^T, \ \tau = \left(\hat{\tau}_t\ \hat{\tau}_\theta\ \hat{\tau}_\phi\right)^T, \\
H(q) &= \begin{bmatrix} 1 & 0 & 0 \\ 0 & \lambda^2\cos^2\phi & 0 \\ 0 & 0 & \lambda^2 \end{bmatrix}, \\
C(q,\dot{q}) &= \left[C_1\ C_2\ C_3\right], \\
B_0 &= \begin{bmatrix} -\frac{2e\sin f}{k} & 0 & 0 \\ 0 & -\frac{2e\sin f}{k} & 0 \\ 0 & 0 & -\frac{2e\sin f}{k} \end{bmatrix}, \\
G(q) &= \begin{bmatrix} -\lambda\cos^2\phi + \frac{\lambda}{k}(1 - 3\cos^2\theta\cos^2\phi) + \rho \\ \frac{3}{k}\lambda^2\cos\theta\sin\theta\cos^2\phi - \frac{2e\sin f}{k} \\ (\lambda^2 + \frac{3}{k}\lambda^2\cos^2\theta)\sin\phi\cos\phi \end{bmatrix}, \\
C_1 &= \begin{bmatrix} 0 \\ (2\lambda + \lambda\dot{\theta})\cos^2\phi \\ \lambda\dot{\phi} \end{bmatrix}, \\
C_2 &= \begin{bmatrix} -(\lambda\dot{\theta} + 2\lambda)\cos^2\phi \\ \lambda\dot{\lambda}\cos^2\phi - \lambda^2\dot{\phi}\sin\phi\cos\phi \\ \lambda^2(\dot{\theta} + 2)\sin\phi\cos\phi \end{bmatrix},
\end{aligned}$$

$$C_3 = \begin{bmatrix} -\lambda\dot{\phi} \\ -(\dot{\theta}+2)\lambda^2 \sin\phi\cos\phi \\ \lambda\dot{\lambda} \end{bmatrix}.$$

It is obvious that $\dot{H}(q) - 2C(q, \dot{q})$ is skew-symmetric, leading to

$$q^T(\dot{H}(q) - 2C(q, \dot{q}))q = 0. \tag{11.6}$$

Remark 11.2 B_0 represents a diagonal non-positive definite matrix related to elliptical orbit and it is clear that B_0 is determined by orbital eccentricity e and true anomaly f. According to the definition of ellipse and celestial kinematics, e is limited in a region $[0, e_0]$ where $e_0 < 1$. Similarly, $\sin f$ is bounded. Thus, we can prove that B_0 is bounded too and denote its upper bound as φ_2 which satisfies $\|B_0\| < \varphi_2$.

Remark 11.3 Denote the maximum dimensionless tether length ratio as $\lambda_{\max}$. Similar to Remark 11.1, the dimensionless tether length cannot be zero. Once $\lambda = 0$ sets up, it means that two satellites of the STS come into collision with each other, which is not permitted during the deployment mission. Even worse, the dimensionless conversion is singular if $\lambda = 0$ is established. To avoid the problems mentioned above, denotation $\lambda_{\min}$ is brought in, such that

$$0 < \lambda_{\min} \le \lambda \le \lambda_{\max}. \tag{11.7}$$

11.3 Fractional-Order Sliding Mode Control

In this section, the control law for STS taking advantage of FO SMC is presented based on the dynamic model established in the previous section. Some requisite preliminaries and properties are shown at first for the subsequent derivation.

11.3.1 Preliminaries

The general torques in the Euler-Lagrange dynamic equation expressed by (11.5) can be parameterized as the expression $Z(q, \dot{q}, \dot{q}, \ddot{q})\Theta = \tau$, in which $Z(q, \dot{q}, \dot{q}, \ddot{q})$ is a function made up of known nonlinear functions and Θ is a dimensionless vector constituting unknown but constant parameters [10]. Substituting a nominal reference $\dot{q}_r$ into (11.5), one has:

$$H(q)\ddot{q}_r + C(q, \dot{q})\dot{q}_r + B_0\dot{q}_r + G(q) = Z_r\Theta, \tag{11.8}$$

where $Z_r = Z_r(q, \dot{q}, \dot{q}_r, \ddot{q}_r)$. Construct the open-loop error dynamics S_r as follows:

$$S_r = \dot{q} - \dot{q}_r. \tag{11.9}$$

Considering (11.5) and (11.8) simultaneously, one gets:

$$H(q)\dot{S}_r + C(q,\dot{q})S_r + B_0 S_r = \tau + \tau_d - Z_r\Theta. \tag{11.10}$$

Property 11.4 ([11]) *One can always find positive scalars $\delta_i (i = 0, \ldots, 5)$ such that*

$$\begin{aligned}
\|H(q)\| &\geq \lambda_{\min}(H(q)) > \delta_0 > 0,\\
\|H(q)\| &\leq \lambda_{\max}(H(q)) < \delta_1 < \infty,\\
\|C(q,\dot{q})\| &\leq \delta_2\|\dot{q}\|,\\
\|G(q)\| &\leq \delta_3,\\
\|\dot{q}_r\| &\leq \delta_4 + \alpha\|\Delta q\| + \beta\|\sigma\|,\\
\|\ddot{q}_r\| &\leq \delta_5 + \alpha\|\Delta\dot{q}\|,
\end{aligned} \tag{11.11}$$

where $\lambda_{\min}(\cdot)$ and $\lambda_{\max}(\cdot)$ stand for the minimum and maximum eigenvalues of the matrix, respectively. We assume that all the states of the STS are bounded, which always holds in practice. In addition, $Z_r\Theta$ is considered bounded such that $Z_r\Theta \leq \eta(t)$ because a bounded function $\eta(t) = f(\Delta q, \Delta\dot{q}, \sigma, \delta_i, t)$ which depends on system states can always be found. We define φ_0 as the bound of $\eta(t)$ such that $|\eta(t)| \leq \varphi_0$, which represents the uncertainty of the system.

In order to deal with the saturation phenomenon, several definitions and comments are brought in.

Definition 11.5 The quasi-saturated function is defined as

$$\text{Qsat}(x) = \begin{cases} \text{sign}(x), & |x| \geq \pi/2,\\ \sin(x), & |x| < \pi/2. \end{cases} \tag{11.12}$$

Remark 11.6 Function $y = \text{Qsat}(x)$ is strictly bounded, or to be more precise, is saturated. In addition, according to Definition 11.5, function $y = \text{Qsat}(x)$ satisfies the equation $|\text{Qsat}(x)| \leq |x|, \forall x$.

Remark 11.7 Utilizing the properties of function $y = \text{Qsat}(x)$ and $y = \arcsin(x)$, one has

$$|\text{Qsat}(x) + \Delta h| \leq |x + \arcsin(\Delta h)|, \tag{11.13}$$

$$\text{sign}(\text{Qsat}(x) + \Delta h) = \text{sign}(x + \arcsin(\Delta h)), \forall x, |\Delta h| < 1. \tag{11.14}$$

Definition 11.8 ([5, 15]) Define a bounded function formed as follows:

$$\tau_{out}(x) = \begin{cases} x_{ub}, & x > x_{ub},\\ x, & x_{lb} \leq x \leq x_{ub},\\ x_{lb}, & x < x_{lb}, \end{cases} \tag{11.15}$$

where x_{lb} and x_{ub} represent the minimum and maximum value of $\tau_{out}(x)$, respectively, and satisfy $x_{lb} < 0 < x_{ub}$.

Remark 11.9 According to the dimensionless transformation described by (11.4), the generalized tension and torques in the channel of in-plane angle and out-of-plane angle are limited, which means that they are in the form of (11.15). Suppose $x = [x_1, x_2, \ldots, x_n]^T$. Then we define the following equations to specify function $\tau_{out}(x)$, arcsin(x) and sign(x) who use vectors as input parameters

$$\tau_{out}(x) = \begin{pmatrix} \tau_{out}(x_1) \\ \tau_{out}(x_2) \\ \cdots \\ \tau_{out}(x_n) \end{pmatrix},$$
$$\arcsin(x) = \left(\arcsin(x_1), \arcsin(x_2), \ldots, \arcsin(x_n)\right)^T,$$
$$\text{sign}(x) = \left(\text{sign}(x_1), \text{sign}(x_2), \ldots, \text{sign}(x_n)\right)^T.$$

Remark 11.10 According to (11.15), the following relations hold

$$\begin{aligned} |\tau_{out}(x)| &\le |x|, \\ \text{sign}(\tau_{out}(x)) &= \text{sign}(x). \end{aligned} \tag{11.16}$$

Introducing a new one dimension parameter J, one can rewrite (11.15) as shown below:

$$\tau_{out}(x) = Jx, \tag{11.17}$$

where J is a nonnegative scalar and the value of J is related to x. J is in the range of [0,1]. Furthermore, from (11.16), we can see that if the input variable x is bounded between $x_{\min}$ and $x_{\max}$, $x_{\min} < 0 < x_{\max}$, there always exists the lower bound of J such that

$$J_{lb} = \min(x_{lb}/x_{\min}, x_{ub}/x_{\max}). \tag{11.18}$$

Assume $x = [x_1, x_2, \ldots, x_n]^T$ and all its elements are bounded. Following equations can be obtained using (11.16)

$$\begin{aligned} \tau_{out}(x_1) &\ge J_{lb1}x_1, \\ \tau_{out}(x_2) &\ge J_{lb2}x_2, \\ &\vdots \\ \tau_{out}(x_n) &\ge J_{lbn}x_n, \end{aligned} \tag{11.19}$$

where $J_{lb1}, J_{lb2}, \ldots J_{lbn}$ are all positive numbers. Equation (11.19) can be denoted in the following matrix form

$$\tau_{out}(x) \geq \begin{bmatrix} J_{lb1} & 0 & 0 \\ 0 & \ddots & 0 \\ 0 & 0 & J_{lbn} \end{bmatrix} x. \tag{11.20}$$

For briefness, denote that

$$J_m = \begin{bmatrix} J_{lb1} & 0 & 0 \\ 0 & \ddots & 0 \\ 0 & 0 & J_{lbn} \end{bmatrix}. \tag{11.21}$$

The aforementioned definitions and remarks present some characteristics of saturated functions, which work in the proof of stability. Before presenting the control law, for clarity, it is necessary to declare the FO operators and some routine lemmas in advance to support the derivation.

Definition 11.11 ([3]) The Riemann-Liouville integral operator is formed as

$${}_{t_0}I_t^{\nu} f(t) = \frac{1}{\Gamma(\nu)} \int_{t_0}^{t} \frac{f(\varsigma)}{(t-\varsigma)^{1-\nu}} d\varsigma. \tag{11.22}$$

Definition 11.12 ([3]) The Riemann-Liouville derivative can be expressed by

$${}_{t_0}D_t^{\nu} f(t) = \frac{1}{\Gamma(m-\nu)} \frac{d^m}{dt^m} \int_{t_0}^{t} \frac{f(\varsigma)}{(t-\varsigma)^{\nu-m+1}} d\varsigma, \tag{11.23}$$

where $\Gamma(z) = \int_0^{\infty} t^{z-1} e^{-t} dt$ is gamma function, m is an integer such that $m-1 < \nu < m$.

Definition 11.13 ([10]) The Laplace Transform of the Riemann-Liouville derivative can be calculated by the following expression

$$\mathscr{L}\{{}_{t_0}D_t^{\nu} f(t)\} = s^{\nu} F(s) - \sum_{k=0}^{m-1} s^k {}_{t_0}D_t^{\nu-k-1} f(t)|_{t=t_0}. \tag{11.24}$$

Lemma 11.14 ([10]) *Consider a FO differential equation shown below*

$$\dot{x} + \alpha_{r-1} x^{((r-1)/r)} + \cdots + \alpha_0 = 0. \tag{11.25}$$

where r is a positive integer. Utilizing Definition 11.13 and transformation $|s| = |w|^r$, one has

$$w^r + \alpha_{r-1} w^{r-1} + \cdots + \alpha_0 = 0. \tag{11.26}$$

Then, the system is stable if

$$|\arg(w)| \geq \pi/2r. \tag{11.27}$$

For better understanding, the relation between root location and state stability in the complex plane is depicted in Fig. 1.1.

In the following, we define the exponential of matrix which is utilized in the control law.

Definition 11.15 The exponential of matrix $K \in R^{n\times n}$ is defined as

$$e^K = \begin{bmatrix} e^{K_{11}} & \dots & e^{K_{1n}} \\ \vdots & & \vdots \\ e^{K_{n1}} & \dots & e^{K_{nn}} \end{bmatrix}. \tag{11.28}$$

When the STS is orbiting in an elliptical orbit, there are external disturbances caused by gravity gradient, aerodynamic drag, and the irregularity of the earth. If not taken into consideration, the external disturbances can cause negative effects on the controllers. To make our simulations closer to reality, an assumption on disturbances is given below.

Assumption 11.5 According to [7], the magnitude of the total disturbance torque is in the range of 1×10^{-5} and its period equals the orbital period T_O. T_O can be obtained by following equations

$$\begin{aligned} T_O &= 2\pi/\Omega, \\ \Omega &= \sqrt{\mu_e/a^3}. \end{aligned} \tag{11.29}$$

Hence, external disturbance after dimensionless transformation is acquired

$$\begin{aligned} \tau_d = [&1.745 \times 10^{-4} \sin f, 4.985 \times 10^{-8} \sin f, \cdots \\ &\cdots 4.985 \times 10^{-8} \sin f]^T. \end{aligned} \tag{11.30}$$

where the orbital parameters are shown in Case 1 in Sect. 4. However, the magnitude of τ_d is rather small. To examine the disturbance-resisting ability of our controllers, we employ τ_d with greater amplitude as

$$\tau_d = [5 \times 10^{-2} \sin f, 5 \times 10^{-2} \sin f, 5 \times 10^{-2} \sin f]^T. \tag{11.31}$$

In addition, the disturbance is assumed to be bounded such that $\|\hat{\tau}_d\| \leq \varphi_3 < \infty$, where φ_3 stands for the bound of the disturbance and it is chosen as $\varphi_3 = 0.1$.

11.3.2 Fractional-Order Adaptive Sliding Mode Control Subject to Input Limitation

Compared with conventional sliding mode control schemes or conventional dynamic sliding PID control, FO SMC subject to input limitation can achieve less stabilization time with smaller overshoot. For improving the performance of the sliding mode dynamics, the FO sliding surface with the nature of memory and heritage is introduced in this subsection. Furthermore, this solution can tackle the input limitation problem.

Based on the above knowledge, the FO adaptive sliding mode control with constrained input can be realized by the subsequent theorem.

Theorem 11.16 (FO adaptive SMC subject to input limitation) *For stabilizing the dynamics of the deployment of the STS in (11.10), construct the fractional-order adaptive SMC input*

$$\tau = -\tau_{out}(\zeta\varphi_1\vartheta_1 \operatorname{sign}(S_{r1}) + \zeta\varphi_2\vartheta_2 S_{r2}), \tag{11.32}$$

where

$$\begin{aligned} S_{r1} &= S_{q1} + \gamma \int_{t_0}^{t} S_{q1}(\varsigma)d\varsigma, \\ S_{q1} &= \text{Qsat}(\Delta\dot{q} + \alpha_0\Delta q + \alpha_1 D^{\nu}\Delta q), \end{aligned} \tag{11.33}$$

in which, $\vartheta_1(0) > 0$, $\vartheta_2(0) > 0$, and the derivatives of ϑ_1 and ϑ_2 can be calculated by:

$$\begin{aligned} \dot{\vartheta}_1 &= \zeta\varphi_1 S_{r2}^T K_1 e^{-2K_1^{-1}\vartheta_1^{-1}} \operatorname{sign}(S_{r2}) \\ &\quad - K_1 e^{K_1^{-1}\vartheta_1^{-1}}(\vartheta_1 - K_1^{-1}) S_{r2}^T \operatorname{sign}(S_{r2}), \\ \dot{\vartheta}_2 &= \zeta\varphi_2 S_{r2}^T K_1 e^{-2K_1^{-1}\vartheta_2^{-1}} S_{r2} \\ &\quad - K_1 e^{K_1^{-1}\vartheta_1^{-1}}(\vartheta_2 - K_1^{-1}) S_{r2}^T (S_{r2}), \end{aligned} \tag{11.34}$$

where ν is the order of the fractional term $\alpha_1 D^{\nu}\Delta q$. Typically, ν is selected as $\nu = 0.5$ such that the dynamics are asymptotically stable. Δq is the system states error and can be expressed by $\Delta q = q - q_d$ where q_d is the reference trajectory. ζ is a design parameter specifying the dynamics of deployment, which is constrained by $\zeta > 1$. φ_1 stands for the bound of lumped uncertainty satisfying $\varphi_1 > \varphi_0 + \varphi_2 + \varphi_3$ and φ_2 is the upper bound of B_0 according to Remark 11.3. γ represents a positive definite diagonal coefficient matrix varying over time. Moreover, S_{q1} is bounded according to Definition 11.5, or to be more exactly, saturated. Therefore, one can have bounded S_{r_1} when the system states are well controlled.

Remark 11.17 In this control law, we adopt two adaptive items ϑ_1 and ϑ_2. ϑ_1 and ϑ_2 are delicately constructed so that they can eliminate the side-effect of input limitation and perturbations generated by elliptical orbits in the process of stability analysis, respectively. On the other hand, from the derivatives of the adaptive terms, we can see that they will remain stable at constant values when $S_{r2} = 0$. As a result, they can be used to reflect whether S_{r2} converges to zero.

Proof Construct a dynamic coordinate change S_{r2} with the system state error Δq and inverse sine function $\arcsin(x)$

$$\begin{aligned} S_{r2} &= S_{q2} + \arcsin(\gamma \int_{t_0}^{t} S_{q1}(\varsigma)d\varsigma), \\ S_{q2} &= \Delta\dot{q} + \alpha_0 \Delta q + \alpha_1 D^{\nu} \Delta q. \end{aligned} \tag{11.35}$$

Until now, we have established two sets of sliding mode manifolds, that is, manifold S_r which includes S_{r1} and S_{r2}, and sub-manifold S_q which includes S_{q1} and S_{q2}. It is kind of complicated to construct so many similar symbols, but they are all helpful during the analysis of Lyapunov stability. The sliding surface $S_{r2} = 0$ is first reached under the designed control input τ and then based on the relationship between S_{r2} and S_{q2}, one can prove the asymptotic stability of S_{q2}. Finally, after reaching the sliding surface $S_{q2} = 0$, we can discuss the stability of the decayed system $\Delta\dot{q} = -\alpha_0 \Delta q - \alpha_1 D^{\nu} \Delta q$. This is the main clue of the following proof.

As we know, the domain of the definition of function $\arcsin(x)$ is [–1,1]. To meet this demand, we can define γ as follows

$$\gamma = \begin{bmatrix} \gamma_1 & 0 & 0 \\ 0 & \gamma_2 & 0 \\ 0 & 0 & \gamma_3 \end{bmatrix}. \tag{11.36}$$

We can denote S_{q1} as $[S_{q1}^1, S_{q1}^2, S_{q1}^3]^T$ according to its definition. Then, for each parameter γ_i, the equation $\gamma_i \int_{t_0}^{t} S_{q1}^i(\varsigma)d\varsigma \in [-1, 1]$ must be fulfilled, where $i = 1, 2, 3$. Hence, we design the adjustment law of γ_i as follows

$$\gamma_i = \begin{cases} \gamma_{i0}, & \gamma_{i0}(|\int_{t_0}^{t} S_{q1}^i(\varsigma)d\varsigma|) \le 1, \\ c_i(|\int_{t_0}^{t} S_{q1}^i(\varsigma)d\varsigma|)^{-1}, & \gamma_{i0}(|\int_{t_0}^{t} S_{q1}^i(\varsigma)d\varsigma|) > 1, \end{cases} \tag{11.37}$$

where γ_{i0} stands for the value of γ_i at the previous moment and c_i is a positive coefficient which satisfies $0 < c_i \le 1$.

According to Remark 11.7, it is apparent that S_{r2} and S_{r1} satisfy:

$$S_{r2}^T \operatorname{sign}(S_{r2}) \ge S_{r1}^T \operatorname{sign}(S_{r1}), \tag{11.38}$$

$$\operatorname{sign}(S_{r2}) = \operatorname{sign}(S_{r1}). \tag{11.39}$$

By introducing the limited input and substituting the dynamic coordinate change S_{r2} in (11.10) for S_r, one can get the below equation:

$$
\begin{aligned}
H(q)\dot{S}_{r2} &= -\tau_{out}(\zeta\varphi\vartheta\,\mathrm{sign}(S_{r1}) + \zeta\varphi_2\vartheta_2 S_{r2}) \\
&\quad +\tau_d - C(q,\dot{q})S_{r2} - B_0 S_{r2} - Z_{r2}\Theta.
\end{aligned} \tag{11.40}
$$

Define a positive definite Lyapunov function

$$
V = \frac{1}{2}S_{r2}^T H(q)S_{r2} + \frac{1}{2}\tilde{\vartheta}_1^{\,2} + \frac{1}{2}\tilde{\vartheta}_2^{\,2}, \tag{11.41}
$$

where $\tilde{\vartheta}_1 = \vartheta_1 e^{K_1^{-1}\vartheta_1^{-1}}$, $\tilde{\vartheta}_2 = \vartheta_2 e^{K_1^{-1}\vartheta_2^{-1}}$, and the time derivative of Lyapunov function V is expressed

$$
\begin{aligned}
\dot{V} &= S_{r2}^T H(q)\dot{S}_{r2} + \frac{1}{2}S_{r2}^T \dot{H}(q)S_{r2} + \tilde{\vartheta}_1\dot{\tilde{\vartheta}}_1 + \tilde{\vartheta}_2\dot{\tilde{\vartheta}}_2 \\
&= -S_{r2}^T\tau_{out}(\zeta\varphi_1\vartheta_1\,\mathrm{sign}(S_{r1}) + \zeta\varphi_2\vartheta_2 S_{r2}) \\
&\quad - S_{r2}^T B_0 S_{r2} + \frac{1}{2}S_{r2}^T\dot{H}(q)S_{r2} - S_{r2}^T Z_{r2}\Theta \\
&\quad - S_{r2}^T C(q,\dot{q})S_{r2} + \tilde{\vartheta}_1\dot{\tilde{\vartheta}}_1 + \tilde{\vartheta}_2\dot{\tilde{\vartheta}}_2 + S_{r2}^T\tau_d \\
&= -S_{r2}^T\tau_{out}(\zeta\varphi_1\vartheta_1\,\mathrm{sign}(S_{r1}) + \zeta\varphi_2\vartheta_2 S_{r2}) \\
&\quad + \frac{1}{2}S_{r2}^T(\dot{H}(q) - 2C(q,\dot{q}))S_{r2} - S_{r2}^T B_0 S_{r2} \\
&\quad - S_{r2}^T Z_{r2}\Theta + \tilde{\vartheta}_1\dot{\tilde{\vartheta}}_1 + \tilde{\vartheta}_2\dot{\tilde{\vartheta}}_2 + S_{r2}^T\tau_d.
\end{aligned} \tag{11.42}
$$

According to (11.6), it has

$$
\begin{aligned}
\dot{V} &= -S_{r2}^T\tau_{out}(\zeta\varphi_1\vartheta_1\,\mathrm{sign}(S_{r1}) + \zeta\varphi_2\vartheta_2 S_{r2}) \\
&\quad - S_{r2}^T Z_{r2}\Theta - S_{r2}^T B_0 S_{r2} + \tilde{\vartheta}_1\dot{\tilde{\vartheta}}_1 + \tilde{\vartheta}_2\dot{\tilde{\vartheta}}_2 + S_{r2}^T\tau_d,
\end{aligned} \tag{11.43}
$$

$\tilde{\vartheta}_1\dot{\tilde{\vartheta}}_1$ and $\tilde{\vartheta}_2\dot{\tilde{\vartheta}}_2$ can be derived as

$$
\begin{aligned}
\tilde{\vartheta}_1\dot{\tilde{\vartheta}}_1 &= \zeta\varphi_1 S_{r2}^T K_1(\vartheta_1 - K_1^{-1})\,\mathrm{sign}(S_{r2}) \\
&\quad - K_1 e^{3K_1^{-1}\vartheta_1^{-1}}(\vartheta_1 - K_1^{-1})^2 S_{r2}^T\,\mathrm{sign}(S_{r2}), \\
\tilde{\vartheta}_2\dot{\tilde{\vartheta}}_2 &= \zeta\varphi_2 S_{r2}^T K_1(\vartheta_2 - K_1^{-1})S_{r2} \\
&\quad - K_1 e^{3K_1^{-1}\vartheta_2^{-1}}(\vartheta_2 - K_1^{-1})^2 S_{r2}^T S_{r2}.
\end{aligned}
$$

Then, (11.42) can be simplified as follows

$$
\begin{aligned}
\dot{V} &\le -S_{r2}^T\tau_{out}(\zeta\varphi_1\vartheta_1\,\mathrm{sign}(S_{r1}) + \zeta\varphi_2\vartheta_2 S_{r2}) \\
&\quad - S_{r2}^T B_0 S_{r2} + \zeta\varphi_1 S_{r2}^T K_1(\vartheta_1 - K_1^{-1})\,\mathrm{sign}(S_{r2}) \\
&\quad - S_{r2}^T Z_{r2}\Theta + \zeta\varphi_2 S_{r2}^T K_1(\vartheta_2 - K_1^{-1})S_{r2} + S_{r2}^T\tau_d.
\end{aligned} \tag{11.44}
$$

According to (11.39) and (11.17), one has

$$\begin{aligned}\dot{V} \le & -S_{r2}^T\tau_{out}(\zeta\varphi_1\vartheta_1 \operatorname{sign}(S_{r2}) + \zeta\varphi_2\vartheta_2 S_{r2}) + S_{r2}^T\tau_d \\ & + K_1 S_{r2}^T(\zeta\varphi_1\vartheta_1 \operatorname{sign}(S_{r1}) + \zeta\varphi_2\vartheta_2 S_{r2}) - S_{r2}^T B_0 S_{r2} \\ & - S_{r2}^T Z_{r2}\Theta - \zeta\varphi_1 S_{r2}^T \operatorname{sign}(S_{r2}) - \zeta\varphi_2 S_{r2}^T S_{r2}.\end{aligned} \tag{11.45}$$

Since that the command input $\zeta\varphi_1\vartheta_1 \operatorname{sign}(S_{r1}) + \zeta\varphi_2\vartheta_2 S_{r2}$ is bounded, then we can choose a positive scalar K_1 such that

$$\begin{aligned}& - S_{r2}^T\tau_{out}(\zeta\varphi_1\vartheta_1 \operatorname{sign}(S_{r2}) + \zeta\varphi_2\vartheta_2 S_{r2}) \\ & \le K_1 S_{r2}^T(\zeta\varphi_1\vartheta_1 \operatorname{sign}(S_{r1}) + \zeta\varphi_2\vartheta_2 S_{r2}),\end{aligned} \tag{11.46}$$

Rewrite (11.45) as

$$\begin{aligned}\dot{V} \le & S_{r2}^T(-B_0 - \zeta\varphi_2)S_{r2} - \zeta\varphi_1 S_{r2}^T \operatorname{sign}(S_{r2}) \\ & - S_{r2}^T Z_{r2}\Theta + S_{r2}^T\tau_d.\end{aligned} \tag{11.47}$$

According to Remark 11.3, (11.47) can be simplified as

$$\begin{aligned}\dot{V} \le & S_{r2}^T\varphi_2(1-\zeta)S_{r2} - \zeta\varphi_1 S_{r2}^T \operatorname{sign}(S_{r2}) \\ & - S_{r2}^T Z_{r2}\Theta + S_{r2}^T\tau_d.\end{aligned} \tag{11.48}$$

If $\zeta > 1$, one has $1 - \zeta < 0$, so

$$\begin{aligned}\dot{V} & \le -\varphi_1\zeta S_{r2}^T \operatorname{sign}(S_{r2}) - S_{r2}^T Z_{r2}\Theta + S_{r2}^T\tau_d \\ & \le -\varphi_1\zeta S_{r2}^T \operatorname{sign}(S_{r2}) + \varphi_0 S_{r2}^T \operatorname{sign}(S_{r2}) + S_{r2}^T\varphi_3 \operatorname{sign}(S_{r2}) \\ & \le -\varphi_1\zeta S_{r2}^T \operatorname{sign}(S_{r2}) + (\varphi_0 + \varphi_3) S_{r2}^T \operatorname{sign}(S_{r2}) \\ & \le \varphi_1(1-\zeta) S_{r2}^T \operatorname{sign}(S_{r2}) \\ & \le 0.\end{aligned} \tag{11.49}$$

Owing to the boundedness property of the expression above, a large enough ζ can be chosen such that $\dot{V} \le 0$. According to Barbalat's theorem [4], it means that S_{r2} asymptotically converges to zero with a set of suitable design parameters. So the first-level sliding surface $S_{r2} = 0$ exists.

Using $S_{r2} = 0$, one has

$$S_{q2} = -\arcsin(\gamma \int_{t_0}^{t} S_{q1}(\varsigma)d\varsigma). \tag{11.50}$$

Select the Lyapunov function candidate consisting of S_{q_2}

$$V_1 = \frac{1}{2} S_{q2}^T S_{q_2}. \tag{11.51}$$

The time derivative is

$$\begin{aligned}
\dot{V}_1 &= S_{q2}^T \dot{S}_{q_2} \\
&= -\frac{S_{q2}^T \gamma S_{q1}}{\sqrt{1 - \|\gamma \int_{t_0}^t S_{q1}(\varsigma) d\varsigma\|^2}} \\
&\le -\frac{S_{q1}^T \gamma S_{q1}}{\sqrt{1 - \|\gamma \int_{t_0}^t S_{q1}(\varsigma) d\varsigma\|^2}} \le 0,
\end{aligned} \tag{11.52}$$

$\dot{V}_1 = 0$ if $S_{q1} = 0$, namely, $S_{q2} = 0$. Therefore, we prove the asymptotic stability of (11.50), that is, $S_{q2} \to 0$ as $t \to \infty$. Based on this result, one has

$$\Delta \dot{q} = -\alpha_0 \Delta q - \alpha_1 D^{\upsilon} \Delta q. \tag{11.53}$$

Until now, from the point of view of the sliding mode control, we have explained the reaching phase by the above contents. Moreover, the sliding phase handles the whole dynamics (11.10), in other words, the stability of (11.53) results in the asymptotic stability of the whole system. Selecting different design parameters α_0 and α_1 corresponds to a critically damped response, over-damped response, or under-damped response. Furthermore, suitable parameters can guarantee the asymptotic stability of (11.33). According to Definition 11.13, Laplace transform of (11.53) gives rise to

$$s + \alpha_1 s^{0.5} + \alpha_0 = 0. \tag{11.54}$$

Substituting $s = w^2$ yields

$$w^2 + \alpha_1 w + \alpha_0 = 0, \tag{11.55}$$

which implies $\arg(s) = 2\arg(w)$. Hence, we can conclude that (11.53) is stable if

$$|\arg(w)| > \pi/4. \tag{11.56}$$

Thus, the parameters can be decided. This completes the proof. ■

Remark 11.18 If $\alpha_1 = 0$, there is no FO differentiation on the sliding surface. Therefore, we can call the decayed control law a traditional integer-order adaptive sliding mode control. Similarly to the previous process, we can prove that $\Delta \dot{q} = -\alpha_0 \Delta q$. By constructing the Lyapunov function, the asymptotic stability of Δq can be proved easily.

11.4 Simulations

In this section, with the purpose of illustrating the effectiveness of the proposed methods and evaluating the performance, the simulations on the dynamics of the STS subject to input limitation are carried out and compared among three different control schemes. The three control methods are FO adaptive sliding mode control subject to input limitation, traditional integer-order adaptive sliding mode control subject to input limitation, and FO SMC proposed by [9], and for briefness, denote them as FDA-SMC, DA-SMC, and FO-SMC, respectively. It should be noted that the first two control laws make use of signum functions, which can cause a response of the unmodeled high-order dynamics, decaying the performance of the control system. To deal with the above possible problems, the boundary layer technique is employed and uniformly ultimately boundedness has been proved according to [1, 2, 6, 8, 16].

Some basic information about the simulation is given out as follows. To make this simulation more accurate and practical, the parameters in the YES2 mission are adopted. According to YES2 mission [14], the earth's radius is 6738 km and the mass of the mother satellite and subsatellite are 6530 kg and 12 kg, respectively. In addition, the subsatellite is deployed with a speed of 2.58 m/s and the destination is 3.5 km away from the initial position (3 m) in the vertical direction [14]. For the sake of better comparison, simulations are carried out under two different orbital eccentricities whose orbital parameters are given in the following subsections, with $e = 0.0027$ and $e = 0.17$, respectively. We take the moment of deployment as the start time of simulation and the STS system is assumed to be located at the perigee at that time. Input limitation is a crucial obstacle to stabilizing the STS in reality, which is solved in this chapter. Thus, we assume that the dimensionless thruster torques for the control of in-plane angle and out-of-plane angle satisfy $|\hat{\tau}_\theta| \leq 5$ and $|\hat{\tau}_\phi| \leq 5$. Accordingly, the dimensionless tension $\hat{\tau}_t$ is saturated in the region $[-2.45, 2.45]$ while ρ is chosen to be 2.55. To illustrate the merits of the FO operator effectively, we employ completely the same parameters for the first two control laws with $\alpha_0 = 1.75$, $\zeta = 3$, and $\varphi_1 = 2$. Other shared design parameters are given below.

$$K_1 = 0.05, \gamma_0 = \begin{bmatrix} 1.2 & 0 & 0 \\ 0 & 3 & 0 \\ 0 & 0 & 0.9 \end{bmatrix}.$$

11.4.1 Case 1: e = 0.0027

Assume that the STS follows an elliptical orbit around the earth with a perigee attitude of 249 km and an apogee attitude of 285 km [14]. Under this setup, the orbit eccentricity $e = 0.0027$. Hence, using celestial kinematics, one can calculate the orbital angular speed at the perigee and get $\dot{f} = 1.17 \times 10^{-3}$ rad/s. With the

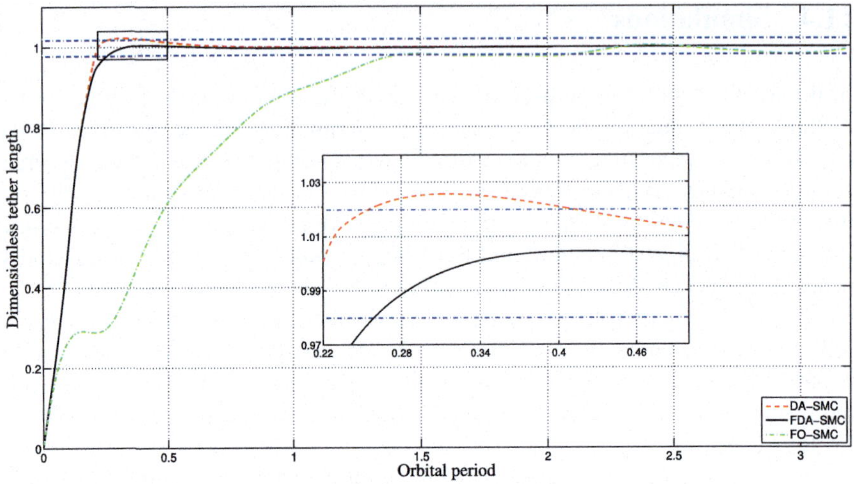

Fig. 11.2 Tether length of the deployment ($e = 0.0027$)

help of dimensionless transformation, initial deployment position, and speed, $\lambda = 0.0009$ and $\dot{\lambda} = 0.7$ are obtained. Apart from that, to show the ability to regulate the in-plane and out-of-plane angles, we simply set $\theta = 0.1$ and $\phi = 0.1$ at the start time, which is non-zero. In general, the initial dimensionless system states are $q_0 = [0.0009, 0.1, 0.1]^T$ and $\dot{q}_0 = [0.7, 0, 0]^T$. The control aims to stabilize the STS at the states $q_d = [1, 0, 0]^T$. The remaining design parameters are given below.

$$\varphi_2 = \frac{2e}{1-e} = 0.0054, \alpha_1 = 0.5. \tag{11.57}$$

At present, we can give the simulation results in curves from Fig. 11.2 to Fig. 11.6. Figure 11.2 shows the dynamics of dimensionless tether length while Fig. 11.3 illustrates the liberation rate of tether length. In-plane angle and out-of-plane angle are depicted in Figs. 11.4 and 11.5, respectively. Finally, the dimensionless tether tension of the above methods is presented in Fig. 11.6.

When it comes to tether length, it is obvious from Fig. 11.2 that tether length regulated by DA-SMC has undesirable overshoot. To some extent, we can reduce overshoot by extending the settling time, just like the tether liberation of FO-SMC, which is a solution with obvious drawbacks. By introducing FO derivative, overshoot and settling time are dramatically reduced as shown in Fig. 11.2. We can see from the zoom-in area in Fig. 11.2 that DA-SMC has a maximum overshoot of about 2% whereas there is no overshoot in FDA-SMC. Furthermore, according to the 2% criterion of adjustment time, tether length controlled by FDA-SMC settles at about 0.28 orbital periods while that of DA-SMC finishes with 0.43 orbital periods. Conversely, the FO-SMC method needs about 1.4 orbital periods to reach the destination and the tether length does not remain steady at $\lambda = 1$ and vibrates slightly. In general, the

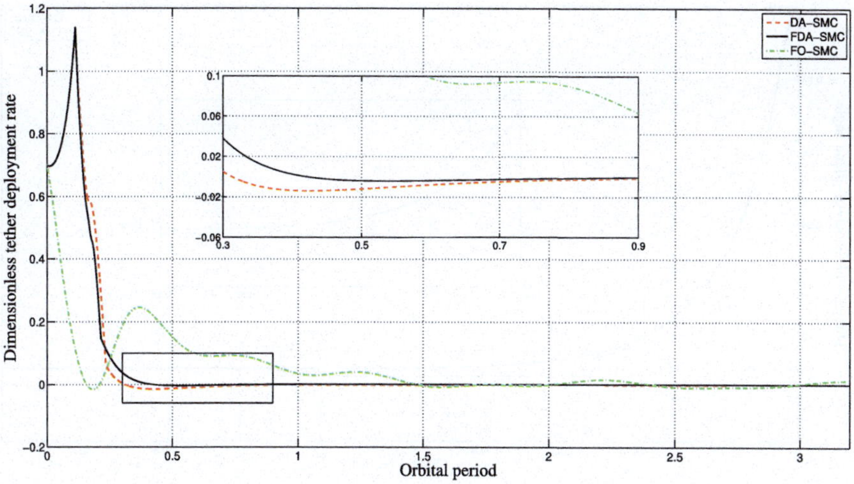

Fig. 11.3 Tether length rate of the deployment ($e = 0.0027$)

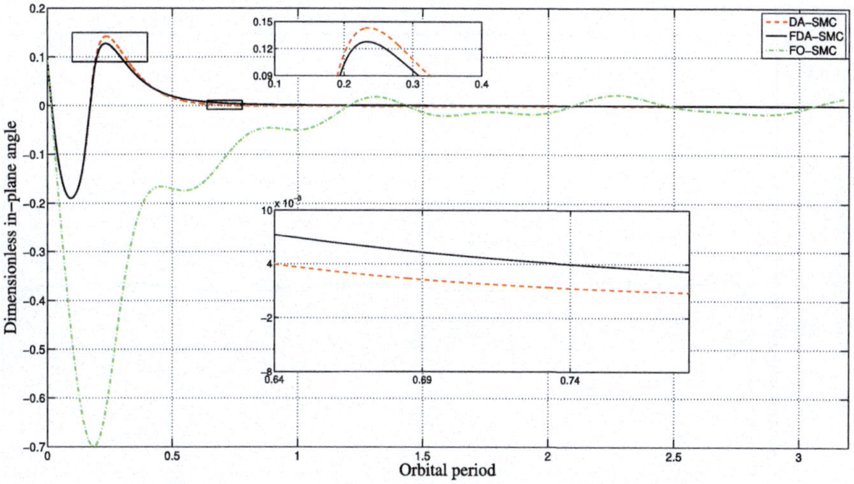

Fig. 11.4 In-plane angle of the deployment ($e = 0.0027$)

FDA-SMC scheme is more advantageous than DA-SMC and FO-SMC both in terms of overshoot and settling time of deployment. Figure 11.3 shows the dimensionless tether liberation rate of FDA-SMC, DA-SMC, and FO-SMC, respectively. Tether liberates significantly faster under the control of FDA-SMC and DA-SMC compared to FO-SMC. In addition, from the zoom-in area, we can find that the tether deployment rate of FDA-SMC is always nonnegative while that of DA-SMC turns negative after 0.38 orbital periods, which indicates that there is no overshoot in FDA-SMC while overshoot exists in DA-SMC. Moreover, the performance of FO-SMC is still poor

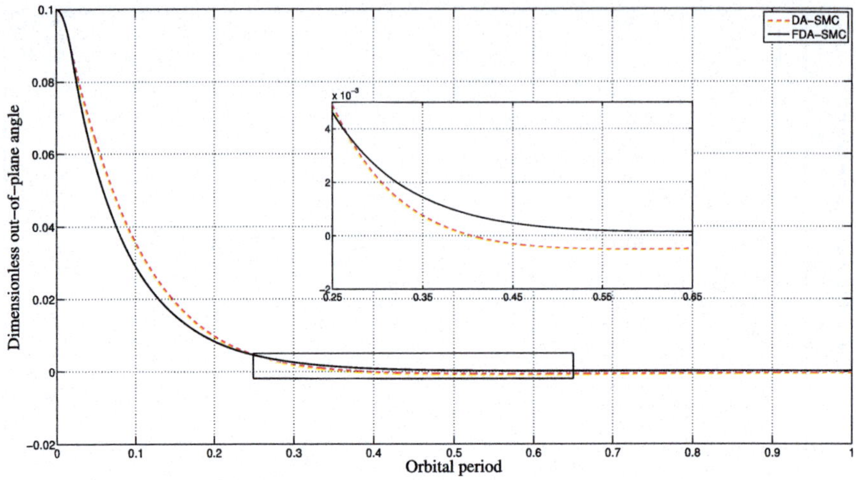

Fig. 11.5 Out-of-plane angle of the deployment ($e = 0.0027$)

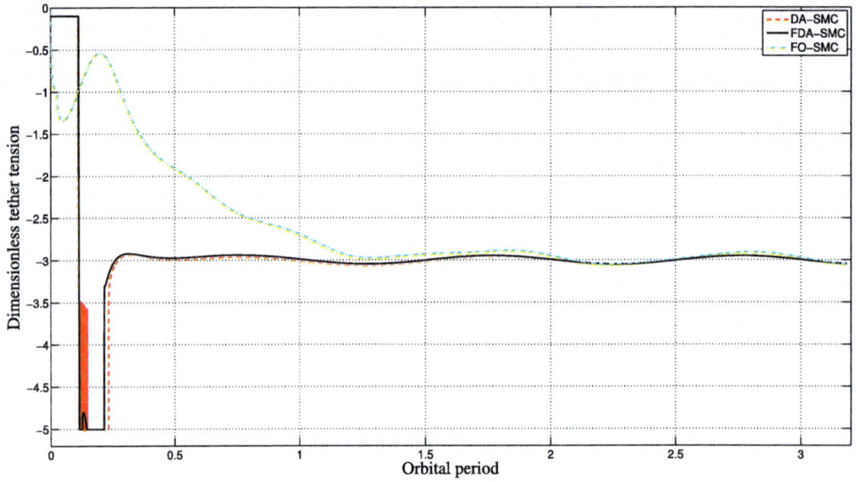

Fig. 11.6 Tether tension using FDA-SMC ($e = 0.0027$)

and its tether rate fluctuates around zero after 1.5 orbital periods, which means that the tether length fluctuates too.

Figures 11.4 and 11.5 depict the in-plane and out-of-plane angles with respect to the orbital period, which is called swinging dynamics. We can see from Fig. 11.4 that, when controlling in-plane angle, FDA-SMC, and DA-SMC perform considerably better than FO-SMC in terms of swing amplitude, showing their superiority to the control law in [9]. Furthermore, from the zoom-in area, it is clear that the swinging amplitude of FDA-SMC is a little smaller. When it comes to settling time, FDA-SMC and DA-SMC methods settle the in-plane angle with almost the same stable time(0.7 orbital periods). However, similarly, FO-SMC continues to regulate the in-plane angle to zero for quite a long time and oscillation. Only the curves of FDA-SMC and DA-SMC are given in Fig. 11.5 since FO-SMC cannot govern out-of-plane dynamics. In Fig. 11.5 the out-of-plane angle of FDA-SMC achieves less overshoot and less settling time compared with those of DA-SMC. In both two figures, angles controlled by FDA-SMC converge faster than the other control method, namely, the FDA-SMC can deal with swinging dynamics better.

Figure 11.6 illustrates the dimensionless tether tension controlled by FDA-SMC, DA-SMC, and FO-SMC. It is clear that saturation phenomena appear and the tether tension is constrained in a proper practical range during the whole deployment process with the help of dimensionless transformation and designed controllers, confirming that input limitation is well solved by the proposed control schemes. Although [9] did not consider input limitation, the tether tension of FO-SMC is located in a proper region.

We can conclude from the pictures and analysis that both FDA-SMC and DA-SMC can regulate the deployment process in elliptical orbits with ideal performance regarding tether length, tether length rate, attitude angles, and tether tension. FO-SMC, which was designed on an oversimplified mathematical model, cannot achieve the desired control effectiveness. Furthermore, the deployment performance of FDA-SMC has a fascinating improvement in overshoot and settling time when compared to DA-SMC, which shows the superiority of fractional control.

11.4.2 Case 2: e = 0.17

In this case, the perigee attitude and the apogee attitude of the elliptical orbit are 300 km and 3000 km respectively, so it is easy to get the orbit eccentricity $e = 0.17$. Utilizing the same calculation process, the initial dimensionless system states are presented as $q_0 = [0.0009, 0.1, 0.1]^T$ and $\dot{q}_0 = [0.6, 0, 0]^T$. The goal is still to deploy the subsatellite to the dimensionless states $q_0 = [1, 0, 0]^T$ and the remaining design parameters are shown as follows.

$$\varphi_2 = \frac{2e}{1-e} = 0.41, \alpha_1 = 0.5.$$

The simulation results are exhibited in Fig. 11.7 to Fig. 11.10. From Fig. 11.7 to Fig. 11.10, the two control schemes, FDA-SMC and DA-SMC, maintain almost the same control performance in comparison with that in the previous case where $e = 0.0027$. This means that our methods can be applied without decaying effectiveness even if the orbital eccentricity is large enough. By contrast, when e becomes larger, FO-SMC fails to regulate and stabilize the systems states as shown in Fig. 11.7. In conclusion, FDA-SMC and DA-SMC presented in this chapter perform much better

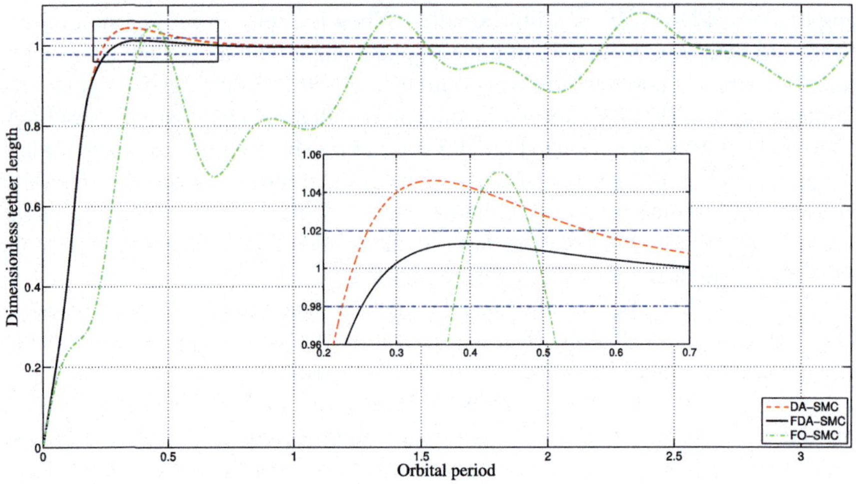

Fig. 11.7 Tether length of the deployment ($e = 0.17$)

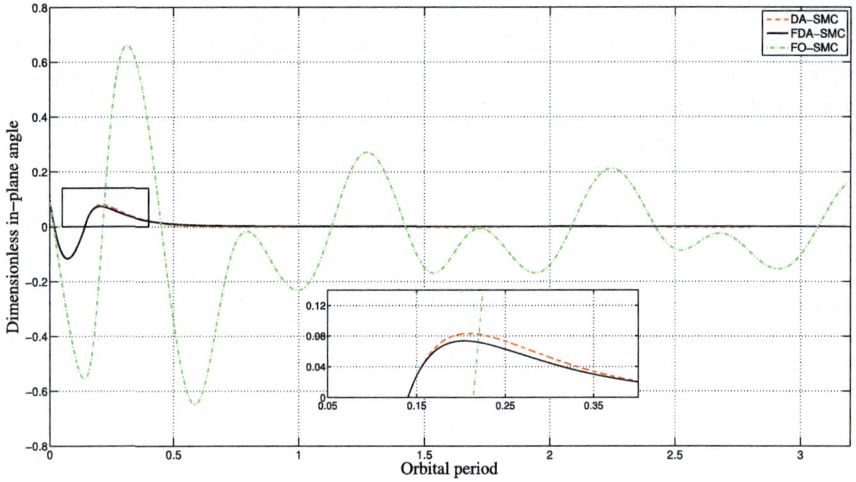

Fig. 11.8 In-plane angle of the deployment ($e = 0.17$)

than FO-SMC, and among the two methods, FDA-SMC achieves a smaller overshoot and shorter settling time (Figs. 11.8 and 11.9).

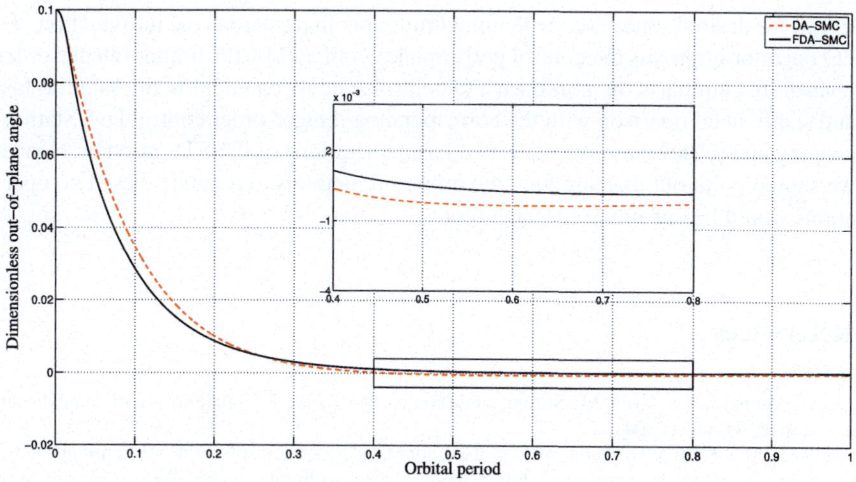

Fig. 11.9 Out-of-plane angle of the deployment ($e = 0.17$)

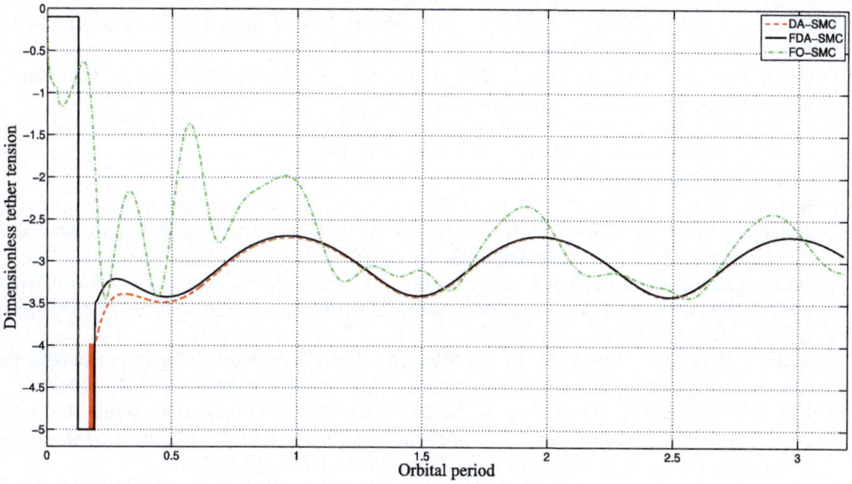

Fig. 11.10 Tether tension using FDA-SMC ($e = 0.17$)

11.5 Conclusion

A novel control scheme based on FO adaptive SMC has been proposed for the deployment mission of STS with consideration of input limitation. With nonlinear dynamics and input limitation considered, this control law is more applicable in the real world, which can achieve more precise control performance especially smaller overshoot and shorter settling time. Simulations show the effectiveness of the proposed control law which can regulate the tether length, in-plane angle, and out-of-plane angle to follow the desired states even with input limitation. In particular, the introduction of a FO operator improves the control performance compared with similar integer-order control. In summary, the nonlinear FO control scheme can deploy the satellite fast and stably in comparison with the corresponding integer-order control law. Station-keeping and retrieval are the other two basic operations of the STS mission. Hence, we should point out that applying the proposed methods to regulate these two operations would be our future research topic.

References

1. Al-Muthairi, N.F., Zribi, M.: Sliding mode control of a magnetic levitation system. Math. Probl. Eng. **2**, 93–107 (2004)
2. Baik, I.C., Kim, K.H., Youn, M.J.: Robust nonlinear speed control of PM synchronous motor using boundary layer integral sliding mode control technique. IEEE Trans. Control. Syst. Technol. **8**(1), 47–54 (2000)
3. Bandyopadhyay, B., Kamal, S.: Stabilization and Control of Fractional Order Systems: a Sliding Mode Approach. Springer, Switzerland (2015)
4. Barbalat, I., et al.: Systemes déquations différentielles doscillations non linéaires. Rev. Math. Pures Appl **4**(2), 267–270 (1959)
5. Du, J., Hu, X., Krstić, M., Sun, Y.: Robust dynamic positioning of ships with disturbances under input saturation. Automatica **73**, 207–214 (2016)
6. Hwang, C.L.: Sliding mode control using time-varying switching gain and boundary layer for electrohydraulic position and differential pressure control. IEE Proc. Control. Theory Appl. **143**(4), 325–332 (1996)
7. Inamori, T., Kawashima, R., Saisutjarit, P., Sako, N., Ohsaki, H.: Magnetic plasma deorbit system for nano-and micro-satellites using magnetic torquer interference with space plasma in low earth orbit. Acta Astronaut. **112**, 192–199 (2015)
8. Jin, M., Lee, J., Chang, P., Choi, C.: Practical nonsingular terminal sliding-mode control of robot manipulators for high-accuracy tracking control. IEEE Trans. Ind. Electron. **56**(9), 3593–3601 (2009)
9. Kang, J., Zhu, Z.H., Wang, W., Li, A., Wang, C.: Fractional order sliding mode control for tethered satellite deployment with disturbances. Adv. Space Res. **59**(1), 263–273 (2017)
10. Munoz-Vazquez, A.J., Parra-Vega, V., Sanchez, A.: Free-model fractional-order absolutely continuous sliding mode control for Euler-Lagrange systems. In: 2014 IEEE 53rd Annual Conference on Decision and Control (CDC), pp. 6933–6938. IEEE (2014)
11. Parra-Vega, V., Arimoto, S., Liu, Y., Hirzinger, G., Akella, P.: Dynamic sliding PID control for tracking of robot manipulators: theory and experiments. IEEE Trans. Robot. Autom. **19**(6), 967–976 (2003)
12. Wen, H., Zhu, Z.H., Jin, D., Hu, H.: Exponentially convergent velocity observer for an electrodynamic tether in an elliptical orbit. J. Guid. Control. Dyn. **39**(5), 1113–1118 (2016)
13. Williams, P., Blanksby, C., Trivailo, P.: Tethered planetary capture maneuvers. J. Spacecr. Rocket. **41**(4), 603–613 (2004)
14. Williams, P., Hyslop, A., Stelzer, M., Kruijff, M.: YES2 optimal trajectories in presence of eccentricity and aerodynamic drag. Acta Astronaut. **64**(7–8), 745–769 (2009)
15. Xu, S., Sun, G., Sun, W.: Takagi-Sugeno fuzzy model based robust dissipative control for uncertain flexible spacecraft with saturated time-delay input. ISA Trans. **66**, 105–121 (2017)
16. Yu, X., Kaynak, O.: Sliding-mode control with soft computing: a survey. IEEE Trans. Ind. Electron. **56**(9), 3275–3285 (2009)
17. Zhao, G., Sun, L., Huang, H.: Thrust control of tethered satellite with a short constant tether in orbital maneuvering. Proc. Inst. Mech. Eng. Part G J. Aerosp. Eng. **228**(14), 2569–2586 (2014)

Chapter 12
Conclusion and Further Work

This monograph is concluded in this chapter, following which some future directions concerning FOSMC are discussed.

12.1 Conclusion

By utilizing SMC, FO calculus, fuzzy systems, and adaptive control, this monograph proposed some effective FO control methods and FOSMC schemes, based on which the problems of the control of linear motors and the deployment of space-tethered systems have been solved. The proposed control schemes can effectively deal with parameter uncertainties, friction, and discrete signals. The performance of the system such as control precision, robustness, and overshoots has been greatly improved. Specifically, the conclusions of this monograph can be drawn as follows.

1. Based on the FO constitutive law of viscoelastic material, a generic FO dynamics has been derived for general rigid-flexible coupling viscoelastic structures in a uniform framework ranging from purely elastic solids to purely viscous fluids. A new FO transfer function of the coupling structures has been derived by separating the dynamic responses of the rigid and flexible components in a distinguished and coupled form: the transfer function of the rigid component and the fractional order transfer function of the flexible component.
2. By designing FO tension control laws, the problems of the deployment of space-tethered systems as well as fast and fast tethered satellite retrieval have been solved. It is shown that the stable region of classic integer-order tension control is the subset of the stable region of the fractional-order tension control. The proposed FO control law can deploy the tethered satellite in a shorter time (1/3 of the integer order control law) stably without the overshoot of tether length due to the fractional order system having a larger stable region and a historical

G. Sun et al., *Fractional-Order Sliding Mode Control: Methodologies and Applications*, Studies in Systems, Decision and Control 532,
https://doi.org/10.1007/978-3-031-60847-6_12

memory effect. The developed FO control law has the memory of previous states and thus controls the tether retrieval more smoothly while maintaining the retrieving speed.

3. Different kinds of continuous and discrete FOSMC strategies have been designed for a linear motor. For a linear motor, an FO sliding mode surface has been designed, and the sliding dynamics have been also analyzed when the linear motor is subject to varying load, unknown uncertainties, and external disturbance. Additionally, the adaptive algorithm has been integrated into the proposed FOSMC to achieve high-precision and fast-response performance with robustness against uncertainties. By conducting simulations and experiments, the proposed control scheme outperforms existing related control strategies, and the performance of the linear motor is drastically improved.
4. For the deployment of space-tethered systems, suitable sliding mode surface functions including FO and integer order items have been constructed in a unified framework. The corresponding sliding dynamics have been analyzed. Novel and effective FOSMC strategies have been proposed for the deployment of space-tethered systems. In the design process, the under-actuated dynamics, time-varying disturbance, as well as input limitation have been addressed to guarantee the feasibility of the proposed control schemes.

12.2 Further Work

In this monograph, some FOSMC schemes have been proposed, and their applications have been discussed. Although the performance of systems can be greatly improved by utilizing the proposed control strategies, there still exist some problems to be addressed, some of which are discussed below.

1. Along with the wide application of smart devices and communication networks in engineering, the issues including data packet dropouts, data packet disorder, network-induced delay, and cyber attack are not addressed. The performance of the system can be degraded or even deteriorated when those issues occur. It is thus of great importance and interest to investigate how to design FO control schemes to guarantee system performance under a complicated network environment.
2. New theoretical methods of fractional calculus combined with the advanced sliding mode control can be further studied, such as fractional-order adaptive sliding mode, fractional-order fuzzy sliding mode, neural-network-based fractional-order sliding mode, and so on. Thus, the control effect and application prospect of fractional-order sliding mode control are effectively improved, and an effective theoretical framework of fractional-order sliding mode control is formed.
3. From the perspective of practical engineering, the discrete-time FOSMC can be paid more attention, which investigates the practical stability of the FO control system. According to the characteristics of the engineering system, the practical implementation length of the historical memory effect for fractional calculus

should be adjusted, to further explore the practical application and engineering value of FOSMC.

4. The implementation of fractional calculus on the motion control of linear motors will take into intelligent approaches to enhance the parameter policy of order selection in the future. Based on the existing results, the deep learning algorithm seems to be an accessible solution to identify the appropriate combination of fractional orders and control parameters, to gain a satisfactory control performance, which can consider the dynamic response and steady-state precision, simultaneously. Moreover, it is believed that this method can help researchers find a new model to describe the behavior of friction, which can be expressed by a fractional-order model, during the motion of a linear motor. If possible, the counter-tracking can benefit from the deep learning method synthesized with the FO algorithm.

SPRINGER NATURE

GPSR Compliance

The European Union's (EU) General Product Safety Regulation (GPSR) is a set of rules that requires consumer products to be safe and our obligations to ensure this.

If you have any concerns about our products, you can contact us on ProductSafety@springernature.com

In case Publisher is established outside the EU, the EU authorized representative is:

Springer Nature Customer Service Center GmbH
Europaplatz 3
69115 Heidelberg, Germany

The manufacturer's authorised representative in the EU is Springer Nature Customer Service Centre GmbH, Europaplatz 3, 69115 Heidelberg, Germany. If you have any concerns regarding our products, please contact ProductSafety@springernature.com

Printed and bound by CPI Group (UK) Ltd, Croydon, CR0 4YY
30/09/2026
02225586-0006